IET ENERGY ENGINEERING SERIES 212

Battery State Estimation

Other volumes in this series:

Battery State Estimation

Methods and models

Edited by
Shunli Wang

The Institution of Engineering and Technology

Published by The Institution of Engineering and Technology, London, United Kingdom

The Institution of Engineering and Technology is registered as a Charity in England & Wales (no. 211014) and Scotland (no. SC038698).

© The Institution of Engineering and Technology 2021

First published 2021

The Institution of Engineering and Technology
Michael Faraday House
Six Hills Way, Stevenage
Herts, SG1 2AY, United Kingdom

www.theiet.org

British Library Cataloguing in Publication Data
A catalogue record for this product is available from the British Library

ISBN 978-1-83953-529-1 (hardback)
ISBN 978-1-83953-530-7 (PDF)

Typeset in India by MPS Limited

Contents

About the editor

Shunli Wang is a professor at Southwest University of Science and Technology, China, where he heads the New Energy Measurement and Control Research Team. His research focuses on modeling and state estimation research for batteries and multiple generation battery systems. He holds 30 patents, has published more than 100 papers, and won several awards.

Foreword

This book mainly focuses on the accurate state estimation of the lithium-ion battery through experiments by introducing the application of adaptive algorithms. The state-estimation-based management systems are the brains of the battery packs, which are responsible for managing outputs, fast charging, and safe discharging, providing timely and reliable notifications of the battery status. Several modeling methods are investigated for the accurate state estimation of lithium-ion batteries. As it is difficult to estimate the battery state accurately, numerous algorithms and techniques are employed. These battery modeling strategies are very important for the battery system management, which determines the parameters to be identified and how it can be done. Consequently, various battery models are analyzed to simplify the circuitry used in the battery management system, aiming to estimate the state of charge, state of power, state of safety, and state of health with efficient performance, reliable and timely information.

Equivalent circuit models are designed to estimate the battery state, according to which experimental approaches are adopted. Data from complex pulse-current tests are used for parameterization. The battery is modeled and simulated with the results and calculations of the experimental data. Improved algorithms are also proposed and analyzed in the quest for accurate state estimation. The idea of improved algorithms is to update the statistical noise covariance parameters, further enhancing the battery state estimation performance. These algorithms reduce the interference of system noise effectively.

This work also employs extended methods alongside the traditional algorithms to improve the effectiveness of the adaptive battery state estimation. Results and computations from the experiment and simulation are compared with that from the improved algorithms, which illustrate that the improved algorithms could present good convergence speed. As the adaptive algorithms are stable with high precision accuracy, they can be effectively used for the accurate battery state estimation.

Preface

The lithium-ion batteries have gradually become the preferred power source for new energy supply working conditions, which have the advantages of high energy, small size, and rechargeability. The internal structure is complex, so its state is affected by various complex factors such as current, self-discharge effect, internal temperature, environmental temperature, and battery aging. It makes the state value difficult to be estimated accurately as the accurate state estimation is a symbolic indication of the energy control technique. Through the accurate residual capacity, the battery application strategy can be planned to realize its optimal operation.

The parameter extraction is picked in the equivalent circuit modeling, and expanded by its internal resistance. The resistance–capacitance values have been calculated mathematically. Furthermore, the hybrid pulse power characterization experiments are conducted to realize the parameter identification and battery state estimation. Then, the iterative calculation is performed by the open-circuit voltage and extended Kalman filter. In the extended calculation process, the iterative prediction and correction strategies are introduced to reduce the initial error.

The research motivation and objective of the battery state estimation methods are elaborately discussed as well as the mathematical analysis of estimation approaches. The critical factors of the state estimation approach are also discussed. The coordinate transformation and iterative estimation are presented, according to which various battery modeling types are introduced in detail together with its mathematical analysis. The data collection tactics are extensively incorporated together with parameter identification and experimental verification.

The whole content is reviewed by Prof. Zonghai Chen, who provides many insightful and constructive opinions on the publication of this book. It has received scientific support from Mianyang Quality Supervision & Inspection Institute, Sichuan Huatai Electric Co., Ltd, Shenzhen Yakeyuan Technology Co., Ltd, and Mianyang Weibo Electronics Co., Ltd.

As battery modeling involves a wide range of aspects, please feel free to contact the authors with the link https://www.researchgate.net/lab/DTlab-Shunli-Wang for effective responses. It is hoped that this book can be served as a communication platform to establish contact with readers and promote the development progress of the battery modeling technologies.

List of contributors

Amdadul Haque	Southwest University of Science and Technology, Mianyang 621010, China
Carlos Fernandez	Robert Gordon University, Aberdeen AB10-7GJ, United Kingdom
Chunmei Yu	Southwest University of Science and Technology, Mianyang 621010, China
Dan Deng	Southwest University of Science and Technology, Mianyang 621010, China
Daniel-Ioan Stroe	Aalborg University, Pontoppidanstraede 111 9220 Aalborg East, Denmark
Ji Wu	Hefei University of Technology, Hefei 230027, China
Jialu Qiao	Southwest University of Science and Technology, Mianyang 621010, China
Jinhao Meng	Sichuan University, Chengdu 610065, China
Junhan Huang	Southwest University of Science and Technology, Mianyang 621010, China
Kailong Liu	WMG, University of Warwick, Coventry, CV4 7AL, United Kingdom
Lei Chen	Southwest University of Science and Technology, Mianyang 621010, China
Lili Xia	Southwest University of Science and Technology, Mianyang 621010, China
Long Zhou	University of Shanghai for Science and Technology, Shanghai 200093, China
Mingfang He	Southwest University of Science and Technology, Mianyang 621010, China
Monirul Islam	Southwest University of Science and Technology, Mianyang 621010, China
Peng Yu	Southwest University of Science and Technology, Mianyang 621010, China
Pu Ren	Southwest University of Science and Technology, Mianyang 621010, China
Shunli Wang	Southwest University of Science and Technology, Mianyang 621010, China
Siyu Jin	Aalborg University, Pontoppidanstraede 111 9220 Aalborg East, Denmark
Weihao Shi	Southwest University of Science and Technology, Mianyang 621010, China

Wenhua Xu	Southwest University of Science and Technology, Mianyang 621010, China
Xiao Yang	Southwest University of Science and Technology, Mianyang 621010, China
Xiaoxia Li	Southwest University of Science and Technology, Mianyang 621010, China
Yanxin Xie	Southwest University of Science and Technology, Mianyang 621010, China
Yongcun Fan	Southwest University of Science and Technology, Mianyang 621010, China
Yunlong Shang	Shandong University, Jinan 250100, China
Yujie Wang	University of Science and Technology of China, Hefei 230027, China
Chuangshi Qi	Southwest University of Science and Technology, Mianyang 621010, China

Chapter 1

Introduction

Abstract

With the development of society, energy security and the environmental pollution caused by it has become a key issue that all sectors of society are concerned about and urgently need to be resolved. The deepening of the world energy crisis has led to the emergence of new energy industries as well as the increasing awareness of environmental protection. Among them, lithium-ion batteries have developed rapidly in the field of new energy due to their higher energy density and longer cyclic life. This chapter briefly introduces the use scenarios and market conditions of lithium-ion batteries, the common methods of lithium-ion battery state estimation, and their research significance. Then, the development status of the battery management system is briefly described. At the same time, the estimation methods of various state parameters are introduced for lithium-ion batteries, which lays a foundation for subsequent research. The research conclusions and further research plans are discussed, the content is reviewed, and the system state estimation is emphasized to achieve the purpose of safety protection and lifetime guarantee.

Keywords: Lithium-ion battery; Energy crisis; Development history; System state estimation; Filtering algorithm; Equivalent circuit model; Safety protection

1.1 State of the art

Currently, lithium-ion batteries are important new energy sources in the twenty-first century, and research in the area of improving and enhancing the technical performance of lithium-ion batteries through various technologies and methods is the key to tapping greater potential. Lithium-ion batteries are used in portable products, such as laptops, notebooks, mobile phones, tablets, portable digital assistants, cameras, rechargeable lamps, toy cars, and toys that use rechargeable batteries and robust products such as the electric vehicle, electric motorcycles, balance cards, scooters, and smart grids, in all aspects of life and all sectors, agriculture, health, education, industry and in wireless sensor networks and recently very popular. The key to improving, controlling, monitoring, and managing the lithium-ion batteries is the battery management system (BMS) [1], as well as the estimation of the state of charge (SOC), state of health (SOH), state of power, and other battery parameters are very important research fields, ensuring the safety and reliability of electronic devices that use the battery as a power source.

The global attention in the direction of the batteries used in electric vehicles (EVs) is gaining much popularity as they are the sustainable mode of transportation and the alternatives of the International Combustion Engine-based vehicles [2]. The EVs effectively overcome the fossil fuel crisis and environmental pollution, considering the major hurdle for the automobile sector [3]. As the only power supply in pure EVs, the capacity of the battery pack is of great importance. In this case, the high specific energy lithium-ion batteries are widely used in EVs with superiority [4]. Lithium-ion batteries have increased in popularity. Due to their advantages such as lightweight, fast charging speed, high energy density, low self-discharge, and long lifespan [5]. The usable energy classified as SOC is the same as the available energy of the vehicle fuel gauge operated by a combustion engine.

BMS is responsible for accurately measuring the status of the battery, ensuring safe operation, and prolonging the battery life [6]. The accurate SOC estimation of the lithium-ion battery is a very challenging task because of its high time-variant, nonlinearity, and complex electrochemical system [6]. An improved Thevenin equivalent circuit model is proposed, designed, and implemented through experimentation and simulation. The model is achieved by adding an extra RC branch to the Thevenin model, making it a second-order resistor capacitor. The second-order Thevenin model has good accuracy, stability, robustness and is very effective for SOC estimation [7]. This model is used to study and record parameters and estimate the relationship between voltage, current, SOC, and charge–discharge characteristics.

Experimental data results and simulated results are compared and analyzed to further appreciate the effectiveness of the improved adaptive extended Kalman filtering (AEKF) algorithm used [8]. The improved AEKF algorithm is employed in this work to accurately estimate the SOC of the lithium-ion batteries. The algorithm estimates the SOC dynamically but easily causes divergence due to the uncertainty of the battery model adopted and to what extent the system noise is taken care of. The AEKF algorithm is used for the SOC estimation of the lithium-ion battery model designed to obtain a more accurate result for the research. It is worth noting that the proposed SOC estimation using the AEKF algorithm is more accurate and reliable comparatively than the extended Kalman filtering (EKF) algorithm.

The safe operation and efficient energy management strategy are essential for the BMS of batteries, which are based on the SOC estimation accuracy [9]. However, the evaluation of battery SOC is challenging in uncertain and complex EVs environments [10]. Therefore, to manage the lithium-ion more efficiently and improve battery performance, it is necessary to obtain the inner state parameters and accurately make an accurate SOC [11]. SOC is not directly measurable and requires a particular type of method to be estimated but often uses some specific models such as the empirical model, the equivalent circuit model, and the electrochemical model [12]. Deterioration of battery cells leading to reduced performance is a problem that limits battery life.

Moreover, electrical equivalent circuit (EEC) modeling has been investigated specifically for applications such as BMS development and vehicle power management control [13]. A good battery model can estimate the battery storage power details of the battery and the voltage response to the load [14]. The EECs can describe these characteristics of lithium-ion batteries [15]. In some cases, modeling the side reactions is required in terms of battery losses, which can also be realized with the EECs [16]. A novel approach for identifying parameters requires a high-order equivalent circuit proposed in this literature with improved mathematical analysis. The EKF algorithm is proposed to estimate the SOC in real-time solid data [5]. The validation results show that the proposed methods have good performance in estimation accuracy and uncertainty of the parameters.

Electrical vehicles powered by lithium-ion batteries are more environmentally friendly than gasoline-powered ones, and this kind of environmentally friendly energy is gradually widely used in various fields. In such high demand, the battery state is significant because it can be estimated. To extend the driving distance, the influencing factors should be measured in real time for the long-term use to avoid the measurement error accumulation of the voltage, resistance, and temperature [17]. The highly nonlinear characteristic also brings lots of difficulties to the battery state estimation. Because the power batteries operating characteristics are highly nonlinear, it is necessary to establish an accurate battery model to estimate the accurate state of the battery [18].

To solve this issue, several algorithms are proposed and used to estimate the battery state, including the current integration, open-circuit voltage (OCV), Kalman filtering (KF) algorithms, and neural network (NN) [19]. The current integral method neglects the influence of the battery self-discharging current rate, aging degree, and charge–discharge current rate on the battery state. Long-term use will lead to the accumulation and expansion of measurement errors, so it is necessary to introduce relevant correction factors to correct the accumulated errors. However, to realize the SOC estimation accurately, the following factors are determined and the core research is objective, including accurate SOC estimation, understanding the dynamic characteristics of battery behavior, studying internal reactions and temperature, and monitoring and controlling charge–discharge flow. An effective BMS is necessary for power lithium-ion batteries.

The above objects are analyzed in various ways to determine a ternary lithium-ion battery for EVs as the research objective. The entire process is carried out by the offline process, and the process is mainly established in the laboratory environment and uses room temperature.

To perform and validate the research objective, the above contents help to carry out the necessary procedures. The Ampere-hour (Ah) commonly known as the coulomb-counting method is to perform the first stage. Then, a similar model can be selected as well as the state-space equation and so on. The high-order improved equivalent circuit model is used as the significant model to get the parameters. Furthermore, these parameters are used for SOC estimation, which is commonly defined as the percentage of the maximum possible charge present inside a rechargeable battery from 0% to 100%. The detailed contents are briefly introduced in Figure 1.1.

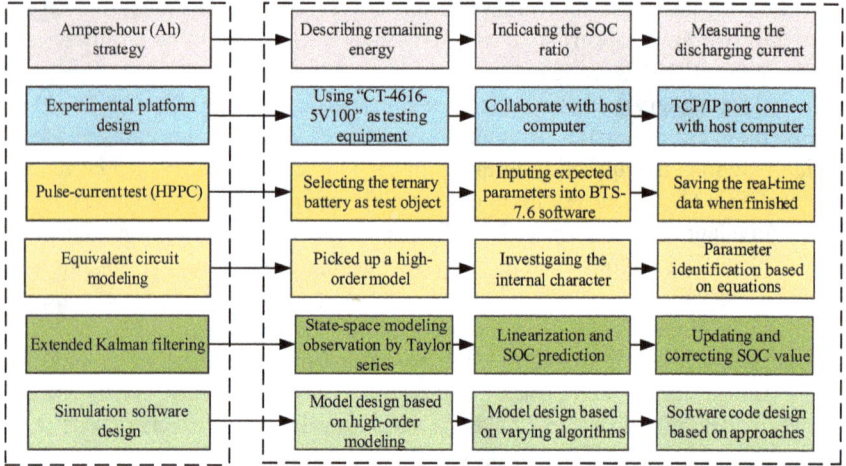

Figure 1.1   Major working procedure for SOC estimation approach

Figure 1.1 shows the working process step by step that has been carried out. Various experimental methods are integrated, such as the hybrid charge–discharge, constant-voltage charging with different multipliers, constant discharge, and cyclic discharge shelved experiments. It is used to study the lithium-ion battery operating characteristics and analyze the characteristics of the battery response. Besides, its long-term use accuracy in estimating the state of the same battery group can be greatly reduced. The complex effect of factors such as battery temperature, self-discharging current rate, and the varying degree of battery aging have been described elaborately with its internal equation.

1.2   Application requirements

The accurate SOC estimation of lithium-ion batteries is one of the most important issues for BMS, especially in EVs and most electronic devices, because it is a critical factor to solve the key issues of monitoring and safety concerns of the batteries. The methods and techniques used for SOC estimation have a great effect on the outcomes and result of analysis. The results greatly depend on appropriate battery modeling and estimation algorithms. The internal resistance and OCV of the battery are influenced by temperature and current detection precision. Therefore, the changes of these parameters are likely to cause a drifting noise in current measurements leading to errors in SOC estimation. Researchers have over the years been using several methods and techniques combining them in the quest to estimate state accurately and improve upon such methods and techniques successfully. This research focuses on the use of an improved second-order Thevenin equivalent circuit model and the application of improved algorithms to improve estimation accuracy and to encourage the use of adaptive algorithms in the estimation process.

EVs and hybrid electric vehicles (HEVs), owing to the benefits of reduced emission and energy conservation, are recognized worldwide as one of the significant growth paths of automotive industries. Lithium-ion batteries are the necessary power source for EVs due to their high energy and energy density. Therefore, it uses rechargeable batteries to maintain a permanent energy supply— various types of rechargeable industrial batteries, such as lead-acid, Ni–MH, Ni–Cd, and lithium-ion. However, lithium-ion battery production seems to be more suitable than other batteries. In such cases, high power density, high voltages, critical load/unload cycles, and safety are superior. The battery charging status must also be controlled explicitly for battery SOC estimation. The BMS configured algorithms to allow better utilization and longer battery life.

The crucial role of the BMS is to control the remaining power. Accuracy in the measurement of the SOC makes it easy for the battery to control adequate electricity, prevents irreversible damage to the internal battery device, and ensures maximum use. Many SOC estimation approaches are available for electrochemical accounting, model based, and black-box often referred to as data oriented. It aims to develop a solution that honors precision and versatility as a target for the deployment of lithium-ion batteries on the BMS. In this idea, an effective SOC estimation algorithm is based on a coulomb-counting algorithm, an accounting approach. It is using a piecewise linear interaction mapping between SOC and OCV to solve this downside approach and increase its accuracy. Experimental data is analyzed and used for parameterization, curve fitting, and simulation of the battery model, while it is used to perfect the estimation and to minimize error. Lithium-ion batteries have become the preferred power source for a lot of electronic devices and require special handling to maintain good SOH and optimum performance.

The ability and capability of BMS are very paramount to estimate the battery state parameters accurately and effectively. Research to the improvement of the function, reliability, and performance of the lithium-ion battery technology is important, and any breakthrough in the area would go a long way to improve upon the technology. As SOC estimation is one of the most important states to be identified in a battery, any advancement towards improving estimation accuracy through the invention of improved estimation methods, enhanced performance, operability of the battery, and extending battery lifespan and safety is vital. Research on lithium-ion batteries and their related parameterization is important across the world today to improve especially on electronic devices, and more specifically, the electric vehicle, the HEVs, smart grids, and UAVs.

1.3 Research methodology

To achieve the objectives of the research and meet the required standards, several methods and techniques are employed including experiments, tests, and simulations.

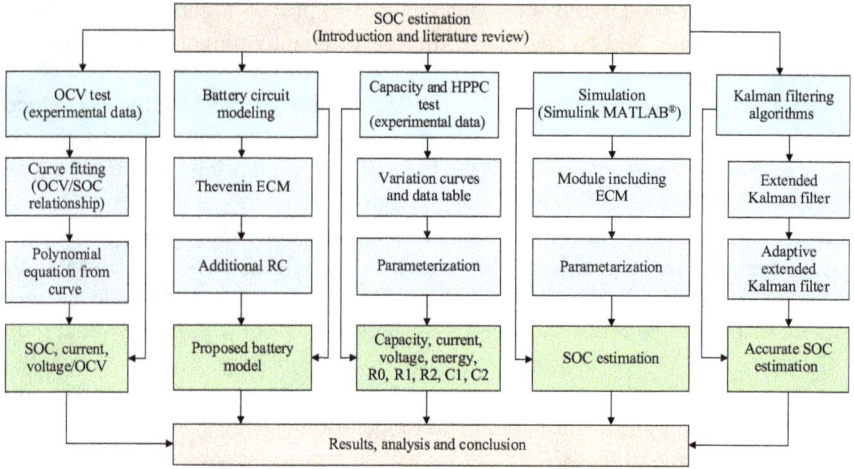

Figure 1.2   Adaptive research approach

The OCV test, capacity test, and hybrid pulse-power characterization test are employed for parameterization. The data from the tests is used as the experimental result for the figures and analyzes made in this work. To further improve the SOC estimation and make it current and significant, an improved AEKF algorithm is proposed and implemented in the simulation.

An experimental research design is a type of research that is used to manipulate, control, test, and understand the causal processes of a system model. Process methodology in research is convenient in the study of mechanisms and functionalities of simple to complex systems. The methodology of experimental research is the use of purposeful abstract terms to represent a real object by representing group components and interactions that allow qualitative and quantitative descriptions of the object. These components and interactions can be constructed and described by text and block diagrams and built into a model with other distinctive modules that operate together to achieve a unique output.

The research content comprises tests, simulations, and algorithms that need to be performed to achieve the objectives and present appropriate results. An overview of the research content from the introduction through the experimental stage and simulation to the use of the AEKF algorithm is shown in Figure 1.2.

1.4   Research status and direction

The common methods used in SOC estimation of the lithium-ion batteries are the OCV method, Ah integral NN method, and KF method. The OCV method uses the one-to-one correspondence between the OCV voltage and SOC to obtain its value by obtaining the battery OCV, realizing the purpose of accurate state estimation. It is a more traditional method to calculate SOC by the integration of current in time.

The NN method estimates the SOC of the battery by processing the amount of real-time input and output data of the battery. The KF method obtains the optimal solution in the sense of minimum variance through continuous iterative operations. Because of the complex electrochemical reaction inside the battery in use, it often shows strong nonlinear characteristics.

Also, some defects of the traditional algorithm itself make the above methods cannot accurately estimate the SOC value of the battery when dealing with the nonlinear systems, which often have the problems of low estimation accuracy and large error. In recent years, researchers put forward some improved algorithms based on traditional algorithms. The complex internal structure of the lithium-ion batteries makes it show strong nonlinear characteristics in the use process, which puts forward new requirements for the traditional SOC estimation methods of the battery. At the same time, the dependence of its estimation on the equivalent model makes the model selection and construction very important. Therefore, the EEC and SOC estimation of the battery still need to be further developed under various conditions.

Research concerning BMS from a global perspective includes those which display an entire BMS design adopting a distributed structure to reach better scalability and portability [20,21]. Different approaches to designing a BMS depend on the functionalities desired for the specific application, but most of them focus on key functions such as SOC estimation [22] and the balancing process [23]. The improvement is towards the design of intelligent BMS for electric and HEVs in artificial intelligence applied for the battery state estimation. Therefore, SOC estimation has drawn the attention of many researchers, and many different methods have been proposed [24,25]. The OCV method, a full charge detector/dynamic load observer, and robust extended Kalman filtering algorithm are combined. It is difficult to determine the specific approach when such methods are used. However, based on the classification [25], two categories are used, namely the direct method and the indirect method, as well as several sub-categories that summarize the trend of SOC estimation and appropriate adjustments.

A Thevenin EEC is used in for every single cell in an array of more than 90 series-connected cells to identify the internal resistance of each cell. Two different branches are used in a Thevenin model for charge–discharge that is connected in series n times to represent n cells in a series. According to [26,27], there are three different EECs of lithium-ion batteries widely adopted because of their excellent dynamic performance. It shows that the second-order EEC is the most accurate and has the best dynamic performance and also the most complex. Also, the discretization equations of each of the three mentioned models are presented and used in combination with the EKF method to estimate the SOC. The Thevenin and second-order EEC modeling methods [27] are used for the SOC estimation. The difference between these models is the way of the SOC equation and calculation procedure. The parameters of the second-order EEC can be calculated with different datasets depending on the scenario where the model is going to be used [28], a comparison between continuous-time and discrete-time equations of the second-order EEC is made and concludes that discrete-time identification methods are less robust due to undesired sensitivity issues in the transformation of discrete domain parameters.

The application of the EKF algorithm in SOC estimation is also presented in [29]. It is common to find a combination of the CC and OCV methods with the KF method to estimate the SOC value [30]. The EKF method [31] is also introduced for the iterative calculation in combination with CC and OCV parameters. Another common improvement to the KF algorithm for SOC estimation is the unscented Kalman filtering (UKF) algorithm, which is used in [31,32] to improve the estimation accuracy. The UKF algorithm is implemented in [33] to estimate SOC using an improved EEC with a resistance and a capacitor correction factor. This is done first of all to measure the effect of different current rates and the SOC estimation on the battery internal resistance and second to identify the impact of different current rates and temperatures on the battery capacity [34]. The work presented in [35] uses multiple modeling approaches combined with the EKF algorithm to estimate the SOC value of the battery. The improved EKF algorithm is implemented in [36] to be more robust to uncertainties in the system, measurement equations, and noise covariance. An SOC estimation approach that uses an improvement in the measurement noise treatment is proposed in [37], and an adaptive Kalman filtering algorithm that can reduce the estimation error is established by correcting the covariance matrix error in the depicted EKF method. To deal with the variation of battery parameters due to temperature changes [38], an online approach is proposed for SOC estimation and parameter updating using a dual square root UKF based on unit spherical unscented transform.

To obtain a more accurate and reliable SOC, an improved Thevenin equivalent model is proposed and its parameters are identified. Likewise, an AEKF algorithm is designed for the SOC estimation of the lithium-ion battery model. The use of the AEKF algorithm in this research is to estimate the SOC and eliminate or reduce errors accurately and diligently. The use of the AEKF algorithm is an innovation in this work coupled with the second-order RC and the other minor features that would lead to the successful implementation.

This research is centered on accurate state estimation. The focus is specifically the use of experimental data for parameterization, the modeling, and the implementation of the circuit for simulation analysis. The study also empathizes with the implementation of the proposed improved adaptive algorithm for accurate state estimation. The introductory aspect of the dissertation is made up of the background which outlines the importance of the research and reasons for the choice of the direction. The problem statement puts forth the main issue to be tackled, and its purpose is to reiterate the main goal and objectives of the research. The research method gives an insight into the methods and techniques used by the researcher to achieve the main goal and finally emphasizes some literature and key issues as far as the research area is concerned.

Theoretical and mathematical analysis deals with the introduction of the lithium-ion batteries specifically and related important areas connected to the study. Battery modeling techniques and experiments are conducted for parameter identification, which is introduced into the SOC estimation methods. The mathematical equations for parameterization and the proposed mathematical algorithms are realized for the battery state estimation. Model building and realization introduces the different battery modeling techniques in the area and puts across the proposed circuit model adopted for the research and underscores the choice of the model and its effect on the SOC estimation.

Experimental verification includes detailed expression, how each testing procedure is performed with a detailed list of steps and flowcharts. The data collection process for parameter identification results from the conducted experiments presents the results from mathematical calculations and simulations. The comparison and verification of the results are also discussed. Conclusion and further research plan present reflections on the work in its entirety review content and emphasizes the effects of the experiments and designs implemented in the work. A concluding statement on the overall performance of the method used, the difficulties encountered, and some suggestions for further research in the area are also opined.

1.5   Chapter summary

This chapter introduces different types of lithium-ion batteries as well as their advantages and disadvantages for applicable occasions. Meanwhile, the working characteristics are analyzed. It provides a basis to choose batteries for different working environments. This chapter briefly introduces the concept, development background, application fields, and working principles of lithium-ion batteries. Meanwhile, the development trend is analyzed. It lays the foundation for the research of lithium-ion batteries.

Acknowledgment

The work is supported by the National Natural Science Foundation of China (No. 61801407), Sichuan Science and Technology Program (No. 2019YFG0427), China Scholarship Council (No. 201908515099), and Fund of Robot Technology Used for Special Environment Key Laboratory of Sichuan Province (No. 18kftk03).

Chapter 2

Mechanism and influencing factors of lithium-ion batteries

Abstract

This chapter introduces the operating mechanism, influencing factors, key indicators, and some mainstream state estimation methods of lithium-ion batteries. First, understanding the main composition and internal working principle of lithium-ion batteries is the prerequisite and basis for other work. Second, internal resistance, open-circuit voltage (OCV), terminal voltage, current thermal energy, capacity variation, and temperature characteristics are the main features of the battery, which can be further used to accurately characterize the battery state. The state of charge, state of health, state of power, depth of discharge, and cyclic life are the key indicators of state estimation, while the discharging test, Ampere-hour integral, OCV, internal resistance, and Kalman filtering are basic state estimation strategies. Based on the Kalman filter, there are many improved algorithms, such as the unscented Kalman filter and adaptive Kalman filter, which have achieved good application effects. Besides, other algorithms such as neural networks, support vector machines, and some improvement strategies are also introduced. The advancement of these algorithms has made an important contribution to improving the whole-life-cycle state estimation effect of lithium-ion batteries.

Keywords: Operating mechanism; Battery characteristics; Influencing factor; State of charge; State estimation; Kalman filtering; Cyclic life; Temperature; Open-circuit voltage

2.1 Introduction

With the development of society, the shortage of primary energy and increasing attention to environmental pollution have promoted the rapid growth of lithium-ion batteries. The invention of new energy and environmental protection are important research topics, and the fact that lithium-ion batteries are relatively environmentally friendly in use has explained the gradual replacement of other relatively polluting batteries.

To improve the function of electronic equipment, especially electric vehicles, hybrid electric vehicles, smart grids, and UAVs that all use lithium-ion batteries [39].

The research on lithium-ion batteries and their related parameterization is of global importance. There are several journals and periodicals with published articles and projects on several important theories and ideas related to lithium-ion batteries and their parameter estimation, and the number is increasing daily [40]. According to the global lithium-ion battery market, it has been growing at a compound annual growth rate of 10.6% since 2016 and is estimated to reach $56 billion by the year 2024, which means that the demand for lithium-ion batteries is set to be more than double by this year. This implies that the increasing demand would lead to more improvements and has a lot of broad prospects for more research in the field, as shown in Figure 2.1.

The battery model studies the relationship between the external characteristics and the internal state of the battery by establishing a mathematical model. Discrete-time and state-space forms are also used for the state of charge (SOC) estimation. The present literature makes mention of electrical equivalent circuit (EEC) models as being widely used as a foundation for model-based estimation and control [41,42]. Generally, the equivalent circuit models are selected for the lithium-ion battery modeling, including the Rint model, Thevenin model, RC model, and PNGV model [43,44]. These models use RC loops of different orders to model the polarization characteristics of batteries [45]. Among them, the Thevenin model is widely used, but as all its components may change with the state of the battery, its accuracy is not high [46].

For more accurate parameter identification [47,48], a new design using charge–discharge is being developed, and SOC estimation methods are combined [49,50] as a means of estimating SOC in the presence of unknown or time-varying battery parameters. Research in this field either assumes the available precise SOC–OCV [51,52] or a constant during discharge. The RC parameters are determined by analyzing the transient state of the battery voltage response [25] under certain excitations, such as constant current or pulse current experiments. The voltage source in an EEC typically represents the open-circuit voltage (OCV), which depends on the SOC [53]. The relationship between SOC and OCV can be

CAGR 10.6% (2016–2024)

56

2015   2016   2017   2018   2019   2020   2021   2022   2023   2024

Figure 2.1   Global lithium-ion battery market size and forecast

identified by charging or discharging the battery with a small current. Parameter identification based on current–voltage data is addressed [48] by a method that simplifies the problem of solving a set of high-order polynomial equations to solving several linear equations and a single variable polynomial equation.

2.2 Operating mechanism

In the global high-tech growth, the lithium-ion battery industry is an excellent direction. The benefits of lithium-ion batteries are high capacity, high conversion rate, long lifespan, and no emission. Because of its good electrochemical durability, high fuel density, long service life, and no maintenance, it is often used in electric vehicles and various power storage systems. Lithium-ion batteries have been widely used at present. It primarily encompasses five fields, including transportation, power storage, mobile communication, modern energy storage, and military aerospace. It will substitute oil with electricity, minimize emissions of greenhouse gas, and store electricity in the power grid in the unnecessary conditions of electric vehicles.

2.2.1 Brief introduction

The battery converts chemical energy into electric energy, which is irreversible in the primary battery and rechargeable in the secondary battery. Lithium-ion batteries can be divided into three different categories, such as lithium metal, lithium-ion, and lithium-ion polymer. Lithium metal batteries are primary batteries, while lithium-ion and lithium-ion polymer batteries are both rechargeables. The lithium-ion battery system is a complex system that integrates chemical, electrical, and mechanical characteristics, so various characteristics must be considered in the design. In particular, the safety and life attenuation characteristics contained in the chemical characteristics of the battery cell cannot be directly evaluated, nor can they be easily predicted in a short time. Therefore, the battery technology, group technology, and battery management system (BMS) technology should be implemented when designing a battery system, as well as battery protection. Consequently, the reliability and lifespan of the battery should be considered.

The lithium-ion battery is an advanced battery technology that uses lithium ions as the key component of its electrochemical reaction. During the discharge cycle, lithium atoms in the anode are ionized and separated from their electrons. The lithium ions are small enough to pass through the micro-permeable separator between the anode and the cathode. This separator recombines with its electrons and is neutralized electrically. A class of organic compounds known as ether is used as the electrolyte of lithium-ion batteries. The most common combination of materials called electrodes is that of lithium cobalt oxide (cathode) and graphite (anode), which is most commonly found in portable electronic devices such as cellphones and laptops. Other cathode materials include lithium manganese oxide and lithium iron phosphate. Lithium manganese oxide is commonly used in hybrid and electric vehicles. Due to the small size of lithium ions, the batteries can have a very high voltage and charge storage per unit mass and unit volume. Lithium is the lightest of all metals, which has the greatest electrochemical potential and provides the largest specific energy per weight. Rechargeable batteries with lithium metal on the anode provide extraordinarily high energy densities.

Lithium-ion batteries are facing competition from numerous alternative battery technologies, most of which are in the development stage [54]. One such alternative called a saltwater drive battery is developed by the energy equation. They are composed of saltwater, manganese oxide, and cotton, which are made by using abundant, non-toxic materials, and modern low-cost manufacturing techniques [55]. Because of this, they are the only cradle-to-cradle-certified battery in the world. The batteries mentioned in it are all lithium-ion batteries. The cathode substrate of a lithium-ion battery is lithium-alloy metal oxide [56]. Its negative electrode component is graphite. The essence of charge and discharge is the electrochemical reaction in lithium-ion batteries. The essential response and working process of lithium-ion batteries are shown in Figure 2.2.

Figure 2.2 shows that the lithium-ion metal oxide positive electrode material is emitted during the battery charging phase [57]. The lithium cobalt metal oxide is transferred through the electrolyte and the separator, and the carbon substrate is incorporated into the negative electrode-coated frame. The positive electrode enters a state of prosperity rich in lithium-ion at this stage. On the contrary, the negative electrode enters a lithium-ion lean state [58]. During the discharge process, lithium-ion can

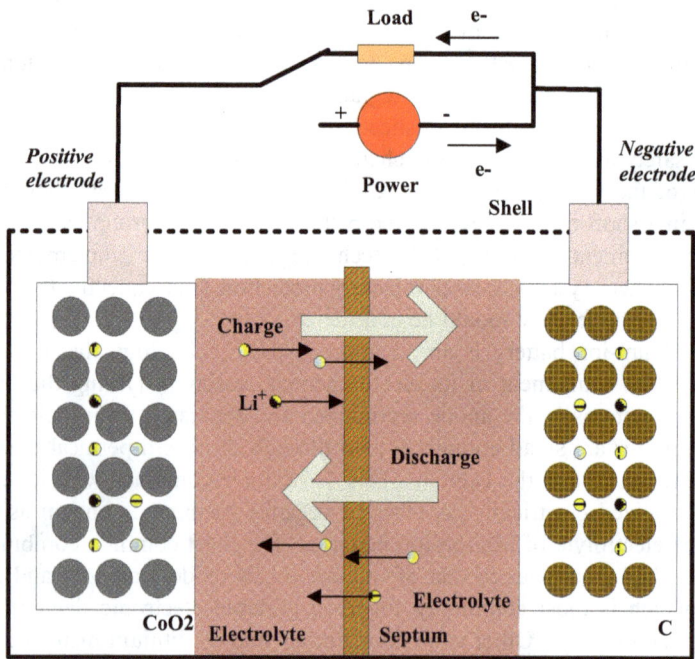

Figure 2.2   Schematic structure of a lithium-ion battery

be released from the negative electrode carbon material layer. The carbon content coating would be deposited in the negative electrode crystal lattice since going through the separator and the electrolyte [59]. The positive electrode is developing towards a lithium-rich state. The negative electrode is developing towards a lithium-poor state. The chemical quality of the electrode continues to recover. The complete equation for the reaction is achieved by positive and negative lithium-ion electrodes. Each battery is shown as follows:

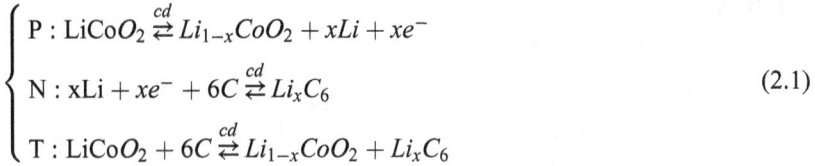

$$
\begin{cases}
P : LiCoO_2 \overset{cd}{\rightleftarrows} Li_{1-x}CoO_2 + xLi + xe^- \\[2mm]
N : xLi + xe^- + 6C \overset{cd}{\rightleftarrows} Li_xC_6 \\[2mm]
T : LiCoO_2 + 6C \overset{cd}{\rightleftarrows} Li_{1-x}CoO_2 + Li_xC_6
\end{cases}
\tag{2.1}
$$

In (2.1), it can be known that the internal resistance, polarization, aging, and other factors are added in the process of the lithium-ion battery. The embedding and unpacking going through the separator and traveling in the electrolyte can impact the calculation of SOC [60]. In the process of model development, these considerations and SOC estimation need to be solved [61].

2.2.2   Battery composition

Lithium-ion batteries generate electricity through the chemical reaction of lithium. The battery consists of four key components. The anode, which determines the capacity and voltage of the battery, is the source of the lithium ions. The cathode allows current to flow through an external circuit, and when the battery is charged, lithium ions are stored in the cathode. The electrolyte is composed of salts, solvents, and additives and acts as the conduit for lithium ions between the cathode and the anode. The separator is a physical barrier separating the cathode and anode. Common cathode materials include lithium cobalt oxide (lithium cobaltate), lithium manganese oxide (spinel or lithium manganate), lithium iron phosphate, lithium nickel manganese cobalt (NMC), and lithium nickel cobalt aluminum oxide (NCA), as shown in Table 2.1.

Table 2.1   Most common lithium-ion battery compositions

Type	Cathode	Anode
Lithium cobalt oxide (LCO)	$LiCoO_2$	Graphite/hard carbon (LiC_6)
Lithium manganese oxide (LMO)	$LiMn_2O_4$	Graphite/hard carbon (LiC_6)
Lithium iron phosphate (LFP)	$LiFePO_4$	Graphite/hard carbon (LiC_6)
Lithium nickel manganese cobalt oxide (NMC)	$LiNiMnCoO_2$	Graphite/hard carbon (LiC_6)
Lithium nickel cobalt aluminum oxide (NCA)	$LiNiCoAlO_2$	Graphite/hard carbon (LiC_6)
Lithium titanate (LTO)	$LiMn_2O_4/$ $LiNiMnCoO_2$	$Li_4Ti_5O_{12}$

During the charge–discharge process of the battery, while the battery is discharging and providing an electric current, the anode releases lithium ions to the cathode, resulting in electron flow from one side to the other, and when charging, lithium ions are released by the cathode and received by the anode.

2.2.3   Working principle

The internal chemical reaction of lithium-ion batteries is a basic oxidation–reduction (OXRED) reaction, which is also the operation theory of the battery in the actual process of converting available energy into heat energy through the chemical reaction. The charge–discharge process of the battery is a chemical reaction involving the movement of lithium ions. The positively charged lithium atom undergoes an oxidation reaction when the battery is charged, which loses electrons and becomes a lithium ion.

Many lithium ions formed by the oxidation reaction of the positive electrode migrate from the positive electrode to the carbon layer of the negative electrode of the battery. At one end, the capacity of the battery is related to the number of lithium ions generated in the positive electrode reaction, and at the other end, the capacity of the battery is related to the number of lithium ions exchanged through the electrolyte in the negative electrode. During discharge, the negative electrode undergoes an oxidation reaction, and the lithium ions trapped in the negative electrode carbon layer appear and move back to the positive electrode. The higher the discharge power, the more lithium ions return to the positive electrode.

More lithium ions are loaded into the anode when the loading power is high. During charging, lithium ions move from the positive electrode to the negative electrode through the electrolyte. The positive and negative reactions of the electrode and the whole reaction equation are as follows. The positive electrode reaction is

$$\text{LiM}_xO_y = \text{Li}_{(1-x)}\text{M}_xO_y + x\text{Li}^+ + xe^- \tag{2.2}$$

Electrons flow from the positive electrode to the negative electrode, which takes a longer path around the external circuit. The electrons and ions combine at the negative electrode and when the battery is fully charged, the ions stop moving. The negative electrode reaction is represented in the following equation:

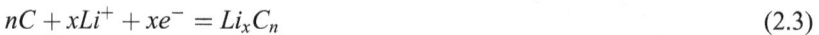

$$nC + xLi^+ + xe^- = Li_xC_n \tag{2.3}$$

During discharging, the ions flow back through the electrolyte from the negative electrode to the positive electrode. Electrons flow from the negative electrode to the positive electrode through the external circuit. When the ions and electrons combine at the positive electrode, lithium is deposited there. The battery is fully discharged and needs charging when all the ions move back to the positive electrode. The total battery response is shown in the following equation:

$$LiM_xO_y + nC = Li_{(1-x)}M_xO_y + Li_xC_n \tag{2.4}$$

Table 2.2   *Life cycle feature of the depth of discharge*

Depth of discharge	Discharge cycles
100	300–600
80	400–900
60	600–900
40	1,000–3,000
20	2,000–9,000
10	6,000–15,000

2.2.4   Cycling lifespan

The electrode design can also be strengthened. The capacity of lithium-ion batteries can be increased by testing the properties and scale of materials used for construction. The chemical reaction on the electrode takes place inside the lithium-ion batteries. The internal resistance of the electrodes can reduce the heat generated during use. The cell potential and lifespan can also be increased [62]. The electrode size should be minimized to improve performance. It has been discovered that there are limitations to the reduction, such that as if the electrode thickness had fallen below. Then the batteries could no longer meet the necessary energy requirements. For thinner content sizes, there are additional trade-offs, including rising processing costs. The electrode scale can be included as the primary design element in the overall design presented in Table 2.2.

In the design of electrodes, another development field has been formed. By changing the materials made, the ion journeys may be optimized to maximize overall conductivity and thermal capacity on the transportation distance or route. In certain situations, the additional advantage of these improvements is to improve the mechanical quality and the heat efficiency or overall power of the mechanical characteristics. Gold, graphene, yttrium, and zinc are used in the components applied to the electrodes. Several various influences affect this mechanism, and further research is ongoing.

2.3   Battery characteristics

After understanding the structural principle of the lithium-ion battery, the parameters, and common concepts of the battery that will appear in this work, it is important to reiterate certain characteristics that affect the battery. Basic parameters of the battery include terminal voltage, electromotive force, capacity, internal resistance, SOC, state of health (SOH), state of power, depth of discharge (DOD), cyclic life, and self-discharge rate. The characteristics that affect the parameterization of the lithium-ion battery include the following contents.

2.3.1   State of power

The physical significance of the capacity per unit time of the lithium-ion batteries is the internal chemical reaction that the battery produces energy output. How much work is done in a certain period is represented by the energy released by a chemical reaction, which is measured in Watt (W) [63–65]. Depending on the project, power can be subdivided into actual power and instantaneous power [66–68]. The battery power equation can be expressed as shown in the following equation:

$$P = UI = I(E - IR) = IE - I^2R \qquad (2.5)$$

wherein E is the battery electromotive force, R is the total battery internal resistance, and I is a unit of time for the average current.

2.3.2   Internal resistance

Lithium ions move from one pole to the other pole inside the lithium-ion battery and the factors that hinder the movement of ions constitute the internal resistance of the battery. Due to the internal resistance of the battery, the terminal voltage of the battery is lower than OCV in the discharge state. In the charging state, the terminal voltage of the battery is higher than the OCV value [69–71]. The essence of electric current is the directional movement of electric charge.

During the movement, the electrons will be resisted by the material itself and the magnetic field. Internal resistance refers to the resistance of current flowing through the battery when the battery is working, which is characterized by internal resistance. Internal resistance is one of the important parameters of the lithium-ion battery [72–74]. The internal resistance directly affects the working voltage, working current, and capacity of the batteries. The internal resistance of the lithium-ion battery is not constant. In the process of battery charging and discharging, temperature changes, charge–discharge ratio, charge–discharge time, electrolyte concentration, and the quality of active substances will all cause changes in the internal resistance of the battery.

2.3.3   Open-circuit voltage

The electrolyte, positive electrode, and negative electrode materials undergo aging during the cyclic charging and discharging process of lithium-ion batteries and in their external structure. The period from the battery start state to the battery life end state is the lifespan of the battery. This is the data of the charge–discharge test by the manufacturer under the specified standard test environment. Generally, discharge is stopped when its capacity drops to 80% of the rated capacity. The number of charge–discharge is the total number of continuous charge–discharge cycles of the battery under this condition.

The OCV value can be measured by connecting a multimeter or voltmeter directly to the positive and negative ends of the battery. It is the voltage used in the simulation of the equivalent model in this book. The battery is not an ideal power source, the electrolyte and electrode material have internal resistance, so the OCV value is slightly less than the electromotive force value. Considering the internal resistance, it can be treated as the electromotive force [75–78]. Under actual operating conditions, the instantaneous closed-circuit voltage of the battery changes dynamically, which cannot avoid overcharge and over-discharge, causing damage to the battery system.

2.3.4 Self-discharge current rate

The self-discharge current rate refers to the ratio of the discharge capacity of the battery to its rated capacity in the no-load state and is used to indicate the consumption rate of the battery capacity. Mainly affected by the battery manufacturing process, materials, storage conditions, and other factors, it is generally expressed as the percentage of battery capacity decrease per unit time, and the unit time is month or year. Therefore, the self-discharge current rate is the leakage current rate, which is usually used to test the leakage current in the test open-circuit environment. Due to the low leakage current, the self-discharge current rate of lithium-ion batteries is usually determined by their monthly values. It can also be regarded as the charge retention rate of the battery.

2.3.5 Terminal voltage

The terminal voltage refers to the potential difference between the positive and negative electrodes of the lithium-ion battery, which can be divided into OCV and operating voltage according to the working conditions of the circuit where OCV is the terminal voltage of the lithium-ion battery without load and other power sources [79–83]. At the end of charging, the maximum allowable voltage of the battery is called the charge cut-off voltage, and the lowest allowable voltage after the end of battery discharge is called the discharge cut-off voltage. Beyond this limit, the battery will suffer some irreversible damage, and the cut-off voltage is an important safety indicator.

Working voltage refers to the potential difference between the positive and negative electrodes when the battery is in working condition. Generally, due to the internal resistance of the battery, the working voltage during discharge is lower than the OCV, and the working voltage during charging is higher than the OCV value [84–86]. Like the rated voltage classification, according to the working state, the operating voltage of the lithium-ion battery is also divided into rated voltage, theoretical voltage, OCV, and working voltage with load. The rated voltage shall be directly calibrated by the battery manufacturer before delivery.

2.3.6 Current heat energy

There is no ideal power source in the world. No matter it is the current source or the voltage source, there is some internal resistance. The lithium-ion battery as a power source is no exception. With a large current through the battery for a long time, it flows through the resistance to generate a large amount of heat energy, as shown below:

$$W = PT = I^2RT \tag{2.6}$$

Resistance will lead to battery energy loss and a continuous increase in the battery temperature, while the temperature rise will also lead to a rise in the resistance value, forming a vicious circle, the battery management system will take measures to deal with heat dissipation after monitoring temperature abnormalities [87–89]. Besides, if the battery is charged for a long time, the high large charge–discharge current of the battery will be lower than below certain levels, which will lead to potential safety hazards. It is very important to monitor the current size of the battery in the charge–discharge process to avoid causing damage to the battery.

2.3.7   Capacity variation

There are different levels of power stored in the battery. When the battery is fully charged, its capacity is the amount of electricity it contains. Batteries of the same type are usually evaluated according to the amount of current they can output over time. Battery capacity refers to the total amount of electricity that can be released in the full state of the battery. It is represented by the symbol Q, and the unit of measurement is milli-Ampere-hour (mAh) or Ampere-hour (Ah) [90–92]. The battery capacity types can be divided into rated capacity, theoretical capacity, and actual capacity.

The rated capacity is measured by the battery manufacturer and calibrated directly on the outer surface of the battery. Rated capacity is an important indicator of battery storage capacity and an indicator of battery life. The rated capacity is also known as calibration capacity, according to the relevant provisions of the relevant departments of the state, under certain discharge conditions (temperature, discharge rate, etc.) to ensure the minimum amount of energy that can be released by the battery, rated capacity is the Ah capacity indicated by the licensed manufacturer, which is one of the important parameters of the battery [93–96]. The theoretical capacity is the theoretical value of the total charge of the battery calculated by Faraday's law according to the electrochemical reaction inside the lithium-ion battery when the internal lithium ions are fully involved in the reaction.

The actual capacity is the maximum capacity that a lithium-ion battery can release when purchased in the store under certain discharge conditions. It is usually calculated by the product of the discharge current and the discharge time. The actual capacity is not necessarily the capacity of the new battery but can also be the capacity of the used battery. Currently, the actual capacity is less than the actual capacity of the new battery. The actual capacity is the amount of electricity that the battery can release when operating in the real environment. The factors that affect the actual capacity of lithium-ion batteries are very complicated, including the temperature of the external environment, the materials used to make the battery, and the service life.

2.3.8   Temperature change

The real-time temperature of the lithium-ion battery is a significant aspect that cannot be ignored in practical applications. Battery activity is directly proportional to temperature. As the temperature increases, the exchange rate of lithium ions between electrolytes increases, and the overall activity of the battery also increases, which is reflected in more battery energy output, increased the usable capacity of the battery, and improved battery efficiency [97–102]. Under the condition of long-term time and low-temperature conditions, the activity of anode and cathode material of the battery decreases. The actual capacity is less than the rated capacity. The charge–discharge efficiency of the battery decreases and the use efficiency is reduced.

2.4 Critical indicators for battery state estimation

SOC estimation is an obligatory part of precise state estimation and battery management for lithium-ion batteries. However, some critical influencing factors are battery related to estimating accuracy. This part of several influences is discussed elaborately in the portion.

2.4.1 Description of major parameters

There are unknown safety risks in using lithium-ion batteries. It is essential to consider the critical criteria of lithium-ion batteries to prevent explosions and fires caused by management failure. To protect against unwanted circumstances, it is obligatory to measure some critical action. To observe the critical situation, the core parameters of lithium-ion batteries are necessary to be considered, as shown in Table 2.3.

In Table 2.3, five key parameters are the basic parameters that deeply affect the performance of lithium-ion batteries. The battery capacity is a significant factor for storage capability, which is expressed in mAh. The nominal voltage is the rated voltage determined by the manufacturer. The end of charge voltage can be obtained when the battery charge is almost near to be ended. While the batteries are completely discharged, the end of discharge voltage associates with the lowest voltage permissible. The self-discharge rate is the capacity consumption rate of a battery

Table 2.3   The major parameters of lithium-ion batteries

Parameters name	Parameter definition
Battery capacity	The amount of active material determines the battery capacity, usually expressed in Ah. For example, 1,000 mAh can be converted into a charge of 3,600 C at a current of 1 A for 1 h
Nominal voltage	The electrical potential difference throughout the batteries between positive and negative electrodes is the nominal battery voltage
End of charge voltage	The active material on the polar plate has reached saturation when the rechargeable batteries are fully charged. Then, it begins to charge and the battery voltage will not increase
Discharge end voltage	When the batteries are fully discharged, the end of discharge voltage corresponds to the lowest voltage permissible. Discharge end voltage is compared to the rate of discharge
Self-discharge rate	The amount of the overall power expended when it is not used for any time. Generally, at an average temperature, the self-discharge rate of lithium-ion batteries is 5%–8%

when it is unloaded. It is generally expressed by the ratio of its unit monthly discharge capacity to the rated capacity, and the ratio is 5%–8%.

2.4.2   Temperature effects

As the environmental temperature affects the electrochemical reaction process of lithium-ion batteries, the measurement accuracy of SOC will continue to be compromised. The higher the temperature and the higher the degree of the internal electrochemical reaction and the energy of the battery will also rise [103]. These factors will generate a large amount of high-temperature gas, which will accelerate the aging process of the battery, and affect its life cycle and even damage the battery. Therefore, only by seeking a suitable temperature can the output of the battery be fully utilized, and the batteries can be charged and discharged efficiently [104].

The battery capacity indicates how much energy it can hold, which decreases with the decrease of temperature and increases with the increase of temperature. Therefore, the battery fails on the cold winter morning, even though it worked the previous day perfectly. The batteries work in a cold environment most of the year. In that case, the decreased power of the lithium-ion battery must be taken into consideration. The typical operating temperature of a battery is 25 °C (77 °F). At approximately −30 °C (−22 °F), the battery power will be reduced to 50%. At freezing, the battery power is decreased by 20%. Capacity is improved at higher temperatures at 122 °F. The capacity of the battery will be around 12% higher.

2.4.3   Charge–discharge current rate

The charge–discharge current can be described by the charge rate and discharge rate of lithium-ion batteries, which is represented by C. The charge–discharge rate refers to the ratio of the working current to the rated capacity of the battery during the charging and discharging process or working process. Within the same duration, the charge–discharge amount of the battery will vary with different rates of charge. The battery is charging, discharge mode, the number of cycles, and DOD will affect the aging process of the battery [105,106].

The C-rates is to regulate the charge–discharge rates of the battery. The rated charge–discharge rate of the battery is usually 1 C, which means that the fully charged battery rated at 1 Ah can supply 1 A for 1 hour. The same battery is discharged at 0.5 C should deliver 500 mA for 2 hours, and 2 C should deliver 2 A for 30 min. Losses at accelerated discharge decrease the time of discharge, and these losses also impact the time of loading. The C-rate of 1 C is often referred to as 1-h releases such as the 2-h release of 0.5 C or C/2 and the 5-h release of 0.2 C or C/5. Any high-performance low-stress batteries can be charged and discharged above 1 C.

2.4.4   Aging degree

The aging of the battery is due to the change in its internal structure, the rise of temperature, and the batteries dropping. The constant charging and discharge will lead to a change in the internal structure, influencing the aging degree [107].

During the battery aging process, the SOC value will continue to decrease until the battery fails to work normally. Generally speaking, the battery charging–discharge mode, the number of cycles, and DOD can affect the batteries aging [108].

The everyday use of batteries in HEVs implies a paradigm of accurate battery aging and battery life. The aging of batteries can be divided into two parts, including the aging of calendars and cycle one. Calendar aging refers to the endless amount of lost storage ability. In other words, battery capacity is responsible for the deterioration. Cycle aging is related to the influence of cycles (charge or discharge) for battery use times. It occurs when the battery is either loaded or unloaded. This is directly due to the level, the temperature of the usage pattern, and current battery specifications. As a result, its performance steadily deteriorates during battery life because its electrochemical components are deteriorating and the performance of EVs and fuel efficiency deteriorates.

2.4.5 Self-discharge rate

When the battery is not in use, some slow chemical reactions are also taking place inside it. As the essence of battery charge–discharge is an electrochemical reaction and the electrochemical reaction inside the battery increases with the rise of temperature, when the temperature is high, the intensity of self-discharge inside the battery is large, which will lead to the automatic drop of SOC.

2.5 Basic state estimation strategies

SOC is primarily used to describe the remaining capacity of the battery, which corresponds to the ratio of the remaining capacity and the rated capacity under the same circumstances following the batteries discharge at a given discharge rate. It is typically expressed as the percentage (%). The range of its value is 0–1. When SOC = 0, this means that the battery is completely discharged. When SOC = 1 is used, this means that the battery is fully charged. Traditional battery SOC prediction techniques include the discharge test method, conductance method, OCV method, and Ah integration method. Their foundation is relatively simple, and the implementation process is relatively simple, but the accuracy is often not high or the adaptability is not strong. New SOC prediction methods for the battery include extended Kalman filter (EKF), particle filtering (PF), fuzzy logic, and back-propagation (BP). These methods combine mathematics and computer theory, and the implementation process is complex, but they can achieve a good estimation effect and strong adaptability.

2.5.1 Discharging test

The discharge test method is to discharge the target battery continuously at a constant current until reaching the cut-off voltage of the battery. The time used in the discharge process is multiplied by the discharge current, which is the capacity of the battery. This method is generally used as the calibration method of SOC estimation of battery or in the later maintenance work of battery. It is relatively simple and reliable to use this method without knowing the SOC value of the battery, and the results are relatively accurate, and it is effective for different kinds of batteries at the same time. However, the discharge experimental method has two shortcomings. First, the experiment process of this method requires a lot of time. Second, this method cannot be used to calculate the SOC of the power battery under the working state.

2.5.2 Ah integral method

The Ah integral method, also known as the Coulomb counting method, is to calculate the total change of a certain charge–discharge period when the SOC basic value of a certain time is known in advance to obtain the SOC value of the battery at the special time point. Comparing the estimated value with the actual value, the error is obtained. Along with the time extension, the error increases obviously and cannot take the correct response strategy to the error accumulation problem, which cannot guarantee the stability of the test results. This approach has a certain dependence on the initial value. The Ah integral method is widely used in SOC estimation. Its goal is to focus on the external characteristics of the battery system, without considering the complex relationship between the electrochemical reaction and parameters in the battery. The principle of the Ah integral method is shown as follows:

$$\text{SOC}_t = \text{SOC}_0 - \frac{1}{C} \int_0^t \eta I dt \qquad (2.7)$$

where SOC_0 is the initial electric quantity of the battery, SOC_t is the electric quantity of the battery at time t, C represents the rated capacity of the battery. The charge–discharge current is described by parameter I, and the discharging direction is the positive direction, and η is the coulomb efficiency coefficient, which reflects the internal electric quantity dissipation of the battery during the charge–discharge process.

2.5.3 Open-circuit voltage method

The OCV method estimates SOC according to the approximately linear relationship between SOC and OCV. The actual operation process is described as follows. First, the battery needs to be stationary for a long time to ensure the stability of its OCV and other parameters, and then the SOC value is obtained according to the known approximate linear function relationship. The OCV method has the advantages of a simple method and strong operability. In this way, only after the battery stops working for a while can the OCV be obtained, so it is difficult to monitor SOC in real time. The OCV refers to the battery positive and negative potential difference when the external circuit current is zero and the battery reaches equilibrium after a long time of standing. The OCV of the battery has a relatively fixed mapping relationship with SOC, as shown below:

$$U_{OC} = F(\text{SOC}) \qquad (2.8)$$

In (2.8), the mapping relationship can be expressed in different forms, mainly divided into discrete and continuous forms. Discrete forms, such as the look-up table method, use a series of discrete isolated points to express the one-to-one correspondence between the OCV and SOC that uses a simple piecewise linear method to express the parts between points, which also exposes the coarseness of this method. The continuous form, such as the function method, is to get the OCV–SOC curve through multiple measurements. The continuous relationship expression between discrete points is obtained by function solution or curve-fitting methods.

2.5.4   Internal resistance method

For the internal resistance method, there is a certain functional relationship between SOC and internal resistance. The resistance method can also be called the conductivity method. It can analyze the relationship between the conductivity or internal resistance of the battery and SOC from numerous experimental data through a long-term tracking test of the conductivity or internal resistance to predict the SOC of the battery. Since the conductance method only relies on the internal resistance of the battery to estimate SOC, the measurement accuracy of the internal resistance of the battery directly affects the accuracy of the SOC estimation. Therefore, it is necessary to ensure good contact during the measurement and try to make the contact resistance zero.

In a certain period, the functional relationship between SOC and internal resistance is relatively stable, so the accuracy of the internal resistance method depends entirely on the accuracy of internal resistance measurement. However, the internal resistance of the battery will gradually increase along with the aging process of the battery, and the fixed function relationship between SOC and internal resistance is no longer applicable, which will affect the accurate estimation of SOC.

2.6   Kalman filtering and its extension

2.6.1   Kalman filtering

The Kalman filtering (KF) method is a theory established by the state-space theory in the time domain. It regards the estimated signal as the output parameter of a stochastic linear system under the superposition of white noise. The input–output equation is given by the state and observation equations in the time domain. The KF algorithm is mainly used to estimate the linear time-invariant systems. The recursive linear minimum variance estimation method is used to repair the unobservable state estimation error by using observable output estimation error of the system, which greatly reduces the noise interference in the data stream and improves the estimation accuracy of the new system.

The KF method constructs a set of recursive equations that can describe the characteristics of the battery system and carry out the recursive operation. Then, the system state-space expression can be obtained, including signal and noise. SOC is one of the internal states in which a mathematical method is used that is based on the estimation result of the previous step and the existing measurement data through the optimal regression data processing to get the current optimal estimation results. The advantage of this method is that the dynamic SOC value can be measured with high accuracy, but the KF method requires the high accuracy of the battery model and complex operation.

2.6.2   Extended Kalman filtering

The EKF algorithm is used to estimate the nonlinear system, linearize the nonlinear state-space model, and then use the basic KF algorithm to achieve. The nonlinear system is linearized by expanding the state-space equation by the Taylor series and discarding the higher-order terms. EKF is a mathematical method combined with probability theory. The idea is to estimate the state of the system optimally based on the minimum variance. Its basic principle is described as follows. Combining the state-space model of signal and noise, the state variables are recursively predicted according to the state equation, and the current observation value is used for their correction and update. The real-time state variable estimation is realized through the continuous "prediction-update" process.

When the KF algorithm estimates the SOC value of the battery, it uses the Ah integration method to calculate the SOC value and uses the measured voltage value combined with the observation equation to continuously modify the SOC value obtained by the Ah integration method. Because the battery is a strong nonlinear system in the working process, and the application object of KF is a linear system, it is necessary to linearize the battery to adapt to KF, that is, the EKF algorithm. When it is used to estimate SOC of the battery, SOC is a component of the state vector, the current is used as control quantity in the input parameter, the output is terminal voltage calculated by the equivalent model, system noise and observation noise are Gauss white noise, and their variance is expressed as Q and R. This method is usually based on the state and measurement equation of the system. The prediction state equation includes the Ah integral method of SOC calculation, and the observation equation represents the equivalent model of the lithium-ion batteries. The accuracy of estimating SOC by the EKF algorithm largely depends on the accuracy of the equivalent model, so it is very important to establish an appropriate equivalent model for the battery.

2.6.3   Unscented Kalman filtering

The unscented Kalman filter (UKF) abandons the traditional method of nonlinear functional linearization, adopts the Kalman linear filter framework, and for one-step prediction equations, unscented transform (UT) is used to process the nonlinear transfer of mean and covariance. In this algorithm, the UKF uses a fixed number of points to approximate the probability distribution of nonlinear functions. The algorithm does not ignore the errors caused by higher-order terms and does not calculate the Jacobian matrix repeatedly, which reduces computational complexity.

The key of this algorithm is UT treatment, in which how to sample has an important impact on the estimation effect. The common sampling methods include symmetric sampling, minimum skewness simplex sampling, and hypersphere simplex sampling. The sampling method is generally selected for convenient calculation and a good effect. The UKF algorithm approximates the probability density distribution of the nonlinear function. It uses a series of samples to approximate the posterior probability density of the state, instead of approximating the nonlinear function. It does not ignore the high-order terms, so it has higher calculation accuracy for the statistics of nonlinear distribution.

2.6.4   Dual Kalman filtering

The essence of SOC estimation using the KF is to combine the OCV method and the Ah integration method, which is used to estimate the system state with the Ah integration method and its feedback with the measurable voltage. Since the KF algorithm largely relies on the accuracy of the equivalent model when estimating SOC, the idea of the dual Kalman filter algorithm is to estimate the equivalent model parameters and system state variables alternately with two KF lines and then adjust their feedback according to the measurable voltage to obtain more accurate estimation results.

In the first stage of Kalman filtering, the internal resistance of the battery model parameter is taken as the state quantity, and after the time update, the internal resistance is measured and updated according to the difference between the actual output voltage and the model output voltage. In the second stage of Kalman filtering, SOC and polarization voltage U_p can be used as two-dimensional state vectors for time update and measurement update.

2.6.5   Adaptive extended Kalman filtering

To improve the accuracy of the SOC estimation of the lithium-ion battery, the adaptive extended Kalman filtering (AEKF) algorithm is used to estimate the statistical characteristics of noise online based on the improved models. Comparatively, the SOC estimation accuracy of the AEKF algorithm is significantly higher than that of the EKF algorithm, which effectively reduces the noise interference in the SOC estimation process, and has certain reliability and practicability.

Although it has higher estimation accuracy and stronger robustness and stability than KF, EKF, and UKF algorithms, it is also based on accurate mathematical models and statistical characteristics of system process noise and observation noise. When the environment around the carrier changes or the motion state changes drastically, the statistical characteristics of process noise and observation noise of the system will change greatly. At this time, the accuracy and stability of the conventional UKF will decrease significantly. In the SOC estimation process of the battery by its standard version, when the working current of the battery changes rapidly with time, the negative determination of the covariance may be encountered in the later stage of the operation.

2.6.6   Square root-unscented Kalman filtering

During the calculation process, the covariance P_k of the state variable SOC becomes negative, while the Cholesky decomposition requires that the matrix must be semi-positive qualitative, otherwise the algorithm cannot continue, making the filter invalid because of rounding errors in the numerical calculation. To solve this problem, a new filtering algorithm named square root-unscented Kalman filter (SR-UKF) is applied. The square root of the covariance of the state variable is used instead of covariance to participate in the iterative operation. This method can guarantee the semi-positive qualitative and numerical stability of the covariance matrix of the state variable and overcome the filtering generation scattered. In the UKF algorithm, the costliest operation is to recalculate a new set of the Sigma points each time it is updated.

2.6.7 Cubature Kalman filtering

The cubature Kalman filtering (CKF) method can use the nonlinear system model to deal with nonlinear problems. The algorithm follows the UKF filtering framework, and its core is the spherical radial volume criterion. By selecting $2n$ volume points of equal weight, a nonlinear transformation is then performed on the volume point set and the corresponding predicted value is obtained according to the transformed statistical characteristics. There is no need to linearize the nonlinear model. Moreover, the weight of the sampling point $1/2n$ is always positive, which ensures the stability of the value, and its calculation amount is small and the filtering accuracy is high. The algorithm can be divided into three parts, namely, initialization, time update, and measurement update.

1. The initialization equation is shown as

$$\begin{cases} \widehat{x}_{k+1|k+1} = \widehat{x}_{k+1|k} + K_{k+1}\left[z_{k+1} - \widehat{z}_{k+1|k}\right] \\ P_{k+1|k+1} = P_{k+1|k} - K_{k+1}P_{z_k z_k}K_{k+1}^T \end{cases} \tag{2.9}$$

2. The time update process is shown as follows:
    1. The volume point calculation is shown as

    $$\begin{cases} P_{k+1|k} = S_{k+1|k}S_{k+1|k}^T \\ x_{k+1|k}^i = S_{k+1|k}\varepsilon_i + \widehat{x}_{k+1|k} \end{cases} \tag{2.10}$$

    In (2.10), ε_i represents the i-th weighted volume data point, and the calculation formula is

    $$\varepsilon_i = \begin{cases} \sqrt{n}I, i = 1, 2, \ldots, 2n \\ -\sqrt{n}I, i = n + 1, n + 2, \ldots, 2n \end{cases} \tag{2.11}$$

    2. The propagation volume point equation is shown as

    $$x_{k+1}^i = f\left(x_k^i\right) \tag{2.12}$$

    3. The calculation formula for the predicted value of the state variable and the predicted value of the error covariance is shown as

    $$\begin{cases} \widehat{x}_{k+1|k} = \dfrac{1}{2n}\sum_{i=1}^{2n} x_{k+1|k}^i \\ P_{k+1} = \dfrac{1}{2n}\sum_{i=1}^{n} x_{k+1|k}^i \left(x_{k+1|k}^i\right)^T - \widehat{x}_{k+1|k}\left(\widehat{x}_{k+1|k}\right)^T + Q \end{cases} \tag{2.13}$$

3. The measurement update is shown as follows:
   The volume point calculation formula is

$$\begin{cases} P_{k+1|k} = S_{k+1|k}S_{k+1|k}^T \\ x_{k+1|k}^i = S_{k+1|k}\varepsilon_i + \widehat{x}_{k+1|k}^i \end{cases} \tag{2.14}$$

1. The calculation formula of the propagation volume point is

$$z_{k+1}^i = h\left(x_{k+1|k}^i\right) \tag{2.15}$$

2. The calculation of the measured predicted value is

$$\widehat{z}_{k+1} = \frac{1}{2n}\sum_{i=1}^{2n} z_{k+1}^i \tag{2.16}$$

3. The measurement error covariance and cross-covariance calculation are

$$\begin{cases} P_{k+1}^z = \frac{1}{2n}\sum_{i=1}^{2n} z_{k+1|k}^i \left(z_{k+1}^i\right)^T - \widehat{z}_{k+1}\left(z_{k+1}^i\right)^T + R \\ P_{k+1}^{xz} = \frac{1}{2n}\sum_{i=1}^{n} x_{k+1|k}^i \left(x_{k+1|k}^i\right)^T - \widehat{x}_{k+1|k}\left(\widehat{z}_{k+1}\right)^T \end{cases} \tag{2.17}$$

4. The update equations of Kalman gain, state variables, and error covariance

$$\begin{cases} K_{k+1} = P_{k+1}^{xz}\left(P_{k+1}^z\right)^{-1} \\ \widehat{x}_{k+1} = \widehat{x}_{k+1|k} + K_{k+1}\left(z_{k+1} - \widehat{z}_{k+1}\right) \\ P_{k+1} = P_{k+1|1} - K_{k+1}P_{k+1}^z K_{k+1}^T \end{cases} \tag{2.18}$$

The CKF algorithm uses current as the system input and the Ah integration method as the system prediction module to estimate the SOC value. Then, the one-step prediction of the current SOC is made that is based on the input current at time point k. To obtain the optimal estimated value of SOC, the observation equation is used as the update module to correct the one-step SOC prediction value. Then, the value is used as the output which feeds back to the prediction module [109]. On this basis, the next stage of SOC continues to be predicted. In this algorithm, the Ah integration method is used as the state equation to estimate the current SOC, and the CKF is used for feedback correction to output the optimal value, which makes full use of the simple applicability of the integration method and the characteristics of the recursive KF correction to realize the estimation with high accuracy. It also reduces the complexity of the system.

The CKF algorithm processes nonlinear functions through deterministic sampling points, avoiding the truncation error introduced by the EKF algorithm when expanding the first-order Taylor formula, which can ensure estimation accuracy. Then, the CKF sampling process has a sufficient theoretical basis in a random system and is less affected by parameters, which has obvious advantages over UKF in numerical stability. Moreover, the number of CKF sampling points is less than that in UKF, and the weight is always maintained at a constant value. In contrast, every time UKF completes the filtering process, the weight needs to be recalculated, so CKF is also less than UKF in terms of calculation.

2.7   Intelligent state estimation methods

Other state estimation methods include PF, neural networks (NN), support vector machines (SVM), and so on. In recent years, some new prediction methods have also been proposed, such as the support vector regression (SVR) algorithm based on deoptimization, and improved neural network methods. These new methods are designed to correct the errors of the traditional methods and new methods in the SOC estimation process and improve its accuracy as much as possible. Most of these new methods are only in the stage of theoretical research and experimental simulation. In the actual battery management system, the battery SOC estimation method still adopts the mature and stable traditional method.

2.7.1   State observer

The observer method is designed by applying the modern control theory to the SOC estimation of battery that resembles the KF method. After analyzing the characteristics of the equivalent circuit model of battery, the spatial model expression of battery state including SOC is established. Then, the SOC value is estimated by the observer design method in control theory. This approach requires a good technical understanding and promising outcomes in the field of automatic control and mathematical matrix theory.

2.7.2   Monte Carlo treatment

The Monte Carlo algorithm is an easy-to-implement method to calculate system random numbers [110]. The basic principle is to treat all integration operations as the mathematical expectation of random variables and then use random sampling estimation to obtain the approximate value of the integration operation. It first randomly samples the affected random variables and then brings the data samples into the function.

The implementation of this method is better because when sampling random numbers, it is necessary to solve the problem, solve the random variable that satisfies the corresponding distribution type, and then substitute the function to solve the problem, which can improve its calculation efficiency. This method is not limited by the problem conditions and does not require discretization. This is a direct solution to the problem. Also, the convergence speed of this method is independent of dimensionality, and the error is easy to determine.

2.7.3   Bayesian estimation

Bayesian estimation is a method that uses prior probability and Bayesian theorem to calculate hypothetical probabilities and combines new data obtained in practice to obtain new probabilities [111]. The random variable x and the prior distribution $p(x)$ of the random variable are known. When the observation data is not available, the random variable x can be estimated according to the prior distribution to obtain the data, thereby optimizing the prior probability of the random variable. The data is corrected to obtain the posterior distribution of the random variable x.

Under the premise that the state equation and the observation equation are known, and assuming that the initial value of the probability density is known, the observation value before time k can be obtained. The prediction equation of the state transition probability density function (PDF) is obtained, and the prediction equation is revised. With the rapid development of new energy sources, lithium-ion batteries have been widely used due to their long lifespan, no pollution, no memory effect, and other advantages. The state estimation of the lithium-ion battery is an important parameter of the entire battery system. Battery technology has achieved considerable research results, among which state estimation algorithms include OCV, Ah integration, discharge experiment, and KF-based methods.

The data-driven prediction method based on the Bayesian method can directly predict the remaining useful life (RUL) [112]. In the offline stage, the unsupervised variable selection method is used to find variables that contain degraded behavior information [113]. In the online stage, the method uses the k-nearest neighbor classifier as a RUL predictor variable. The Bayesian framework can be constructed for reliability analysis by integrating bivariate degradation data and life data. Combining the Bayesian method and Markov chain Monte Carlo, the limited bivariate degradation data can be integrated with the life data of other similar methods. Then, reliability assessment evaluation and RUL prediction are carried out. The combination of the Bayesian non-parametric method and Gaussian process regression can predict the capacity attenuation under various usage conditions [114], in which the standardized root means the square error is 4.3%.

The technical novelty is produced by appropriate model migration [115] to solve the battery problem. In the presence of noise measurement and modeling errors, the Bayesian Monte Carlo algorithm is applied to health prediction tasks. Based on 30 cycles of training data, the root-mean-square error (RMSE) of the proposed algorithm is within 2.5%. The deep learning integrated prediction method can also be realized with uncertainty management based on Bayesian model averaging and long short-term memory network [116]. Based on the features extracted by the dynamic Bayesian network (DBN) from the charging process, the battery SOH estimation and RUL prediction can be performed as well. In the case of incomplete battery operating characteristics information, the promotion of the DBN-PF method is suitable for various practical situations.

A migration-based framework can be constructed for battery modeling [117], where the effects of temperature and aging are regarded as uncertain factors. An accurate model of the new battery is established and migrated to the degraded battery through the Bayesian Monte Carlo method. For temperature changes of up to 40 °C and capacity degradation of up to 20%, the typical voltage prediction error can be limited to 20 mV. Based on the accelerated aging model, the velocity aging model can also be established through the migration process, and the migration factor is determined by the Bayesian Monte Carlo method and hierarchical re-sampling technology. The Bayesian simulation evaluation theory is introduced into the updating process of the battery RUL prediction model. Consequently, uncertainty quantification methods are developed, a simulator is used to evaluate the theoretical statistical structure, introducing the deviation function and measurement error into the Bayesian model update in the form of GP. The modular Markov chain Monte Carlo method is used to modify the model with multiple uncertain parameters. Since uncertainty is systematically considered in the inference process, reliable RUL predictions can be provided.

2.7.4   Support vector machine

Corinna Cortes and Vapnik proposed the support vector machine (SVM) algorithm to solve machine learning problems such as small samples, nonlinearity, high-dimensional pattern recognition, and function fitting. It can be used for analysis, pattern recognition, classification, and regression analysis. Based on the VC dimension theory of statistical learning theory and the principle of structural risk minimization, this method can not only solve the linear classification problems but also implicitly map the input to effective high-dimensional feature space for non-linear classification. Therefore, the SVM is a method based on the principle of structural risk minimization considering VC dimensional and statistical theories.

In machine learning, SVMs have supervised learning models with associated learning algorithms, which are used to analyze data for classification and regression analysis. Given a set of training samples, each sample is marked as belonging to one of two categories. The SVM training algorithm builds a model and assigns the new sample to one category or the other, making it a non-probabilistic binary linear classifier. It maps training examples to points in space to maximize the width of the gap between the two categories. Then, the new examples are mapped to the same space and predicted to belong to a category based on which side of the gap they fall.

More formally, the SVM constructs a hyperplane or a set of hyperplanes in a high-dimensional or infinite-dimensional space, which can be used for tasks such as classification, regression, or outlier detection. Intuitively, a good separation is achieved through a hyperplane, which is the largest distance from the nearest training data point in any class. Usually, the larger the margin, the lower the generalization error of the classifier.

Whereas the original problem may be stated in a finite-dimensional space, it often happens that the sets to discriminate are not linearly separable in that space. Therefore, it is proposed that the original finite-dimensional space be mapped to a much higher-dimensional space, presumably to make the separation easier in that space. To keep the computational load reasonable, the mappings used by the SVM scheme is designed to ensure that dot products of pairs of input data vectors may be computed easily in terms of the variables in the original space by defining them in terms of a kernel function selected to suit the problem.

The hyperplane in the higher-dimensional space is defined as a point set whose dot product with a vector in the space is a constant, where such a set of vectors is an orthogonal set of vectors that defines a hyperplane. The hyperplanes describing the vector can be linearly combined with the image parameters of the function vector contained in the database. By selecting the hyperplane, the points mapped to the hyperplane in the feature space are defined by relations. In this way, the sum of kernels above can be used to measure the relative nearness of each test point to the data points originating in one or the other of the sets to be discriminated. Note the fact that the set of points mapped to any hyperplane can be quite convoluted as a result, allowing much more complex discrimination between sets that are not convex at all in the original space.

Also, to perform linear classification, SVMs can use so-called kernel techniques to perform nonlinear classification effectively, implicitly mapping their inputs to high-dimensional feature spaces. When the data is not marked, supervised learning is impossible. An unsupervised learning approach is required, which attempts to find natural clusters of the data to groups, and then map new data to these formed groups. The support vector clustering algorithm, created by Hava Siegelmann and Vladimir Vapnik, applies the statistics of support vectors, developed in the SVM algorithm, to classify unlabeled data. It is one of the commonly used classification algorithms in industry and research applications.

The differential evolution (DE) algorithm is realized [118] to obtain the SVR kernel parameters to predict the RUL of the lithium-ion battery. The error of this method is about 1/99 at the starting point of 80 cycles. Critical features are extracted from the voltage and temperature curves [119] and, based on the critical features, used SVM to build RUL classification and regression models. The RMSE of the model is 0.357%, and the estimated upper and lower errors are 95%, which are 7.87% and 10.75%, respectively. A combination method [120] is realized based on the quantum-behavioral particle swarm optimization and incremental support vector regression, which has an RMSE of 0.0202 Ah and an MAPE of 0.0255%. An SVR-based RUL prediction model is established [121] for lithium-ion batteries under coupling stress using six sets of coupled stress experimental data, in which the relative error of six hundred cycles is within 5%.

2.7.5 Particle filtering

The PF is also known as the sequential Monte Carlo method, which is a Bayesian filtering technology based on the Monte Carlo method. The PF algorithm is a recursive Bayesian filter implemented by the non-parametric Monte Carlo simulation method. It is suitable for any nonlinear system described by the state-space model and the optimal estimation of its accuracy approximation. The particle filtering algorithm can be well adapted to nonlinear and non-Gaussian systems and has no restrictions on measurement noise and process noise. The basic principle of PF is to find a group of random particles (samples) propagating in the state-space to describe the state of the system. The integral operation in Bayesian estimation is processed by the Monte Carlo method, and the minimum mean square error (MSE) estimation of the system state is obtained. When the number of particles tends to infinity, it can approach the system state that obeys any probability distribution.

Compared with other filtering methods, PF does not need to make any assumptions about the system. Theoretically, it can be applied to any stochastic system which can be described by a state-space model. However, two problems limit the further development of the PF. One is a large amount of PF calculation. Another problem is that the degradation of particles leads to the waste of computing resources and the deviation of computing results. Subsequently, with the development of semiconductor technology, the cost-effectiveness of computing resources has become higher and higher. These problems have been effectively solved, and PF has gradually become one of the key technologies. In practical applications, the accuracy of the model can improve the accuracy of SOC estimation. Under the condition of ensuring accuracy, the amount of calculation can be reduced by reducing the number of samples.

The PF algorithm has an important relationship with particles, and prediction is based on particles. This method is superior to other dynamic parameter prediction and tracking methods and can track dynamic parameters more accurately. The probability of random events in particle filtering is very low, which is equivalent to a significant advantage of the algorithm. Bayesian importance sampling puts forward the important concept of probability density, sampling independent and uniformly distributed sampling points from a known distribution. Instead of sampling from the posterior probability density, the difficulty of the sampling method is theoretically solved. The Bayesian importance sampling method requires resampling every time, which increases the complexity of sampling.

After each sampling, the weight of each ion needs to be recalculated. Then, new observations can be obtained, which also increases time and calculation. Sequential importance sampling avoids these troubles when acquiring new observation data. Based on Bayesian importance sampling, the particle mass is calculated by the recursive updating method, which reduces the amount of calculation. The algorithm theory is deduced in detail as follows, and the iterative calculation process is obtained, and described by the following steps.

1. Initialization: N SOC initial particles and particle weight are generated by prior probability.
2. The algorithm loops as follows:
   1. Updating particle weight

$$w_k^i = w_{k-1}^i p\left(U_{L(k)} \big| SOC_k^i\right) = w_{k-1}^i p\left(U_{L(k)} - h\left(SOC_k^i\right)\right), i = 1, 2, \ldots, N$$
(2.19)

   2. Normalizing weight

$$w_k^i = \frac{w_k^i}{\sum\limits_{i=1}^{N} w_k^i}$$
(2.20)

   3. Calculating the least mean square estimation

$$\widehat{SOC}_k \approx \sum_{i=1}^{N} w_k^i SOC_k^i$$
(2.21)

4. The number of effective particles is resampled.
5. The state equation prediction is conducted to predict the unknown parameters.
6. The end condition of the program is judged. If it is not over, at time $k = k + 1$, jump to step 1.

2.7.6 Neural network

The neural network method and the battery neural network model are the same ideas, which are based on the unique characteristics of the neural network. The neural network is a relatively new method. According to the parameters measured during battery charging, a highly nonlinear autonomous learning mechanism is used to realize the neural network model and estimate the SOC value. The more types of input parameters, the more the data, the more accurate the SOC estimation. However, the neural network requires a lot of mathematical operations, which requires high speed and capability of the CPU, and the cost is high. The neural network method is generally more complicated, and the estimation error is largely depending on the training data and training methods.

The principle of using neural networks to estimate the battery SOC is to use battery voltage, current, and other external characteristic parameters as input and train the system through a large number of sample data. When the SOC reaches the required error range, the system is used to estimate the new input SOC. The advantage of using an NN model to estimate the SOC value of the battery is that no specific mathematical model is required.

The neural network is a multilayer feed-forward system. Its main features are as follows. The signal propagates forward, and the error propagates backward. The computing network consists of an input layer, a hidden layer, and an output layer. Given a training set, its input consists of attributes and its output is a one-dimensional real-valued variable. The calculation algorithm generally uses a three-layer network structure, as shown in Figure 2.3.

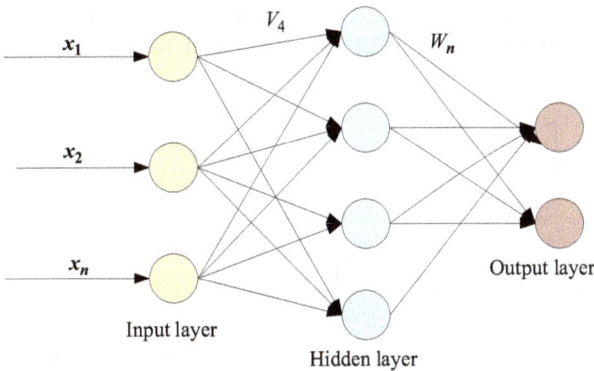

Figure 2.3 Network structure diagram

To reduce the estimation error of the network, it is necessary to adjust its weight coefficient. The function always changes the fastest along the gradual direction. Then, the partial derivative of each parameter is obtained that needs to be adjusted. Then, the partial derivative is used to get the value that the parameter needs to change. The output layer expression is shown as follows:

$$o_k = f(net_k), net_k = \sum_{j=0}^{m} w_{jk} y_i, (k = 1, 2, 3, \ldots, l) \tag{2.22}$$

The hidden layer expression is shown as

$$\begin{cases} y_i = f(net_j)(j = 1, 2, 3, \ldots, m) \\ net_j = \sum_{i=0}^{n} w_{ij} x_i (j = 1, 2, 3, \ldots, l) \end{cases} \tag{2.23}$$

When the output of the network is different from the expected output, the error can be generated that is defined by

$$E = \frac{1}{2}(d - o)^2 = \frac{1}{2} \sum_{k=1}^{l} (d_k - o_k)^2 \tag{2.24}$$

The training process of the calculation algorithm is to continuously modify the weight reduction error of the network. The gradient descent method is used to update the parameters. The relationship between the hidden layer and the input error is shown as

$$E = \frac{1}{2} \sum_{k=1}^{l} (d_k - f(net_k))^2 = \frac{1}{2} \sum_{k=1}^{l} \left[d_k - f\left(\sum_{j=0}^{m} w_{jk} y_i \right) \right]^2 \tag{2.25}$$

If the partial derivative is greater than 0, it should change in the opposite direction of the partial derivative. If the partial derivative is less than 0, the change in this direction can be calculated. The variables are brought into the above equations and they are simplified to get the functional relationship as shown in the following equation:

$$E = \frac{1}{2} \sum_{k=1}^{l} \left[d_k - f\left(\sum_{j=0}^{m} w_{jk} f(net_j) \right) \right]^2$$

$$= \frac{1}{2} \sum_{k=1}^{l} \left\{ d_k - f\left[\sum_{j=0}^{m} w_{jk} f\left(\sum_{i=0}^{n} w_{ij} x_i \right) \right] \right\}^2 \tag{2.26}$$

As can be seen from (2.26), the error of the neural network is closely related to the weight of each layer. At the same time, the learning rate η is set; the learning rate cannot be too fast or too slow. Too fast may cause exceeding the optimal solution. If the learning rate is too slow, it may reduce the efficiency of the algorithm. The parameter expression is shown as follows:

$$
\begin{cases}
\Delta w_{jk} = -\eta \dfrac{\partial E}{\partial w_{jk}} \,(j = 1, 2, 3, \ldots, l) \\[3mm]
\Delta v_{ij} = -\eta \dfrac{\partial E}{\partial w_{ij}} \,(i = 1, 2, 3, \ldots, n; j = 1, 2, 3, \ldots, m)
\end{cases}
\tag{2.27}
$$

where the negative sign indicates a gradient decrease, in which η is a constant. It reflects the learning rate varying from 0 to 1.

An NN model is a series of algorithms that endeavors to recognize potential relationships in a set of data through a process that mimics the way the human brain operates. In this sense, neural networks refer to neuronal systems, whether organic or artificial. NNs can adapt to changing input. Therefore, the network generates the best possible result without redesigning the output criteria. The NN concept originated from artificial intelligence and become popular in the development of trading systems.

In the financial world, neural networks contribute to the development of processes such as time-series forecasting, algorithmic trading, securities classification, credit risk modeling, and the construction of proprietary indicators and price derivatives. The working principle of the neural network likes that of the human brain neural network. The neuron in a neural network is a mathematical function that collects and classifies information according to a specific architecture. The network is analogous to statistical approaches such as curve fitting and regression analysis.

A neural network contains layers of interconnected nodes. Each node is a perceptron and resembles the multiple linear regression. The perceptron feeds the signal generated by multiple linear regression into an activation function that may be nonlinear. In a multi-layered perceptron, the perceptrons are arranged in interconnected layers. The input layer collects input patterns. The output layer has categories or output signals to which input patterns can be mapped. For instance, the patterns may comprise a quantitative list for technical indicators of security. Hidden layers fine tune the input weights until the neural network margin of error is minimal. It is hypothesized that the hidden layer extrapolates the input data to have the salient features of output prediction ability. This describes feature extraction, which implements utility-like statistical techniques such as the principle-component analysis shown in Figure 2.4.

The neural network is widely used in financial operations, enterprise planning, trade, business analytics, and product maintenance. Neural networks have also gained widespread adoption in commercial applications, such as forecasting and marketing research solutions, fraud detection, and risk assessment. A neural network evaluates price data and unearths opportunities for making trade decisions based on the data analysis. These networks can distinguish subtle nonlinear interdependencies and patterns that cannot be distinguished by other technical analysis methods. According to research, the accuracy of neural networks for stock price prediction is different. Some models predict the correct stock prices 50%–60% of

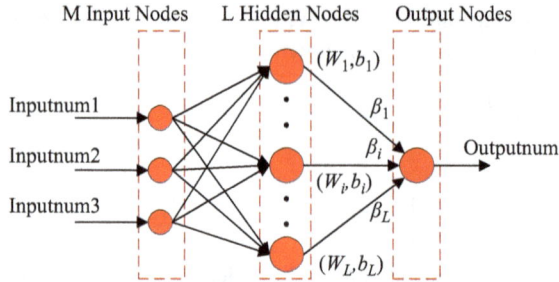

Figure 2.4   Schematic diagrams of the NN training process

the time, while others are accurate in 70% of all instances. Some have posited that investors can get a 10% increase in efficiency from neural networks.

There will always be data sets and task classes that can be better analyzed by using previously developed algorithms. The important thing is not the algorithm but the well-prepared input data on the target index, which ultimately determines the success of the neural network. Wu *et al.* applied the Bat particle filter (Bat-PF) to recursively update the parameters of the NN degradation model. The RUL prediction error of the proposed NN + Bat-PF method is 2 cycles in five hundred prediction cycles, and the PDF width is 35 cycles. She *et al.* proposed a battery aging assessment method based on incremental capacity analysis and radial basis function neural network models. The average prediction error of this method reaches 4.00%, the confidential interval of the derived model is 92%, and the prediction accuracy is 90%. Wu *et al.* proposed an online method to estimate the RUL of lithium-ion batteries using importance sampling (IS) for the feed-forward neural network (FFNN) input selection and setting the number of hidden neurons to 40.

The error of this FFNN + IS method in actual operation is less than 5%. Li *et al.* established an indirect prediction method based on the Ellman neural network and estimated the RUL of lithium-ion batteries under vibration stress. The mean absolute error (MAE) is 0.0243, and the MSE is 0.0278. Tang *et al.* built a model through the input–output slope and bias correction (SBC) method and further integrated the SBC model into a four-layer feed-forward migration NN and trained it via the gradient correlation algorithm. When only the first 30% of the aging trajectory is used for NN training, the RMSE of the prediction is less than 2.5%.

A hybrid neural network (HNN) is proposed by combining convolutional neural networks and long- and short-term memory (LSTM) with the false nearest neighbor method [122], and the accuracy of the proposed method can reach 98.21%. A method is proposed [123] to predict the RUL value based on empirical mode decomposition (EMD), deep neural network (DNN), and LSTM model. Compared with the mixed model of EMD and autoregressive comprehensive moving average model, the average value of RMSE is reduced effectively.

A model framework is proposed for battery SOH monitoring and RUL prediction [124] based on temporal convolutional networks (TCN). For RUL prediction tasks, in the case of offline data, the offline data can be de-noised through EMD and then combined with some online data to fine-tune the model to achieve highly accurate RUL prediction. The average RMSE accuracy of the TCN model is 5% higher than that of traditional networks with different starting points. Compared with other models, the average RUL error is nearly eight cycles higher. An online synthesis method is proposed by combining part of the incremental capacity and artificial neural network (ANN) [125]. The ANN model has only two inputs and its structure is simple. The MAE and RMSE of RUL estimated by the model are less than four cycles and six cycles, respectively. First, a simple model is created from the current battery aging data set to explain the decay of power over time. Then, the basic model is transformed by the structure of the input–output SBC method to capture the degradation of the target unit. To enhance its nonlinear transmission capability, the SBC model is further integrated into a four-layer NN and easily trained via the gradient correlation algorithm.

The NN algorithm has many advantages, but it needs to estimate large numbers of parameters. When there is one hidden layer or multiple hidden layers, and multiple input variables, the number of parameters that need to be accurately estimated explodes. However, each parameter should be estimated consuming some information budget. It guarantees to overfit this model. This has nothing to do with the computational feasibility of the algorithm. If a continuous variable predictive model is established, the logistic regression model is better than the NN. It will use fewer parameters. Herein, only one or two variables are used to fit the model, and then the test set is used to see if other variables outside the intercept cause instability and reduce sample accuracy. Regarding the X variable, the principal component analysis can be used to improve its estimation effect, in which only the previous one or two PCs need to be extracted.

2.7.7   Deep learning

Deep learning architectures, such as deep neural network, deep belief network, recurrent neural network, and convolutional neural network, have been applied in many fields including computer vision, machine vision, speech recognition, natural language processing, audio recognition, social network filtering, machine translation, bioinformatics, drug design, medical image analysis, material inspection, and board game programs, where they have produced results comparable to and in some cases surpassing human expert performance.

ANNs are inspired by information processing and distributed communication nodes in biological systems. ANNs have differences from biological brains. Specifically, neural networks tend to be static and symbolic, and most biological brains are dynamic (plastic) and analog. The adjective "deep" in deep learning refers to the use of multiple layers in the network. Earlier studies have shown that a linear perceptron cannot be a universal classifier, and a network with a non-polynomial activation function with one hidden layer of unbounded width can be a general classifier. Deep learning is a modern variation that is concerned with an unbounded number of layers of bounded size, which permits practical application and optimized implementation while retaining theoretical universality under mild conditions. In deep learning, the layers are also permitted to be heterogeneous and to deviate widely from biologically informed connectionist models, for the sake of efficiency, trainability, and understandability, whence the structured part.

Most modern deep learning models are based on artificial neural networks, especially convolutional neural networks (CNNs), although they can also include proposition equations or latent variables organized hierarchically in-depth generation models, such as the nodes in deep belief networks and deep Boltzmann machines. In deep learning, each level learns to transform its input data into a slightly abstract and compound representation. In the recognition application, the original input can be a matrix of pixels. The first representational layer may extract pixels and encode edges. The second layer may compose and encode the edge arrangement. The third layer may encode a nose and eyes, and the fourth layer may recognize that the image contains a face. Importantly, a deep learning process can learn which features to optimally place in which level on its own. This does not eliminate the need for manual tuning. For example, different layer modes and layer sizes can provide different degrees of abstraction.

The word deep in deep learning refers to the number of layers through data transformation. More precisely, deep learning systems have a substantial credit assignment path (CAP) depth. The CAP is the conversion chain from input to output. CAPs describe the potential causal connections between input and output. For a feed-forward neural network, the depth of the CAPs is that of the network and is the number of hidden layers plus one (as the output layer is also parameterized). For recurrent neural networks, the signal can propagate multiple times through one layer, and its coverage depth may be unlimited. No universally agreed depth threshold distinguishes shallow learning from deep learning, but most researchers believe that the CAP depth involved in deep learning is greater than 2. CAP of depth 2 is a universal approximator, which can simulate any function. Besides, more layers do not add to the function approximator ability of the network. Deep models (CAP > 2) can extract better features than shallow models, so the extra layers help to learn features effectively.

The deep learning architecture can be constructed by using a greedy layer-by-layer method. This helps to sort out these abstractions and pick out which features can improve performance. For supervised learning tasks, it eliminates feature engineering by transforming data into compact intermediate representations similar to principal components and derives layered structures that eliminate redundancy in the representation. The deep learning algorithm can be applied to unsupervised learning tasks. This is an important benefit because unlabeled data is more abundant than the labeled data. Examples of deep structures that can be trained in an unsupervised manner are neural history compressors and deep belief networks.

The empirical mode decomposition algorithm with long- and short-term memory (LSTM) and the Elman neural network are combined [126] to predict the batteries RUL. The relative prediction errors of this Elman–LSTM method are 3.3% and 3.21%, respectively. The LSTM network can also be optimized by particle swarm optimization and attention mechanism to predict battery RUL and

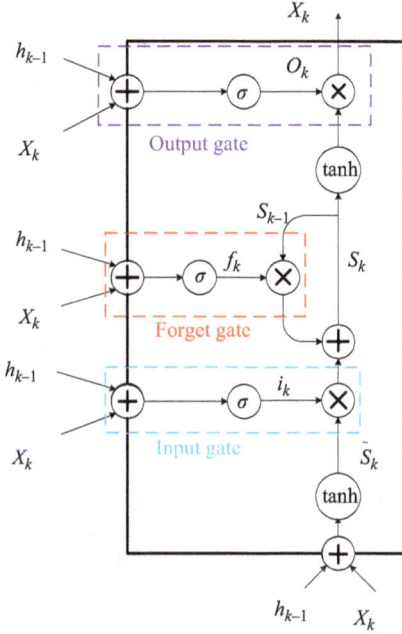

Figure 2.5 LSTM-based RNN structure

realize SOH monitoring. The UKF algorithm is combined with the LSTM network and the NN model to form a data-model fusion to realize the SOH estimation and RUL prediction of lithium-ion batteries. The HNN can be combined with the convolutional neural network and bidirectional long- and short-term memory (Bi-LSTM). Recurrent neural network-long- and short-term memory (RNN-LSTM) can be used to fully select the best subset of statistical features, as shown in Figure 2.5.

The battery RUL predictor can be constructed [127] by using LSTM-RNN. The resilient mean square BP technique is used to adaptively optimize the LSTM-RNN, and the dropout technique is used to solve the overfitting problem. RUL prediction technologies based on LSTM have a good effect, in which the many-to-one structure is used to adapt to various input types. A variant of the LSTM named AST-LSTM is designed [128], which has many-to-one and one-to-one mapping structures. The ARMSE of SOH is estimated to be 0.0216, and the CE of RUL is 0.0831.

2.8 Algorithm improvement strategies

2.8.1 Bayesian importance sampling

Bayesian importance sampling puts forward the concept of probability density importance, sampling from a known distribution to obtain independent and uniformly distributed sampling points. Instead of sampling from the posterior probability density, the difficulty of the sampling method is theoretically solved. The Bayesian importance sampling method requires re-sampling each time it samples, which increases the complexity of sampling.

The weight of each ion needs to be recalculated after each sampling. Then, new observations can be obtained, which also increases time and calculations. Sequential importance sampling avoids these troubles when acquiring new observation data. The recursive update method is used to measure the particle weight based on Bayesian significance sampling, which decreases the measurement size. Bayesian importance sampling puts forward the concept of probability density importance, sampling from a known distribution to get independent and uniformly distributed sampling points. It does not use posterior probability density sampling, which theoretically solves the difficulty of the sampling method in theory.

The Bayesian importance sampling method requires re-sampling each time it samples, which increases the complexity of sampling. The weight of each ion needs to be recalculated after each sampling. Then, new observations can be obtained, which also increases time and calculations. Sequential importance sampling avoids these troubles when acquiring new observation data. Based on Bayesian importance sampling, the recursive update method is used to calculate the particle weight, which reduces the amount of calculation.

2.8.2   Coordinate transformation

One concept is to exchange the coordinate system horizontal and vertical coordinates to transform the OCV into an independent variable. The SOC is also a dependent variable until modifying the fitting curve and polynomial equation through discrete points [129,130]. The curve fitting is carried out to achieve an OCV curve and its polynomial functional relationship. This way is possible since it can be realized without considering the nonlinear curve fitting degree. An open-circuit and closed-circuit voltage difference can be defined as shown in Figure 2.6.

The dynamic nonlinear relationship between the OCV and SOC is presented in Figure 2.6(a). The direct polynomial fitting effect is sufficient because OCV is connected to the internal states of the batteries [131]. The transfer coordinates a significant difference after correction in the same order in Figure 2.6(b) that can be

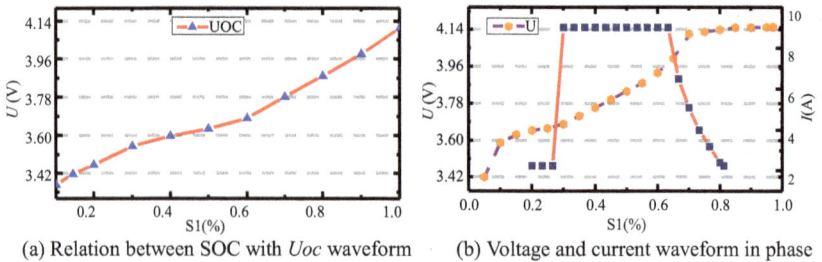

(a) Relation between SOC with U_{oc} waveform   (b) Voltage and current waveform in phase

Figure 2.6   Voltage variation of SOC estimation waveform

seen from the contrast of the fitting effect before and after a coordinated transition. The approach is not extremely feasible for direct coordinate transformation, except if the predicted result has been reached under such conditions. The fitting curve converges and it may only refer to a particular category or even a batch of batteries and not widely right. Moreover, the SOC induces a subsequent change in the OCV in terms of the lithium-ion batteries functional theory, rather than the latter being the first to be calculated.

2.8.3 Binary iteration treatment

The EEC implementation process is the basis for a detailed battery state determination. The model of the battery will adequately define the relationship between the condition and different battery parameters. Meanwhile, the influence on variables such as temperature, age, and other variables should be taken into consideration. As a research goal, a ternary lithium-ion battery is considered. The Kalman filtering algorithm is generalized, and the status calculation can be carried out, as shown in Figure 2.7.

The battery model parameters are tested for different SOC points, and an accurate battery equivalent model is established utilizing a complete composite pulse-power experiment. On this basis, the algorithm is used to approximate the value of the state. To establish the algorithm based on high-order equivalent model circuit used to evaluate the exact battery characterization. As a selectable model-based battery system, the EKF can handle the battery well. The lithium-ion battery carries a nonlinear behavior, which can realize its accurate estimation.

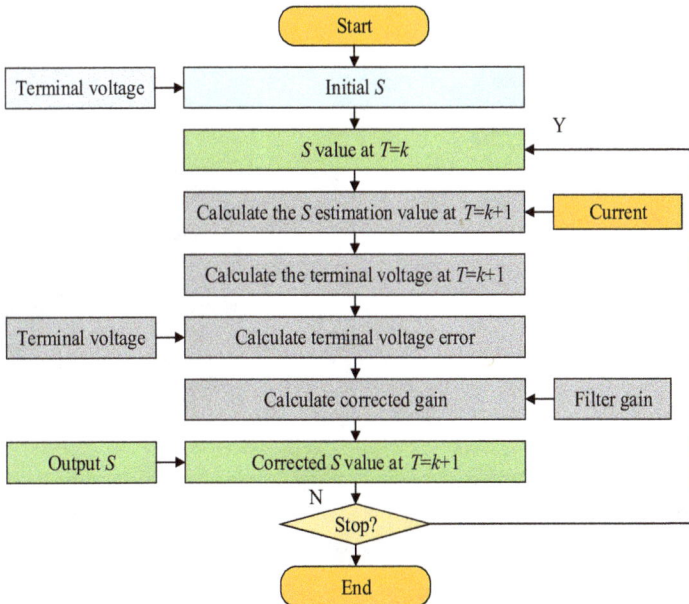

Figure 2.7   Iterative SOC estimation technique

2.9   Chapter summary

Battery modeling and SOC estimation are fundamental aspects to improve the reliability of electric vehicles for the BMS. This chapter outlines operation and mechanism strategies for lithium-ion batteries. After then, several estimation processes and detailed descriptions of these techniques are introduced. This chapter also discusses different mathematical analyzes of the identified model parameters to ensure accuracy, parameterization, and reliability. The accurate parameter identification to determine the battery state is the focus. Furthermore, the EKF algorithm is discussed comprehensively, in which the mathematical expression and model construction are also presented.

Acknowledgment

The work is supported by the National Natural Science Foundation of China (No. 61801407), Sichuan Science and Technology Program (No. 2019YFG0427), China Scholarship Council (No. 201908515099), and Fund of Robot Technology Used for Special Environment Key Laboratory of Sichuan Province (No. 18kftk03).

Chapter 3

Equivalent modeling, improvement, and state-space description

Abstract

It is vital to establish an accurate battery model for the characteristic analysis and performance optimization of the batteries. This chapter introduces several popular modeling strategies of batteries, including Rint, partnership new generation of vehicles, and Thevenin modeling methods. According to the different internal characteristics of various batteries, four kinds of battery modeling methods are studied, including the electrochemical model, mathematical model, thermal-based model, and shepherd model. The advantages and disadvantages of these models are discussed in detail through theoretical analysis and open-circuit voltage characteristics. Meanwhile, several improvement measures are investigated for the Thevenin equivalent circuit modeling. Their benefits and limitations are also compared. Through the demonstration of simulated models, the basis for accurate battery state of charge estimation is constructed. Finally, different parameter identification methods for various battery models are introduced in detail and verified with the Beijing bus dynamic stress test. The results show that the model parameters can be well identified.

Keywords: Equivalent circuit modeling; Battery modeling strategy; Electrochemical model; Mathematical model; Thermal-based model; Shepherd model; Improved Thevenin equivalent modeling; Improved equivalent circuit modeling; Model parameter description

3.1 Introduction

The internal resistance characteristic of the lithium-ion battery is highly nonlinear. Therefore, effective modeling techniques are very important for understanding its dynamic characteristics. Researchers have already designed various kinds of battery modeling techniques, which are essential for their implementation to be realized [132]. Some of these models are interrelated, but they are very different from each other. However, all of them have a different tradeoff between accuracy and complexity [133]. The method proposed in this section is also a model-based method, which includes a very popular battery modeling technology with low complexity and accuracy [134]. Different types of battery modeling techniques are discussed as well as the mathematical state-space expression methods.

3.1.1   Application background

Models can help explain the system, describe the effects of different components, and predict their behavior. The battery models can be typically divided into several categories, which differ with the level of abstraction they assume concerning the physical nature of the battery [135]. The battery models can be generally organized into various categories, such as electrochemical models, mathematical models, and empirical models. Each category of battery model contains several modeling types, which present different characteristics and complexity [136].

The application of lithium-ion batteries produces complex chemical reactions, and the environment is easy to affect the reaction process. As for the battery state calculation, the offline experimental data fitting function is usually used to evaluate the values of different battery model parameters [137]. However, using this method may lead to significant errors in prediction performance. To improve the state estimation accuracy, online detection of model parameters and their real-time correction is crucial. Battery simulation has been widely used in the field of state of charge (SOC) estimation. However, the relationship between resistance–capacitance (RC) and electrical equivalent circuit (EEC) modeling accuracy needs further clarification [138]. Some publications also attempted to clarify the correlation between RC network numbers and EEC modeling accuracy. In addition, the relationship between these various modeling methods is discussed, as well as the opposition and negation of each method [139].

3.1.2   Modeling principle

The battery simulation model can be used to verify the accuracy of parameter setting in the model, so its input is current and its output is the terminal voltage. In the battery management system (BMS), both current and terminal voltage are input [140]. The electrochemical model is complex and difficult to be applied in practical products, which is mainly used to assist in the design and manufacture of batteries [141]. The intelligent mathematical model can be the neural network model, which can theoretically complete its equivalent modeling but needs a large amount of data. Its practical application is restricted by the training, high technical threshold, and long processing time.

The application of lithium-ion batteries is becoming more and more extensive. According to the needs of various power supply applications, researchers have conducted a comprehensive study on battery models, including electrochemical model, thermal model, coupling model, and performance model. Based on the electrochemical theory, the electrochemical model describes the reaction process of the battery in a mathematical way. The thermal model describes the heating and heat transfer process of the battery. The coupling model interacts with the electrochemical process, heat generation, and heat transfer process. This type of modeling describes the external characteristics of the battery during operation.

The lithium-ion battery exhibits polarization and hysteresis effects during application, and this effect is consistent with the characteristics exhibited by the circuit resistance model. This is the basis of the EEC modeling and state-space expression that can be performed on the battery. It is a battery performance model that uses circuit components, such as resistors, capacitors, and voltage sources, to form circuits that simulate the dynamic characteristics of the battery [43,44]. The equivalent circuit model is easy to be analyzed by using circuit and mathematical methods, which has become the most widely used battery model. Typical examples include the Rint model, RC model, Thevenin model, and partnership new generation of vehicles (PNGV) model.

3.1.3   Modeling types and concepts

In the lithium-ion battery SOC estimation process, establishing operating state and properties is very important for the equivalent model when simulating the battery characteristics. The SOC estimation accuracy largely depends on the characterization degree of the dynamic battery factors by equivalent modeling.

Lithium-ion batteries undergo complex chemical reactions during the application, and their reaction process is easily affected by the working environment. As for the battery state estimation, the offline experimental data fitting function is usually used to determine the values of various parameters in the battery model. However, using this method will lead to large errors in the estimation results. To improve the state estimation accuracy, online identification of model parameters and their real-time correction is particularly important. Based on the second-order resistance–capacitance equivalent circuit modeling, this chapter discusses the identification methods and results of model parameters by using the forgetting factor recursive least-square (RLS) algorithm.

The mathematical model is based on theoretical state-space expressions, such as the physical equations and chemical reaction principle. The controlled object is regarded as a black box, and the model is constructed experimentally, recording the variation law of the target object characteristic parameters. Combining mechanism modeling with experimental testing, the model can be established. As the battery is a highly complex nonlinear electrochemical energy storage device, it is difficult to explain the interaction and reaction of the controlled correcting process by mathematical equations. Data modeling generated by experiments, such as neural networks, requires a lot of data input and learning, so hybrid modeling is more commonly used.

To obtain the battery state conveniently and accurately, it is necessary to establish an appropriate battery model for obtaining the effective scheme. (1) The internal resistance equivalent model is used, in which the ideal voltage source is introduced to express the OCV characteristics effectively. Both the internal resistance R_0 and the open-circuit voltage U_{OC} are all functions related to the SOC levels. The charge–discharge direction can be determined only by determining the positive and negative current. (2) The resistance–capacitance equivalent model can be obtained by using the parallel-connected RC circuit that can describe the surface effect of the circuit effectively [41,142–144]. The improved EEC model consists of three resistors and two capacitors, in which C_a is the battery capacity and C_b is the small capacitance produced by its surface effect. (3) The Thevenin equivalent model can be obtained by combining the internal resistance model and the resistance–capacitance model analog circuit, which can take the effects of temperature and polarization into account. (4) According to the battery test manual, the PNGV equivalent model is also proposed. The circuit is connected serially with the capacitance C_b. R_0 is the internal resistance, R_p is the polarization resistance, and C_p is the polarization capacitance. The load current is described by $I(t)$ and U_L is the terminal voltage.

Among the above four commonly used equivalent modeling types, although the internal resistance equivalent model is convenient and fast to measure U_{OC}, it does not consider the transient characteristics of the electrochemical reaction process of the battery, nor can it accurately describe the battery changing process. Although the resistance–capacitance equivalent model can better simulate the dynamic characteristics of the battery and increase the description of the battery surface effect, it ignores the influence of temperature and battery polarization. The PNGV model has high accuracy due to the self-discharge effect, but introducing the series-connected capacitance C_b makes this method prone to cumulative error in long-time simulation. The Thevenin model can overcome the error of the polarization effect. The steps are short, and the principle is clear. It is suitable for the transient power battery charge–discharge analysis. Compared with the PNGV model, general non-linear (GNL) model, and other models, the Thevenin model has a simple structure belonging to the nonlinear low order model. It involves fewer parameters, and its accuracy can meet the requirements of engineering applications.

3.1.4   Model building principle

The equivalent models used exclusively for battery simulation check whether the parameter settings of the device are correct so that its input is current and the output is the terminal voltage. Both current and terminal voltage are taken as input parameters for the BMS. In practical devices, which are mostly used to support the design and produce batteries, the electrochemical model is much more complex and challenging to adapt. The intelligent mathematical model is primarily a neural network model, which is theoretically capable of completing battery modeling but requires a lot of data. Its practical application is restricted by preparation, high technological thresholds, and long working hours. The model focuses on the sample data and the identification method of battery dynamic equipment. The established principle of the model building process is shown in Figure 3.1.

According to different requirements, various battery equivalent models are selected, which have a significant impact on the battery modeling accuracy. The more complex the structure, the higher the precision of the model, and the more difficult it is to be operated. The corresponding hardware specifications have been strengthened. The equivalent circuit with the resistance–capacitance loop may have higher accuracy in the corresponding model. When using the high-order resistance and capacitance equivalent circuit model, the measurement accuracy is high.

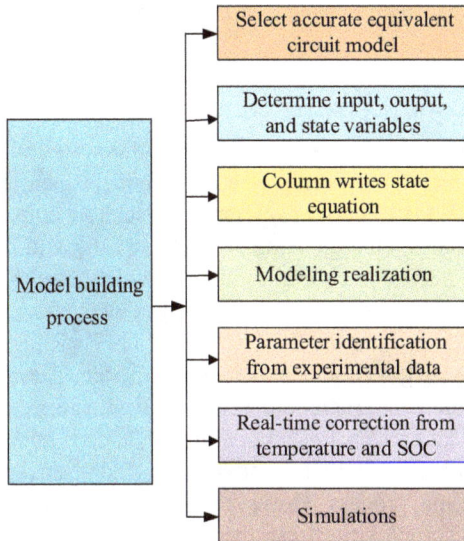

Figure 3.1 Dynamic established model building system

3.1.5 Battery modeling methods

In the BMS, battery modeling based on electrical equivalent treatment is the most commonly used technology. Compared with modeling based on electrochemistry and heat, this model is easier to understand and realize. The core analytical part of this modeling technique is based on the equivalent circuit theory. The model parameters are also well known, such as resistance, capacitance, current, and voltage. The accuracy of the battery model based on electricity largely depends on the voltage and current characteristics. The further classification of these modeling techniques has also been described by different researchers as runtime-based electrical models, Shepherd models, and Thevenin-based electrical models.

Based on the advantages of the lithium-ion battery, it proposes a new battery model that intuitively describes the relationship between the external character-istics of the battery voltage, current, and temperature. Then, internal state quantities are used to establish the mathematical model, such as resistance and electromotive force. A relational expression calculates the amount of internal state indirectly based on the external characteristics of the battery. The battery equivalent modeling is the basis of the state estimation. Its accuracy directly affects the result of state estimation.

At present, battery models can be divided into different types according to research mechanisms and research objects. The most commonly used methods are electrochemical models and equivalent circuit models. The equivalent circuit model uses the ideal voltage source, resistance, capacitance, and other electrical components to form a circuit model for simulating the external characteristics of the power battery. After years of research, equivalent circuit models of power batteries have been gradually formed, such as internal resistance (Rint), PNGV, Thevenin, and GNL models.

3.1.6   Modeling characteristic comparison

In a practical model-based estimation approach, the selection of the appropriate battery models is essential. Several electrical models have been proposed for accurate simulation of the battery characteristics, such as empirical models [145], equivalent circuit models, electrochemical models [146], and data-driven models. In these models, the electronic control module is widely used for simulation of the battery dynamic behavior, comprising a current–voltage response under certain conditions. The EECs include the PNGV model [147], Rint model [148], GNL model, Thevenin model [149], and higher-order model. These models consist of essential circuit components, including sources of voltage, resistances, and capacitances. The classification of battery modeling methods is presented in Figure 3.2.

As discussed in Figure 3.2, various advantages and disadvantages can be found in the battery modeling process. However, these advantages and disadvantages bring suitability and complexity to use the modeling methods. Based on the validation and characterization process inside the reaction of batteries, some advantages and disadvantages are described in Table 3.1.

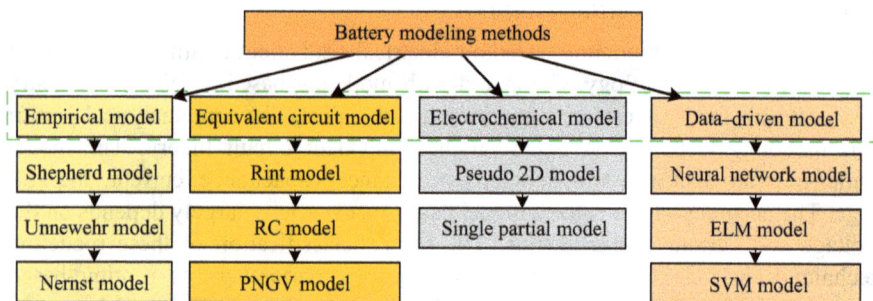

Figure 3.2   The classifications of battery modeling methods

Table 3.1   Advantages and disadvantages of the modeling methods

Modeling methods	Advantages	Disadvantages
Empirical model	Simple expression and computational efficiency	Limited capability of describing the terminal voltage
Equivalent circuit model (EEC)	Easily understand and widely used in SOC estimation	The complex parameter identification process
Electrochemical model	High accuracy of voltage calculation	Required prior knowledge of the batteries
Data-driven model	High accuracy of voltage calculation	Laborious training dataset collection process

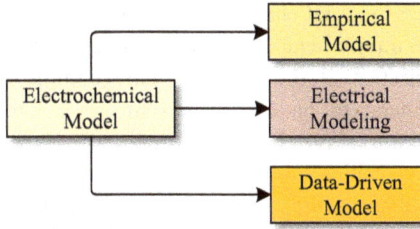

Figure 3.3   Connection procedure of the modeling system

Different battery modeling methods have some basic connections and describe the specific reactions inside the lithium-ion battery. Compared with statistical, electrochemical, and electrical models, the model is easy to use. Usually, all-electric systems have equivalent circuits consisting of passive components including resistors and capacitors, possibly inductors, and voltage sources. Therefore, it is especially suitable for circuit simulators. For many applications, the precision achievable with these models in terms of friction, current, and charge conditions is exactly appropriate. The connections between the four modeling methods are discussed in Figure 3.3.

The equivalent circuit of the battery model can be directly used in the production of the BMS and vehicle power controlling [150]. A good battery model should forecast the battery capacity details and the voltage response to the load. These characteristics of lithium-ion batteries can be represented by an analogous circuit model. In certain situations, it is necessary to model the side effects of battery loss [98], which can also be done by EEC modeling.

3.2   Electrochemical modeling

Electrochemical-based models describe the fundamental characteristics of batteries such as voltage and current. One of the earliest introductions to the electrochemical-based battery model had come from Doyle *et al.* [334]. Their research literature is an important contribution to physical cell designing and quantification of macroscopic battery variables. The main drawback of the electrochemical-based battery modeling methods is that it requires a variety of cell parameters and the calculation process is also very complex. The implementation of these battery modeling methods is conducted in the simulation research on the behavior of electrical terminal terminals, and this modeling method is particularly not suitable for the estimation method of the SOC or the state of health (SOH).

3.2.1   Electrochemical modeling

The original intention of the electrochemical model is to design the battery structure. Through the analysis of macroscopic data onto the battery and the internal microscopic particle activity, the internal state of the battery is obtained and combined with the battery mechanism to generate energy. The model primarily represents the state of the internal chemical reaction to the battery. Electrochemical models can analyze the internal characteristics from a microscopic perspective. To express these characteristics of the battery, many complex time-varying partial differential equations need to be established in the electrochemical model. It usually takes a long time to solve these equations.

The structure of the lithium-ion battery is generally composed of five parts, namely, the positive electrode, negative electrode, electrolyte, diaphragm, shell, and electrode lead. The positive electrode, negative electrode, electrolyte, and diaphragm are closely related to the electrochemical reaction of the battery. Cathode materials can be regarded as spherical lithium compound particles immersed in an electrolyte, and lithium ions can escape and embed from these spherical particles in the charge–discharge process. The membrane separates the anode from the cathode, and the pore tissue allows lithium ions to travel through it, thus providing the effect of ionic conductivity.

As can be known from the perspective of electrochemistry, it is the difference in the distribution of lithium ions in the positive electrode, negative electrode, electrolyte, and membrane region that forms the potential difference. Therefore, the lithium-ion secondary battery can be regarded as the lithium-ion concentration difference battery. During the charging process, lithium ions escape from the positive lithium-ion compound particles that are embedded into the negative lithium-ion compound particles through the electrolyte and membrane, so that the negative electrode is in the lithium-rich state and the positive electrode is in the lithium-poor state. At the same time, the compensation charge of the electron is supplied to the carbon anode from the external circuit to maintain the charge balance.

On the contrary, lithium ions escape from the negative lithium-ion compound particles and are embedded in the positive particles through electrolyte and membrane, and the positive electrode is in a lithium-rich state. Under normal charge–discharge conditions, the interbedding and escape of lithium ions between the layers of carbon materials and the layers of oxides can only cause the change of the layer spacing without damaging the crystal structure, and the chemical structure of the anode material is unchanged. Therefore, the working voltage of the lithium-ion battery is related to the lithium-ion embedded compounds that make up the electrode and the lithium-ion concentration.

The original intention of the electrochemical model is to design the battery structure of the macroscopic data onto the battery and the internal microscopic particle activity analysis. Combined with the basic mechanism of the battery, the energy is generated and the internal state of the battery can be obtained. The model mainly reflects the chemical reaction state of the internal battery. Electrochemical models can analyze the internal characteristics of batteries from a microscopic perspective; however, to express these characteristics of the battery, electrochemical models usually need to establish multiple sets of complex time-varying partial differential equations. Due to the complexity, solving these equations usually takes a lot of time.

The electrochemical model aims to construct the battery structure by analyzing macroscopic battery data and internal microscopic partial operation. The internal state is obtained and coupled to the energy generation phase of the battery [151]. The model is primarily the chemical reaction state of the internal battery. Electrochemical models shall analyze the characteristics of internal batteries from a microscopic point of view to express specific characteristics of the battery [152]. Electrochemical models usually involve many sets of complex time-variable partial differential equations. Generally speaking, these equations take a long time to solve.

The model includes a wide range of cell parameters and sophisticated numerical computational methods. Regardless of the accuracy of these models, the simulation environment cannot accept them. The battery's electrical terminal behavior and charging conditions will be calculated according to the actual calculation time. In the law of electrolysis, the pore-wall flux J_i is related to the divergence of current flow in the electrolyte phase. The pore-wall flux is calculated according to the following equation:

$$J_n(t) = \frac{I(t)}{F \cdot S_n'}, J_p(t) = \frac{I(t)}{F \cdot S_p'} \tag{3.1}$$

where n denotes the negative electrodes and p represents the positive electrodes. Ohm's law shows the distribution of potential between the electrolyte and the active substance. The distribution law of Fick shows the relationship between concentration and diffusion, which is capable of defining the diffusion of both the electrolyte and the electrode. The Butler–Volmer equation shows the effect of the electrode potential for the current of the electrode.

3.2.2 Mathematical Shepherd modeling

In social sciences, natural sciences, and engineering, mathematical models are used to describe physical systems using mathematical concepts and languages. Mathematical models can take many forms, including dynamical systems, statistical models, differential equations, and game-theoretic models. Mathematical models may include logical models, and in many cases, it is so. The quality of science depends on the degree of agreement between the mathematical model established in theory and the results of repeatable experiments. Inconclusive results and concord between theoretical mathematical models and experimental measurements often lead to important advances as better theories are developed and discovered.

Shepherd is one of the oldest battery modeling techniques. It is first described in 1965 and improved by Tremblay *et al.* [153]. This method is also well known as the empirical modeling technique. Shepherd models represent the battery terminal voltage as a mathematical function of SOC and current as below:

$$y_k = E_0 - Ri_k - \frac{K_1}{z_k} \tag{3.2}$$

Equation (3.2) shows the model equation of the Shepherd model, where y_k (V) is the terminal voltage, E_0 (V) is OCV, R (Ω) is the internal resistance, K_1 (Ω) is the polarization resistance, i_k (A) is the instantaneous current, and Z_k is the SOC.

3.2.3   *Electrochemical thermal modeling*

Thermal-based battery modeling is one of the commonly used battery modeling techniques for hybrid and electric vehicle applications. One of the earliest introductions to thermal-based battery molding had come from Pals and Newman *et al.* [154]. Like the electrochemical-based model, it also requires a complex parameter calculation process. In this method, the battery pack needs to be installed in a temperature control chamber to maintain the constant temperature effect of the battery. This method only classifies the environmental ambient temperature effect and considers that the internal temperature will not affect the model behavior.

The existing thermal models of lithium-ion batteries can be divided into the zero-dimensional thermal model, one-dimensional thermal models, two-dimensional thermal models, and three-dimensional thermal models according to their dimensions. The calculation amount of the zero-dimensional thermal model is small, but only considering the battery temperature is not enough, as the inner temperature variation cannot be measured in real time. One-dimensional, two-dimensional, and three-dimensional thermal models can accurately describe the real-time temperature field, but the huge complexity of the corresponding models limits its application of online control management.

According to the principle, the existing thermal models of lithium-ion batteries can be divided into the electrochemical–thermal coupling model, electro-thermal coupling model, and thermal runaway model. The electrochemical–thermal coupling model is a coupling model of electric energy, chemical energy, and thermal energy. It relates the internal chemical reaction of the battery with its external temperature performance, and its theoretical basis is energy conservation. There are two kinds of electrical–thermal coupling models: uniform parameter electro-thermal coupling model and distributed parameter electro-thermal coupling model. The thermal runaway model is a thermal model that adds reaction heat under abnormal working conditions based on the conventional charging and discharging thermal model.

At present, most of the commonly used thermal models in the world are constructed based on heat transfer mechanisms, and the research on heat transfer mechanisms focuses on the heat generation mechanism of batteries. In the field of control, thermal models are often established based on the Bernardi heat generation model and energy conservation theorem or Kirchhoff's law.

3.3   Electrical equivalent modeling

New battery models have been constructed for the advent of lithium-ion batteries to visually characterize the mathematical relationship between external characteristics such as battery voltage, current, and temperature and internal state quantities such as SOC, SOH, state of power (SOP), resistance, and electromotive force. The lithium-ion battery equivalent modeling is the basis of SOC estimation, and its accuracy will directly affect the estimation accuracy. Battery models can be divided into different types according to the research mechanism and research objects.

The most commonly used modeling types can be divided into the electrochemical model and equivalent circuit model. To address the critical behavior of lithium-ion batteries, this section briefly discusses three main battery models. The internal resistance model, the Thevenin model, and the PNGV model presented the benefit of measuring state-space characteristics. However, these models have a low statistical numerical analysis that allows for gaining parameters.

3.3.1   Equivalent circuit modeling

The equivalent circuit model uses basic circuit components to form a specific circuit network to characterize the operating characteristics of circuits. It establishes the relationship between the external characteristics of the battery of operation and its internal state. The model is more intuitive, easy to handle, moderate in computation, suitable for simulation experiments with circuits, and the parameters of the model are easy to identify. It is therefore widely used in practical engineering applications.

The equivalent circuit model uses circuit parts to shape a particular circuit network to describe the circuit's operating characteristics. This model establishes the relationship between the external characteristics of work and the internal state of the battery [155]. The equivalent circuit model is more intuitive, easy to process, and mild in computing. The model parameters are easy to define and are ideal for simulation studies with circuits. Analog circuit models are also commonly used in functional engineering applications.

The design of the battery equivalent model consists of two processes, namely, theoretical and experimental analyses. The theoretical study is based on the interpretation research purpose of internal law, which deduces the complex equation of the law of change in the entity. The experimental research has to capture the input and output signals to the object [156]. The benefit is that the model has high accuracy and can show the battery characteristics evolution in detail. The downside is that there is a high complexity level of the model, which is not useful for its engineering usage.

The logical structure of the circuit model should be defined as follows. The different configuration processes of battery models can be split into a basic electrochemical model, an intelligent mathematical model, and an analog circuit model. The equivalent circuit model is currently commonly used because of its exact physical meaning and necessary mathematical expression [157,158]. Since the battery characteristics, such as current rate, SOP, SOC, and temperature, are nonlinear, the computational complexity is increased if the battery modeling has to take full account of these variables. The general controller cannot comply with the specifications. The modeling establishment critical points can be extracted from its mathematical description and its conceptual structure.

3.3.2   Internal resistance modeling

The Rint model is a more common equivalent circuit battery model, which can be formed by connecting an ideal voltage source U_{OC} in series with an equivalent internal resistance R_0. The voltage source U_{OC} represents the electromotive force of the battery, which can be obtained by measuring the terminal voltage of the battery after the battery is in an idle state and after sufficient standing. The resistance R_0 represents the DC internal resistance of the battery, which can be tested by the battery load. The terminal voltage and the current flowing through the battery loop are calculated as

$$R_0 = (V_{OC} - V_L)/I \tag{3.3}$$

where V_L is the terminal voltage of the model and I is the current flowing through the model. When this model is applied, the accuracy of the model is generally insufficient due to the consideration of the static characteristics of the battery, and the dynamic characteristics exhibited inside the battery are neglected, so it is not suitable for battery modeling. The internal resistance model is a more common battery model, which consists of an ideal voltage source U_{OC} in series with an equivalent internal resistance. The U_{OC} represents the electromotive force of the battery, which can be obtained by measuring the terminal voltage of the battery after the battery is in an ideal state and after sufficient shelved [156]. The resistance R_{int} represents the direct-current battery internal resistance, and the battery can be tested when it relates to the load. The simple Rint battery model is illustrated as shown in Figure 3.4.

The internal resistance R_{int} reflects the energy losses that allow the batteries to heat up. The terminal voltage U_L can only be triggered with the U_{OC} when the circuit is at open-circuit conditions. When the battery relates to the electrical load, the terminal voltage can be obtained as

$$U_L = U_{OC} - (I_L \cdot R_{int}) \tag{3.4}$$

where U_{OC} and R_{int} are considered constant, and the battery capacity is considered infinite. Parameter I_L characterizes the current flowing through the model. When this model is applied, the accuracy of the model is generally insufficient due to the consideration of the shelved characteristics of the battery, and the dynamic characteristics exhibited inside the battery are neglected, so it is not suitable for the battery modeling.

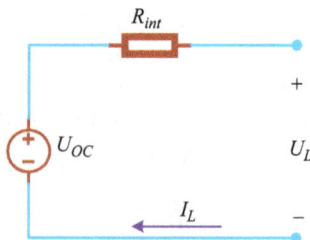

Figure 3.4   Schematic diagrams of Rint model

3.3.3 Resistance–capacitance modeling

The RC model consists of two capacitors and three resistors [159,160], in which the large capacitor C_b describes its ability to store energy. The capacitance C_s describes the surface capacitor and diffusion effect of the battery. R_t is the termination resistance, R_s is the surface resistance, and R_e is the end resistance. The model structure is shown in Figure 3.5.

Taking the voltage V_{Cb} and V_{Cs} on the two capacitors as the state variables, the terminal current I_L as the input variable, and the terminal voltage V_L as the output variable, equations are listed according to Kirchhoff's law, to obtain the state-space equation of the RC model, as shown below:

$$
\begin{cases}
\begin{bmatrix} V_{C_b}* \\ V_{C_s}* \end{bmatrix} = \begin{bmatrix} -\dfrac{1}{(R_e+R_s)C_b} & \dfrac{1}{(R_e+R_s)C_b} \\[3mm] \dfrac{1}{(R_e+R_s)C_s} & -\dfrac{1}{(R_e+R_s)C_s} \end{bmatrix} \begin{bmatrix} V_{C_b} \\ V_{C_s} \end{bmatrix} + \begin{bmatrix} \dfrac{R_s}{(R_e+R_s)C_b} \\[3mm] \dfrac{R_e}{(R_e+R_s)C_s} \end{bmatrix} I_L \\[10mm]
V_L = \begin{bmatrix} \dfrac{R_s}{R_e+R_s} & \dfrac{R_e}{R_e+R_s} \end{bmatrix} \begin{bmatrix} V_{C_b} \\ V_{C_s} \end{bmatrix} + \begin{bmatrix} \dfrac{R_s+R_e}{R_e+R_s} + R_t \end{bmatrix} I_L
\end{cases}
$$

$$(3.5)$$

3.3.4 Electrical modeling effect comparison

Designing a trustworthy battery model that represents the accurate battery internal characteristic is only possible by the Thevenin-based EEC technique [161]. Thevenin-based models are a very straightforward and typical approach to identifying the dynamic nature of the lithium-ion battery. Generally, they contain a voltage source, an ohmic resistance, and an RC network. The RC network illustrates the electrical contact between the inputs and the terminal voltage. Compared with complexity, computational burden, and accuracy, there should be a good trade-off between different models. The basic structure of some widely used

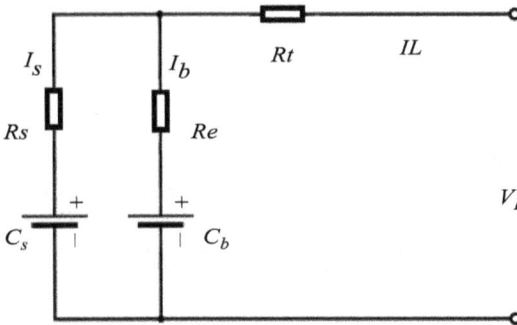

Figure 3.5 RC battery model

electrical equivalent models with a fundamental circuit diagram and OCV equation is described briefly, as shown in Table 3.2.

As can be seen from Table 3.2, the Rint model consists of only a single voltage source U_{OC} and an ohmic resistance R_0. I denotes the emitting current and R_0

Table 3.2   Thevenin-based models and related mathematical expression

Battery models	Mathematical expression
	Known as the Rint model Mathematical equation: $U_t = U_{OC} - IR_0$ Here, U_t (V) is the terminal voltage, U_{OC} (V) is the open-circuit voltage, R_0 (Ω) is the internal resistance, and I (A) is the current
	Known as the first-order Thevenin model Mathematical equation: $U_t = U_{OC} - U_1 - IR_0$ Here U_t (V) is the terminal voltage, U_{OC} (V) is the open-circuit voltage, U_1 (V) is th RC network voltage, R_0 (Ω) is the internal resistance, and I (A) is the current
	Known as the PNGV model Mathematical equation: $U_t = U_{OC} - U_{cap} - U_1 - IR_0$ Here U_t (V) is the terminal voltage, U_{OC} (V) is the open-circuit voltage, U_{cap} (V) is the bulk capacitor voltage, U_1 (V) is the RC network voltage, R_0 (Ω) is the internal resistance, and I (A) is the current
	Known as the GNL model Here, U_t (V) is the terminal voltage, U_{OC} (V) is the open-circuit voltage, U_{cap} (V) is the bulk capacitor voltage, U_1 (V) is the first RC network voltage, U_2 (V) is the second RC network voltage, R_0 (Ω) is the internal resistance, and I (A) is the current

denotes the internal resistance of the battery. In the Rint model, the battery is equivalent to the series of the ideal voltage source U_{OC} and the resistance R_0. This is the simplest representation of Thevenin-based battery models. In the update of the Rint model with an added RC network, the first-order Thevenin model is carried out. It considers the similar characteristics of battery and capacitor. In the model, the ideal voltage source U_{OC} describes the battery OCV, the resistor R_0 is the internal resistance of the battery, and the capacitor C_p and resistor R_p are in parallel to describe the overpotential of the battery.

This updated model can abduct the changing terminal voltage of the battery with more accuracy. However, the divergence effects of OCV reduce in this way. That is why researchers introduced the PNGV model with an additional capacitor C_{cap} (F) connected to series with the RC networks. In the PNGV model, this C_{cap} can express the OCV deviation by the inflation of the discharging current [162]. For this reason, the PNGV model has higher accuracy than the Rint model and the first-order Thevenin model. After all, higher accuracy could be possible through considering the concentration polarization impact by updating the Thevenin model with an additional RC network. This new model is known as the GNL model.

3.3.5 Surface effect modeling

The PNGV model is proposed by the United States for New Generation of Vehicles. This model is also the standard battery equivalent circuit model specified in the PNGV battery experiment manual. The PNGV model has a clear physical meaning. Its concept reflects a relationship of two capacitors and two resistors for a next-generation vehicle comparable cycle models for lithium-ion batteries. The OCV model is distinguished by a more accurate model, as the PNGV similar circuit model requires a battery polarization and internal ohmic resistance. The PNGV electrical battery modeling circuit is shown in Figure 3.6.

The PNGV circuit equivalent model consists of U_{OC}, which is the OCV. R_p is the polarization resistance. C_p is the polarization capacitance. The ohmic resistance is described by R_{int}. U_b is the bulk capacity. The battery terminal voltage is shown as

$$U_L = U_{OC} - U_b - U_1 - I_L R_{int} \tag{3.6}$$

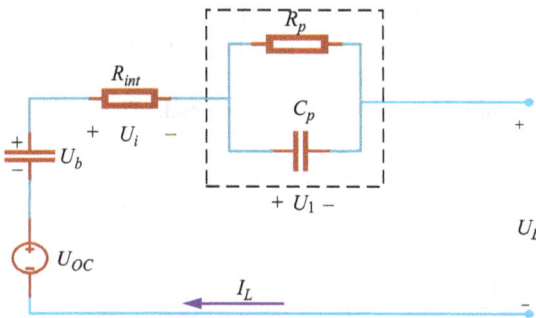

Figure 3.6 Schematic diagrams of the PNGV model

When the battery is drained, the accumulation of current and time induces a difference in the SOC estimation process. The battery OCV also varies, which is expressed in the capacitor voltage shift model. The size of the capacitor characterizes both the battery power and the direct-current battery response to compensate for the limitations of the Thevenin model. To balance the curve of the model with the calculated voltage curve, the updated model incorporates a series of resistance–capacitance circuits to achieve a better degree of alignment in the curve-fitting process. The equation is obtained according to Kirchhoff's law, and the state-space equation of the PNGV model is obtained, as shown below:

$$
\begin{cases}
\begin{bmatrix} V^*_{CO} \\ V^*_{CP} \end{bmatrix} = \begin{bmatrix} 0 & 0 \\ 0 & -\dfrac{1}{R_P C_P} \end{bmatrix} \begin{bmatrix} V_{CO} \\ V_{CP} \end{bmatrix} + \begin{bmatrix} \dfrac{1}{C_O} \\ \dfrac{1}{C_P} \end{bmatrix} I_L \\[20pt]
V_L = \begin{bmatrix} 1 & 1 \end{bmatrix} \begin{bmatrix} V_{CO} \\ V_{CP} \end{bmatrix} + R_O I_L + OCV
\end{cases}
\tag{3.7}
$$

3.4   Improved Thevenin equivalent modeling

3.4.1   Thevenin electrical modeling

The SOC value for the lithium-ion battery is the remaining capacity of the battery and is written mathematically as the ratio of remaining capacity to the maximum available capacity. This can be expressed as

$$
\text{SOC} = \frac{Q_d}{Q_s} \times 100\%
\tag{3.8}
$$

wherein Q_d represents the remaining capacity and Q_s represents the rated capacity of the battery.

The Thevenin model is one of the most used models. Based on the PNGV model, the capacitance C_b representing the OCV change is reduced, and its parameters have the same meaning as the PNGV model. It is generally believed that the OCV in the Thevenin model is varied so that the capacitance C_b in the PNGV model can be replaced by the equivalent of the battery OCV–SOC. Further, the Thevenin battery model is obtained. The voltage V_{Cp} across the capacitor C_P is set as the state variable. The terminal current I_L is set as the input variable, and the terminal voltage V_L is set as the output variable. According to Kirchhoff's law, the state-space equation of the Thevenin model can be obtained, as shown below:

$$
\begin{cases}
V^*_{CP} = -\dfrac{1}{R_P C} V_{CP} + \dfrac{1}{C_P} I_L \\[10pt]
V_L = V_{CP} + R_O I_L + OCV
\end{cases}
\tag{3.9}
$$

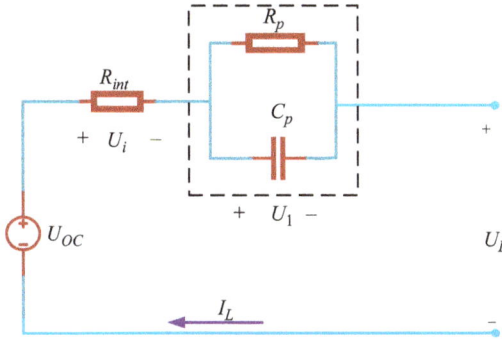

Figure 3.7   Schematic diagrams of the Thevenin model

The Thevenin equivalent circuit model is simple in modeling parameters and easy to meet the basic requirements of batteries. This accounts for its widespread use in various research involving lithium-ion batteries. The Rint model should not consider the transient behavior of the battery. The addition of the parallel resistance–capacitance (RC) branch is obtained, which allows consideration of the short-term transient due to electrolyte polarization. Similar to the Rint model, U_{OC} and R_{int} can be used to describe the OCV and electrode resistance batteries, R_P is the polarization resistance, and C_P is the polarization capacitance [163]. To improve model precision and consider-transient phenomena with various time constants, other RC divisions can be used in the Thevenin model sequence, as shown in Figure 3.7.

The parameterization process of the model becomes more complicated. The model is employed to simulate the battery behavior in one operating condition at a given SOC. In this case, the model parameters can be counted as constants. Otherwise, if wide SOC operating range must be simulated, the parameters can be considered dependent on temperature and SOC. The terminal voltage is shown as

$$U_L = U_{OC} - U_1 - (I_L R_{int}) \tag{3.10}$$

However, the above four parameters are required to parameterize the model completely by using U_{OC}, R_{int}, R_P, and C_P. All of them are SOC dependent. The requirements of the manufacturers do not contain adequate detail for this model to continue with the parameterization process. Any preliminary experiments must also be carried out to the pulse discharge test (PDT). The PDT consists of unloading a fully charged battery with a current pulse of defined amplitude and length. At the end of the pulse, the battery is left to rest in the open-circuit status. At the end of the rest time, another current pulse is added, and the process is repeated before the battery hits the cut-off voltage.

3.4.2   Second-order circuit modeling

An improved Thevenin equivalent circuit model is proposed for experimental and simulation purposes and to model the behavior of lithium-ion batteries. The

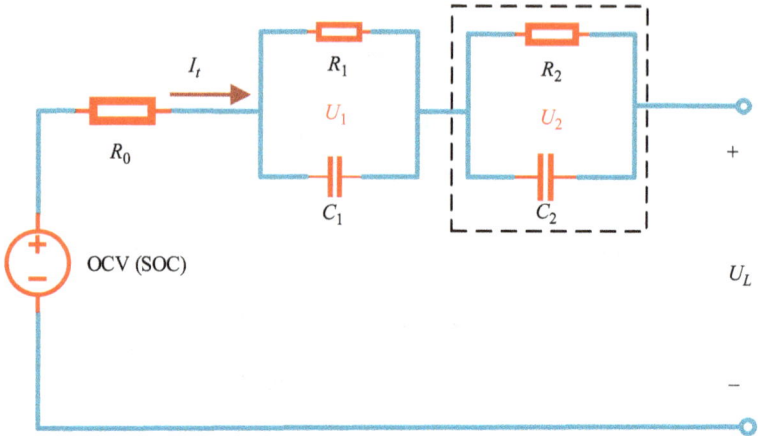

Figure 3.8   The improved Thevenin equivalent circuit model

dynamic system of the Thevenin model describes the fixed parameters in the classical model as variables that vary with the SOC and temperature, enabling a more accurate description of the battery performance. This method is used because the topology generally proposes a tradeoff between battery cell computational requirements and the approximation of voltage precision. The model is made up of a series of DC internal resistance or ohmic resistance, a voltage source, and two RC parallel circuit networks, as shown in Figure 3.8, where OCV (SOC) is the open-circuit voltage, R_0 is the ohmic resistance of the battery, I_t is the charge–discharge current flowing from the voltage source, and U_L represents the terminal voltage of the battery cell. R_1 and R_2 denote the electrochemical polarization resistance and concentration polarization resistance, respectively. C_1 and C_2 represent the electrochemical polarization capacitance and concentration polarization capacitance, respectively.

3.4.3   Dynamic high-order equivalent modeling

A precise model-based estimation approach requires the appropriate and accurate equivalent model. The dynamic high-order RC equivalent model of the battery is crafted here to imitate the original battery behavior. The dynamic high-order RC equivalent model consists of major circuit components such as the DC voltage sources, R_0 internal resistance, and three parallel RC circuits in series. The dynamic high-order circuit model is an enhancement version of the second-order resistance–capacitance equivalent model that obtains high accuracy and precision, as shown in Figure 3.9.

The dynamic high-order equivalent model is composed of OCV represented by E_m, which can characterize the change in voltage response at the time point of battery charging and discharging, a resistor R_0, and the three parallel RC networks connected in series, including R_{p1}–C_{p1}, R_{p2}–C_{p2}, and R_{p3}–C_{p3}. R_0 is the ohmic resistance caused by the accumulation and dissipation of charge in the electrical

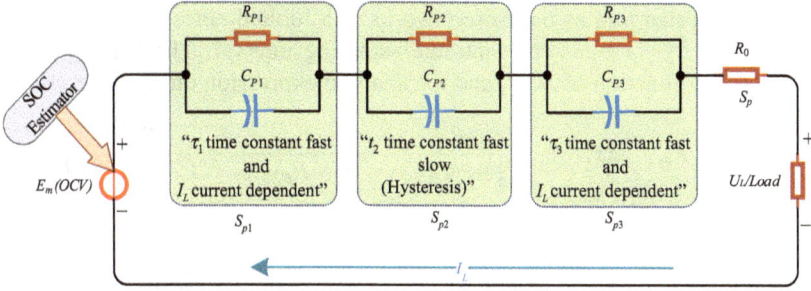

Figure 3.9   Dynamic high-order equivalent circuit structure

triple layer, R_{p1} and C_{p1} are the electrochemical polarization resistance and capacitance, respectively. R_{p2} and C_{p3} are the concentration polarization resistance and capacitance, respectively. The end of the network R_{p3} is the resistance across capacitance C_{p3}, which brings adequate accuracy on the parameter extraction.

S_{p1}, S_{p2}, and S_{p3} denote the polarization capacitance voltage of C_{p3}, respectively. I_L is the load current through the branch, which is supposed to be positive for discharge and negative for the charge. The terminal voltage is formatted as U_L. According to the high-order RC electrical behavior of the battery, the structure model can be governed by

$$\begin{cases} U_F = I_L * R_0 \\ C_{p1}\dfrac{dS_{p1}}{dt} + \dfrac{S_{p1}}{R_{p1}} = I_L \\ C_{p2}\dfrac{dS_{p2}}{dt} + \dfrac{S_{p2}}{R_{p2}} = I_L \\ C_{p3}\dfrac{dS_{p3}}{dt} + \dfrac{S_{p3}}{R_{p3}} = I_L \end{cases} \tag{3.11}$$

For dynamic high-order equivalent circuit modeling, the change of terminal voltage variation can be determined by the polarization circuit id by

$$U_L = E_m(\text{OCV}) - S_p - S_{p1} - S_{p2} - S_{p3} \tag{3.12}$$

Generally, battery SOC is defined as the ratio of the remaining capacity over the nominal available capacity, and the SOC calculated by coulomb counting can be expressed as

$$\text{SOC}\,(k+1) = \text{SOC}\,(k) - \frac{\eta \Delta t I_L(k)}{Q_N} \tag{3.13}$$

where SOC $(k+1)$ and SOC (k) are the SOC values at $(k+1)$-th and k-th sampling time, respectively. η is the coulomb efficiency that is assumed to be 1 at charging and 0.98 at discharging as the battery works in a limited current range. Q_N is the nominal capacity, and Δt represents the sampling interval. The battery terminal voltage U_L is a function of SOC, and the analytic expression on U_L is described as

$$U_L(s) = E_m + I(s)\left(R_0 + \frac{R_{P1}}{1 + R_{P1}C_{P1}} + \frac{R_{P2}}{1 + R_{P2}C_{P2}} + \frac{R_{P3}}{1 + R_{P3}C_{P3}}\right)$$

(3.14)

With a good curve fitting of the parameters, R_0, R_{P1}, R_{P2}, R_{P3}, C_{P1}, C_{P2}, and C_{P3} of the equivalent circuit model. Then, the voltage response curve can be reproduced accurately.

3.4.4   Double internal resistance modeling

The double internal resistance model is based on the Thevenin model, considering the difference between the charge and discharge processes. Great differences exist in battery characteristics during charging and discharging, mainly due to the internal structural characteristics of batteries, as shown in Figure 3.10.

The double internal resistance model uses two different internal resistances to characterize the different structural characteristics of the battery during charging and discharging.

3.4.5   Improved surface effect modeling

To better reflect the dynamic characteristics of the battery and reduce the fitting error, the polarization circuit is expanded that is based on the PNGV equivalent circuit model. The specific improvement method is to change one RC circuit into two RC circuits in the PNGV equivalent circuit model to two series RC circuits. The improved equivalent circuit model is shown in Figure 3.11.

The improved PNGV circuit model is composed of U_{OC}, which is the OCV. R_s and R_L are the polarization resistances. C_s and C_L are the polarization capacitances. The ohmic resistance is represented by R_0. C_b is the bulk capacity.

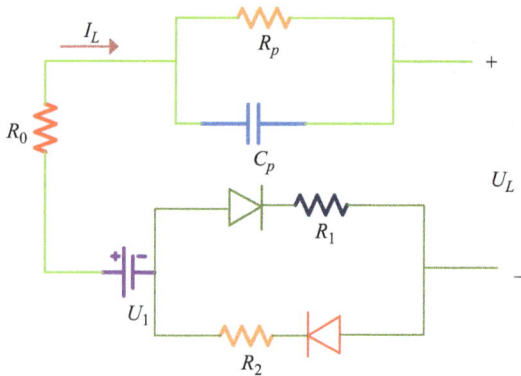

Figure 3.10   Double internal resistance model

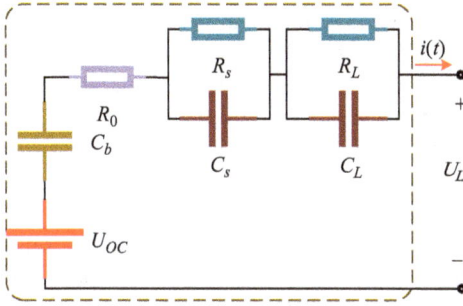

Figure 3.11   Improved PNGV circuit model

3.4.6   State-space description

The state-space representation is a mathematical model of a physical system as a set of input, output, and state variables related by first-order differential equations or difference equations. Concerning the equivalent model and the application of Kirchhoff's law, the mathematical relationship is obtained and written as

$$
\begin{cases}
U_L = U_{OC}(\text{SOC}) - i(t)R_0 - U_1 - U_2 \\
\dfrac{dU_1}{dt} = -\dfrac{U_1}{R_1 C_1} + \dfrac{i}{C_1}, \dfrac{dU_2}{dt} = -\dfrac{U_2}{R_2 C_2} + \dfrac{i}{C_2}
\end{cases}
\tag{3.15}
$$

where [SOC, U_1, U_2] is selected as the state variables which need to be realized. After the discretization and considering the definition of SOC as stated earlier, its state-space equation can be written as

$$
\begin{cases}
\begin{bmatrix} \text{SOC}_{k+1} \\ U_{1,k+1} \\ U_{2,k+1} \end{bmatrix} =
\begin{bmatrix}
1 & 0 & 0 \\
0 & 1 - \dfrac{T}{\tau_1} & 0 \\
0 & 0 & 1 - \dfrac{T}{\tau_2}
\end{bmatrix} \\
U_{L,k+1} = U_{OC}(SOC, k+1) - U_1 - U_2 - IR_0
\end{cases}
\tag{3.16}
$$

In the above equation, the parameters to be identified by the model include ohmic resistance R_0, open-circuit voltage U_{OC}, and polarization internal resistances R_1 and R_2. The polarization capacitors C_1 and C_2 will lead to identifying U_1 and U_2.

3.4.7   Simulation realization

The simulation model of the lithium-ion battery is established after obtaining the required circuit model parameters. The simulation model is mainly composed of the SOC calculation module, the circuit parameter updating module, and the terminal output voltage calculation module. The SOC calculation module is based on the EKF algorithm and the proposed AEKF algorithm. The SOC values are obtained using the codes for calculations based on the algorithms and to prevent the battery from overcharge and over-discharge.

In this work, the influence of temperature change on the output voltage of the lithium-ion battery is ignored. For the time-domain ordinary differential equation of the second-order Thevenin equivalent circuit, the corresponding voltage response equation needs to be discretized before modeling, and the discretized state-space equation of the model is obtained. The logical structure of the circuit model is shown in Figure 3.12.

In the BMS, both current and terminal voltage are inputs, as the battery simulation model is to verify the accuracy of the parameter setting in the model, the current is the input and the terminal voltage is the output. The simulation module can be built and the second-order RC internal circuit is the core part of the whole module, and the circuit structure is directly used to build the module, including an ohmic resistance, two RC parallel structures, a controllable voltage source, and a controllable current source, a voltage and current sensor, and an input and output interface, as shown in Figure 3.13.

Figure 3.13 shows the internal component of the model representing the proposed second-order Thevenin equivalent circuit model. The necessary inputs and outputs are labeled and other components are duly presented in the figure based on the proposed model, as shown in Figure 3.14.

The controllable voltage source and controllable current source are the signal interface in SIMULINK, which can turn the signal into a material port. The external input voltage source is converted into a voltage and current source that the circuit

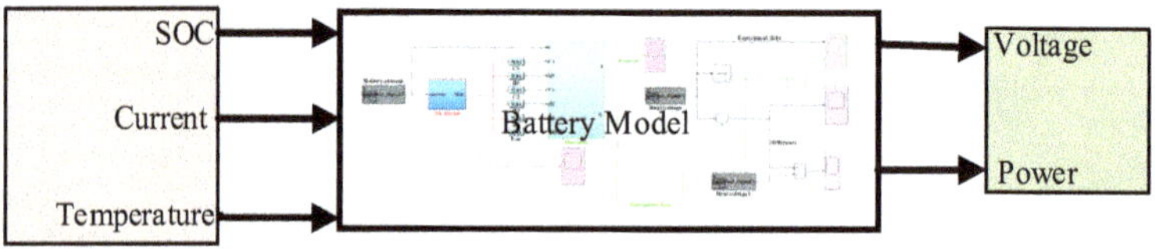

Figure 3.12   Logical structure of the circuit model

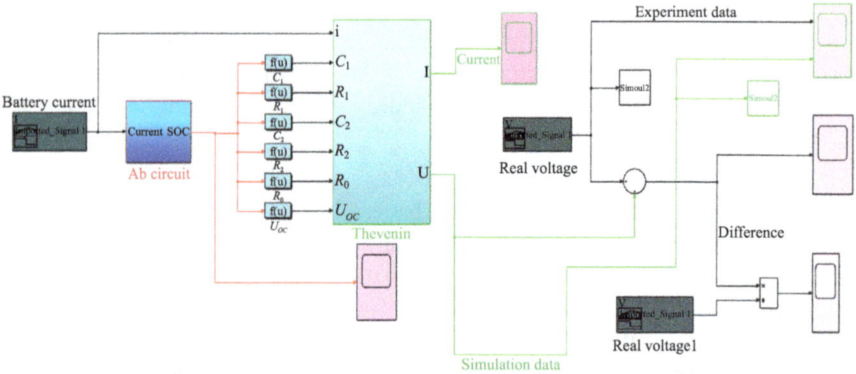

Figure 3.13   The simulation model of the second-order Thevenin equivalent model

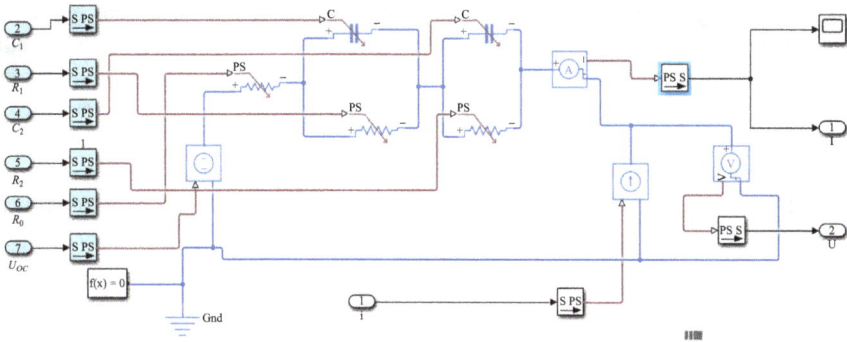

Figure 3.14   The internal circuit of the second-order Thevenin equivalent model

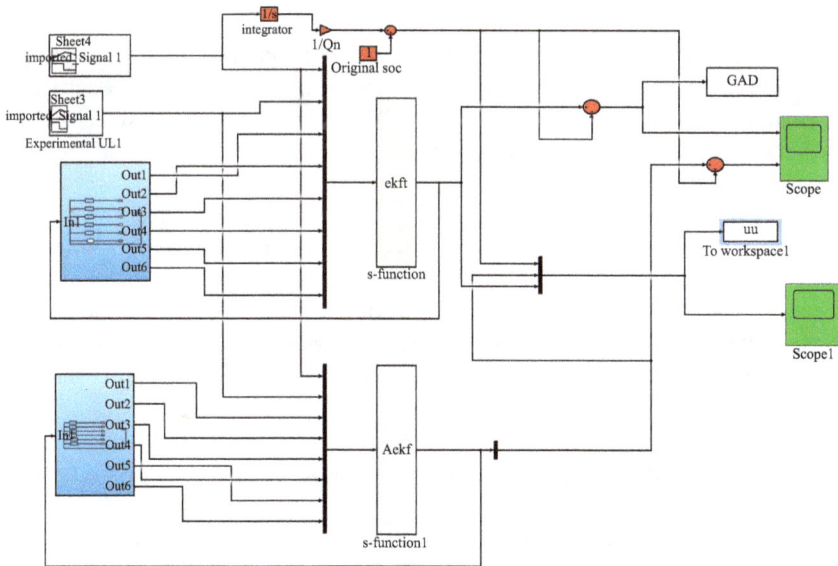

Figure 3.15   Integration of algorithms into the simulation diagram

can connect to. Voltage sensors and current sensors are also signal transducers, converting physical interfaces into signal interfaces. The EKF algorithm and the AEKF algorithm are coded into the simulation model, and the complete diagram is shown in Figure 3.15.

3.5   Improved equivalent circuit modeling

An appropriate equivalent circuit model is selected that is modeled and described, and the input, output, and state variables are determined. Using the Simulink modeling, the parameters are identified through experimental data, and real-time corrections are made according to the internal resistance, capacitance, current, and other variables of the model parameters. The simulation results verify the effectiveness of the method and provide a basis for the accurate estimation of the battery SOC value.

The SOC of the lithium-ion battery is the remaining capacity of the battery, which is mathematically written as the ratio of remaining capacity to the maximum usable capacity. This can be expressed as

$$SOC = \frac{Q_d}{Q_s} \times 100\% \tag{3.17}$$

where Q_d represents the remaining capacity and Q_s represents the rated capacity of the battery.

3.5.1   Runtime electrical modeling

The runtime-based electrical model is an updated version of the usual second-order Thevenin model. The original model of this electrical model was first described by Hagemann *et al.* [164] and it is expanded by Chen *et al.* [165]. Figure 3.16 shows the expanded runtime-based electrical model, where the model is divided into two parts (left and right).

The left side contains a capacitor C_c (Ah), a self-discharge resistor R_{dis} (Ω), and a current control current source I_{bat} (A). On the other hand, the right side of the model is similar to the second-order Thevenin model, where R_0 (Ω) is the internal resistance, R_1 (Ω) and R_2 (Ω) are the polarization resistance, and C_1 (F) and C_2 (F) are the polarization capacitance. In this case, the voltage source V_{oc} (V) is a regulated voltage source. The mathematical derivation of the battery model is shown as

$$C_C = Q_{nom} \cdot f_1(N) \cdot f_2(T) \tag{3.18}$$

Among them, C_c (Ah) is defined as a function of the nominal battery capacity Q_{nom} (Ah) and the correction coefficients $f_1(N)$ and $f_2(T)$. The main implementation of this model is to use the capacity model to calculate the impact of running time on the battery life cycle.

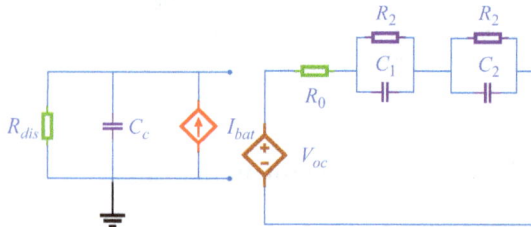

Figure 3.16   Runtime-based electrical battery model

3.5.2 Fractional-order electrical model

On the one hand, the equivalent circuit model has a simple structure, convenient calculation, and easy application in practical engineering. However, since it cannot accurately describe the electrochemical dynamics inside the battery, its accuracy is not as good as the electrochemical model. On the other hand, although the accuracy of the electrochemical model is high, the measurement costs are increased significantly due to the difficulty in calculating the complex mutual coupling in actual work. Because of the above reasons in recent years, scholars have proposed an effective combination of the electrochemical models and equivalent circuit models to improve the accuracy of the model and reduce the calculation cost.

To characterize the dynamic working characteristics and electrochemical reaction characteristics of lithium-ion batteries more accurately and improve the adaptability and robustness of the model, based on the integer-order Thevenin model and the chemical characteristics of the lithium-ion battery, the electro-chemical impedance of the lithium-ion battery is analyzed, and the dispersion effect and charge-transfer effect between the solid electrode/electrolyte interface and the double electron layer effect of lithium-ion batteries are fully considered. The constant-phase element (CPE) and semi-infinite diffusion (Warburg) elements are added, and the fractional-order equivalent circuit model corresponding to the integer-order equivalent circuit model is established, as shown in Figure 3.17.

The parallel circuit is composed of resistors and CPE, which represents the resistance and polarization capacitance of the diaphragm. Unlike the RC polarization loop in the integer-order equivalent circuit model, CPE has an attribute between resistance and capacitance, as shown below:

$$Z_{CPE1} = \frac{1}{CS^{\alpha}}, 0 < \alpha < 1 \tag{3.19}$$

wherein C is the capacitance value and α is the order of the CPE. The Warburg element is a constant-phase element with a fixed fractional order of 0.5, and its properties are like CPE, and the impedance expression is shown as

$$Z_w = \frac{1}{WS^{\beta}}, 0 < \beta < 1 \tag{3.20}$$

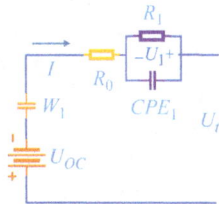

Figure 3.17 Fractional-order equivalent circuit model

where W is the capacitance value and β is the order of the Warburg element. The impedance expression of the CPE element is in the form of fractional calculus. The only discrete form of fractional calculus can be obtained by the Grünwald–Letnikov definition, as shown below:

$$D_t^p f(x) = \lim_{h} \frac{1}{h^p} \sum_{j=0}^{[t/h]} (-1)^j \binom{p}{j} f(t - j * h) \tag{3.21}$$

where $D_t^p f(t)$ represents the first-order integral–differential value of the parameter f (t), t_0 defaults by 0, h is the sampling interval, p is the number of fractional operators, and $[(t - t_0)/h]$ is the historical data point involved in the calculation. The matrix of p and j is the Newtonian binomial coefficient, which can be converted with the gamma function. The definition equation and the conversion relationship between them are

$$\begin{cases} \binom{p}{j} = \dfrac{p!}{j!(p-j)!} \\ \Gamma(x) = \displaystyle\int_0^{\infty} t^{x-1} e^{-t} dt, x > 0 \\ \Gamma(p+1) = p! \end{cases} \tag{3.22}$$

Through the transformation relationship of (3.22), the application field of fractional calculus defined by Grünwald–Letnikov can be extended from the field of positive integers to the field of the real number. Compared with the integer equivalent circuit model, the fractional-order equivalent model can more accurately characterize the electrochemical effects of lithium-ion batteries, improve the model accuracy, and increase the calculation difficulty accordingly.

3.5.3   Improved Thevenin model

Although the first-order Thevenin battery equivalent circuit model can express the state of the battery accurately, it also has some certain defects. The disadvantage of this model is that it cannot distinguish between ohmic resistance and open-circuit voltage (OCV) during charging and discharging. For this reason, an improved Thevenin model is proposed, which uses the unidirectional conductivity of the diode to distinguish between the ohmic resistance and OCV in different states. Therefore, this model can improve the accuracy of the battery model and reduce the response error of the battery in different state conditions. The improved Thevenin equivalent circuit model is shown in Figure 3.18.

Here, diodes D_0 and D_1 are used to divide the open-circuit voltage U_{OC} into the open-circuit voltage U_{OC1} in the discharging stage and the open-circuit voltage U_{OC2} in the charging stage. The ohmic resistance is divided into the ohmic resistance R_1 in the discharging stage and the ohmic resistance R_2 in the charging stage. Therefore, the expression of the circuit model in the discharge stage is

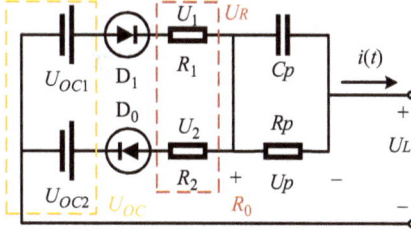

Figure 3.18 Improved Thevenin model

derived from (3.23) and (3.24). The expression of the circuit model in the charging stage is derived from

$$U_L = U_{OC1} - U_P - iR_1 \qquad (3.23)$$

$$U_P = IR_P\left(1 - e^{-t/\tau}\right) \qquad (3.24)$$

$$U_L = U_{OC2} - U_P - iR_2 \qquad (3.25)$$

$$I = \frac{U_P}{R_P} + C_P\frac{dU_P}{dt} \qquad (3.26)$$

where τ represents the time constant for U_p and the current of the circuit will be derived. The battery model is the main battery model that is used in this research work to validate the proposed battery aging effect. These equations are the fundamental equations for this battery model.

The improved equivalent circuit model is proposed for experimental and simulation purposes and is used to simulate the behavior of lithium-ion batteries. The dynamic system of the Thevenin model describes the fixed parameters in the classical model as variables that change with the SOC and temperature, which can more accurately describe the performance of the battery. The model includes a series of direct-current (DC) internal resistance or ohmic resistance, a DC voltage source, and two resistor–capacitor (RC) parallel circuit networks.

R_0 is the ohmic resistance of the battery, $i(t)$ is the charge–discharge current flowing from the voltage source, and U_L represents the terminal voltage of the battery cell. R_1 and R_2 represent the ohmic resistance in the discharging stage and the ohmic resistance in the charging phase, respectively. The open-circuit voltage is indicated in the discharge phase and charge phase, respectively.

3.5.4 State-space description

The state-space representation is a mathematical model of a physical system. It is a set of input, output, and state variables. These variables are connected by first-order differential equations or difference equations. Concerning the application of Kirchhoff's law, the mathematical analysis is conducted as follows:

$$\begin{cases} U_L = U_{OC}(\text{SOC}) - i(t)R_0 - U_1 - U_2 \\ \dfrac{dU_1}{dt} = -\dfrac{U_1}{R_1C_1} + \dfrac{i}{C_1}, \dfrac{dU_2}{dt} = -\dfrac{U_2}{R_2C_2} + \dfrac{i}{C_2} \end{cases} \tag{3.27}$$

Among them, [SOC, U_p] is selected as the state variable that needs to be realized. After the discretization, considering the definition of SOC described above, the state-space equation can be written as

$$\begin{cases} \begin{bmatrix} \text{SOC}_{k+1} \\ U_{p,k+1} \end{bmatrix} = U_{OC}(\text{SOC}) - i(t)R_0 - U_1 - U_2 \\ U_{L,k+1} = U_{OC}(\text{SOC}, k+1) - U_p - iR_0 \end{cases} \tag{3.28}$$

In the above equation, the parameters that need to be identified in the model are ohmic resistance R_0, including ohmic resistance R_1 in the discharging stage and ohmic resistance R_2 in the charging stage, open-circuit voltage U_{OC}, including open-circuit voltage U_{OC1} in the discharge stage, open-circuit voltage U_{OC2} in the charging stage, polarization internal resistance R_p, and polar capacitance C_p, which will lead to the identification of U_p.

3.5.5   Simulation realization

After obtaining the required circuit model parameters, the simulation model of the lithium-ion battery is established. The simulation model is mainly composed of the SOC calculation module, the circuit parameter update module, and the terminal output voltage calculation module. The SOC calculation module is based on the extended Kalman filtering (EKF) algorithm and the adaptive extended Kalman filtering (AEKF) algorithm. The SOC value can be obtained by using an algorithm-based calculation program to prevent the battery from overcharging and over-discharging. In this work, the influence of temperature change is ignored on the output voltage of the lithium-ion battery. For the time-domain ordinary differential equation of the improved equivalent circuit, the corresponding voltage response equation is solved, and discretization is required before modeling to obtain the discretization state-space equation of the model. The logical structure of the circuit model is shown in Figure 3.19.

To verify the accuracy of the model, the battery simulation model with current as input and terminal voltage as output is established. The calculation results are

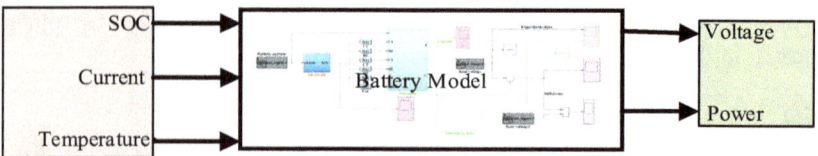

Figure 3.19   Logical structure of the circuit model

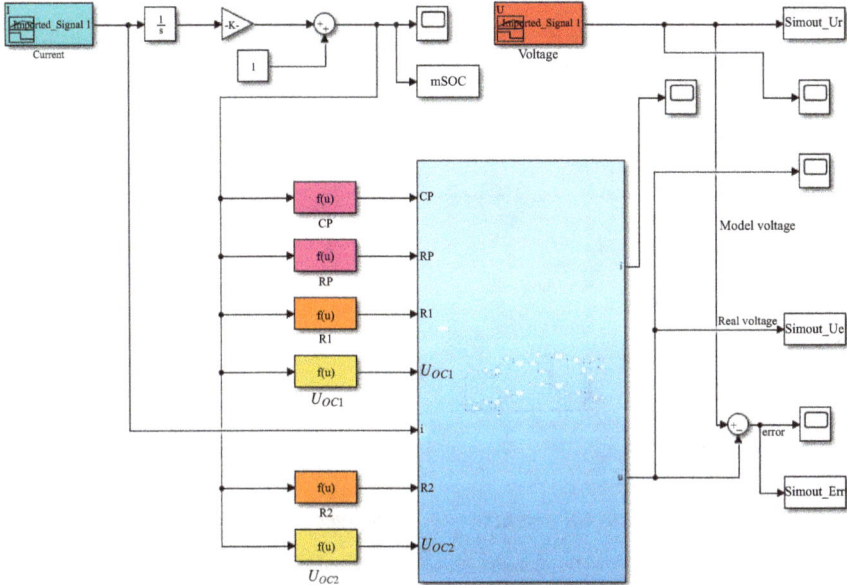

Figure 3.20   The simulation model of the improved Thevenin equivalent model

obtained by introducing experimental data, which verifies the accuracy of the model parameters. The internal circuit of the improved model is the core part of the whole model, and the circuit structure is directly used to build the module, including two ohmic resistances, two diodes, one RC parallel structure, two controllable voltage sources, controllable current sources, voltage and current sensors, and input and output ports, as shown in Figure 3.20.

The internal components of the model represent the proposed improved equivalent circuit model. The necessary inputs and outputs are marked, and other components are appropriately displayed in the figure based on the proposed model, as shown in Figure 3.21.

3.6   High-order model establishment

The high-order battery model is a highly precision lithium-ion battery control model designed for a transient reaction simulation. It is sufficient for high current, phase sort, and complicated load–unload conditions. The internal resistance modeling and the estimation of the state-space characteristics are provided in this section to obtain a zero-input state equation.

3.6.1   High-order electrical modeling

Compared with the Rint model, the change in the high-order model versus the Rint model is to add two RC network circuits to describe the lithium-ion battery

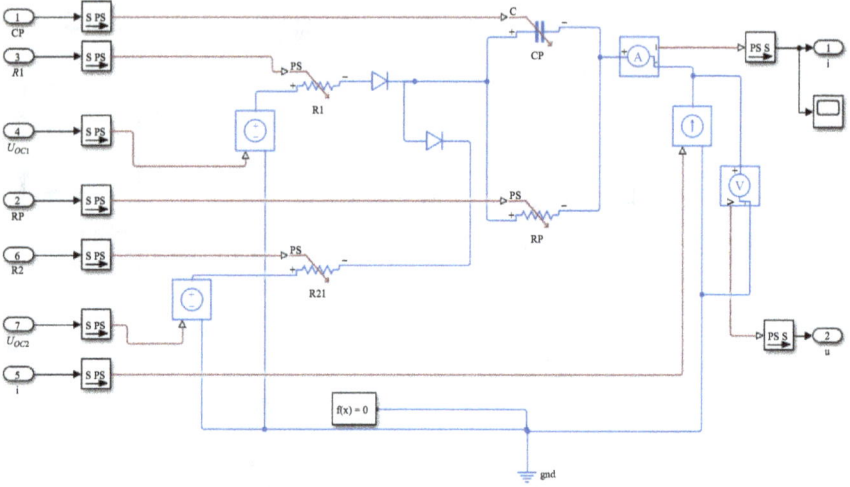

Figure 3.21   The internal circuit of the improved Thevenin equivalent model

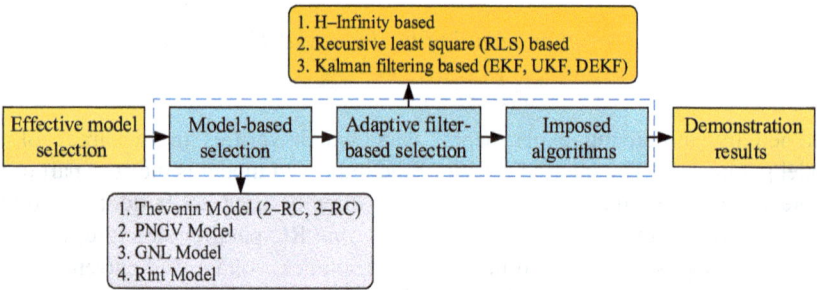

Figure 3.22   Essential steps for selecting appropriate models

operations polarization effects. The high-order model will describe the complex response of the battery. During and after the charge–discharge treatment, the RC circuits will be reflected. As a high-order equivalent circuit model, the battery voltage incremental drift is simple and easy to implement, which can meet the needs of simulation. The model is also used for practical applications. A correct battery model is essential to ensure safe and effective operation [166]. To prevent improper operations such as overloading, overcharging, and high temperatures, the battery behaviors should be predicted under various operating conditions. The battery architecture will also lead to onboard control and continuous maintenance design techniques.

Another crucial component, such as charging state and protection, is the leading battery operating states, but it cannot be calculated directly. Nevertheless, the model-based estimation approach is the other vital use of the model battery seen in Figure 3.22.

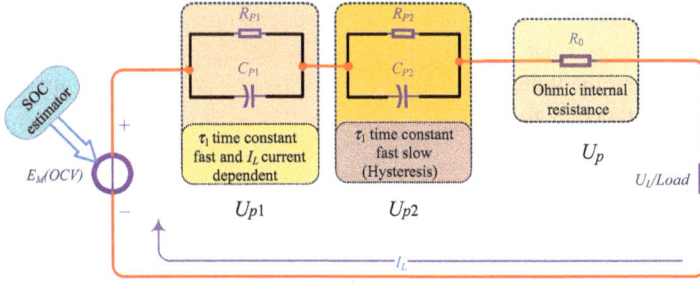

Figure 3.23   Schematic diagrams of the high-order modeling

However, the high-order model is made by the essential circuit components such as voltage sources, resistors, and capacitors [167]. In general, the improved RC model is a high-order equivalent circuit, which increases the precision and structure of the models. In this study, a similar circuit model is used to extend two RC networks, which are composed of a resistor R_0, two RC networks, and the U_{OC} voltage source. The schematic diagram of the high-order model is presented in Figure 3.23.

The configuration of the corresponding circuit model is adequate for a two-state circuit. Within the model, R_0 is the internal ohmic resistance of the lithium-ion battery. R_{p1} and C_{p1} have indicated the electrochemical polarization resistance and capacitance, respectively. Another parameter is R_{p2}, and C_{p2} denotes the resistance and capacitance of the concentration polarization, respectively. The voltage sources can be denoted by U_{OC}, which features a monotonous relationship with the SOC. I_L is the charging current and U_L is the terminal voltage. The battery terminal voltage is

$$U_L = E_m(OCV) - U_p - U_{p1} - U_{p2} \tag{3.29}$$

The measurement equipment used in the hybrid pulse-power characteristic system is not difficult to use, the measurement cost is not high, and higher accuracy can be obtained. However, the application of dynamic integration in a functional environment is challenging, and it is difficult to obtain the posterior probability filtering density with the above methods. Therefore, this approach is limited to the available boosts and needs to be paired with other approaches to solve the underlying probability density. The required posterior probability density is obtained to measure the required parameters.

3.6.2   Ohmic resistance identification

The theory of the internal resistance method can be used in the open-circuit voltage system. There is an individual interaction between SOC and internal resistance. The internal resistance is measured and converted to an SOC value [168]. Since the accuracy of the internal resistance method completely depends on the calculation accuracy of the internal resistance, the functional relationship between SOC and internal resistance is relatively consistent. However, another polarization voltage decrease terminates the fast phase exponentially before the voltage returns to the steady-state voltage value [149,169]. Assuming that the voltage before the current changes is U_1 and the voltage after the current changes is U_2, the rule for measuring the ohmic resistance is as follows:

$$R_0(S) = \frac{(U_1 - U_2)}{I_L} \tag{3.30}$$

According to (3.30), batteries' terminal voltage will suddenly change at the time point of discharge and the moment of stop, both caused by internal ohmic resistance [170]. Therefore, the overview of internal resistance and voltage curve is presented in Figure 3.24.

The method of resistance may also be called a conductivity method. The relationship between conductivity or internal resistance of the lithium-ion batteries and the SOC can be studied from several experimental results, and the SOC can be estimated by testing the long-term conductivity and internal resistance of the lithium-ion battery.

3.6.3   State-space expression

The research carried out in the year developed a model that considered three aspects of the electrical model, thermal model, and degradation model of lithium-ion batteries installed in electric vehicles (EVs) [171]. U_1, U_2, and SOC are state variables, and U_L is vector observation. The equivalent circuit model equation can be deduced as

$$\begin{cases} C_{p1} \dfrac{dU_{p1}}{dt} + \dfrac{U_{p1}}{R_{p1}} = I_L \Rightarrow U_{P1} = \dfrac{I_L}{C_{p1}} - \dfrac{U_{p1}}{C_{p1}R_{p1}} \\[3mm] C_{p2} \dfrac{dU_{p2}}{dt} + \dfrac{U_{p2}}{R_{p2}} = I_L \Rightarrow U_{P2} = \dfrac{I_L}{C_{p2}} - \dfrac{U_{p2}}{C_{p2}R_{p2}} \end{cases} \tag{3.31}$$

The SOC vector can express the OCV, and a nonlinear function relationship can be formed. The analogous circuit model can be discretized using the experience

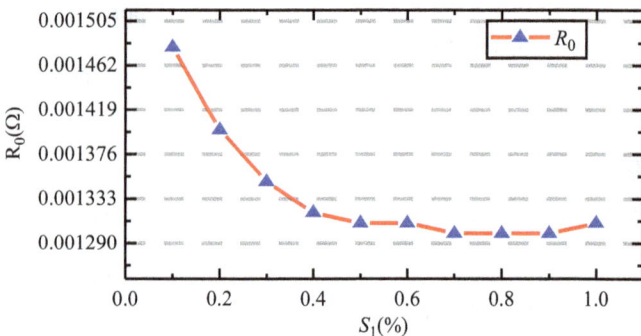

Figure 3.24   Relation between ohmic resistance with SOC

of the current control theory. For the selected chosen high-order equivalent circuits, [SOC U_{p1} U_{p2}] is selected as the state variable. Following the absolute equation description of SOC, its state-space equation is shown as

$$
\left\{
\begin{bmatrix} SOC_{k+1} \\ U_{p,k+1} \\ U_{c,k+1} \end{bmatrix}
=
\begin{bmatrix} 1 & 0 & 0 \\ 0 & e^{(-t/\tau_p)} & 0 \\ 0 & 0 & e^{(-t/\tau_c)} \end{bmatrix}
\begin{bmatrix} SOC_k \\ U_{p,k} \\ U_{c,k} \end{bmatrix}
+
\begin{bmatrix} -t/\eta Q_k \\ R_p\left(1 - e^{(-t/\tau_p)}\right) \\ R_c\left(1 - e^{(-t/\tau_c)}\right) \end{bmatrix}
I(k)
\right.
$$
$$
U_{k+1} = U_{OC}(SOC_{k+1}) - U_{p,k+1} - U_{C,k+1} - iR_0
$$

(3.32)

Equation (3.32) can be substituted for the RLS method, and it is taken as parameters for direct identification. According to the identification results of these parameters, the circuit modeling parameters of R_0, R_{p1}, C_{p1}, R_{p2}, and C_{p2} are derived. These parameters can be obtained by the relationship between the open-circuit voltage and the terminal voltage as follows:

$$
\begin{cases}
U_L = U_{OC} - be^{-et} - ce^{-ft} \\
U_L = \left[U_{OC} - \left(I_L R_{p1} e^{-t/\tau_1}\right) - \left(I_L R_{p2} e^{-t/\tau_2}\right)\right]
\end{cases}
$$

(3.33)

In the equivalent circuit model, these equations are the values of arbitrary parameters. In the process of data analysis and processing, the prosperous data section is derived from the initial experimental data, and then the extracted data segments are analyzed. The parameters of τ_1 and τ_2 can be calculated accordingly, as shown below:

$$
\begin{cases}
\tau_1 = \left[R_{p1}(S) \times C_{p1}(S)\right] \\
\tau_2 = \left[R_{p2}(S) \times C_{p2}(S)\right]
\end{cases}
$$

(3.34)

Assuming that the voltage before the current changes is S_{P1} and the voltage after the current changes is S_{P2}, the ohmic resistance can be considered as a unique factor that causes the voltage drop at that time point. An effective processing method is used to create the relationship between the internal parameters of the equivalent circuit model and the SOC. Also, the values of the remaining parameters can be obtained by integration. However, compared with the coefficient, the values of the parameters can be obtained as

$$
\begin{cases}
R_{p1}(S) = \dfrac{b}{I_L}, C_{p1} = \dfrac{1}{\left[R_{p1}(S) \times e\right]} \\[3mm]
R_{p2}(S) = \dfrac{c}{I_L}, C_{p2} = \dfrac{1}{\left[R_{p2}(S) \times f\right]}
\end{cases}
$$

(3.35)

The step-by-step variation of the terminal voltage is consistent with the pure resistance characteristic for the instantaneous disappearance of the current. The ohmic resistance can be considered as a unique factor that causes the voltage drop at the time point. The working procedure for the identification of a high-order model is shown in Figure 3.25.

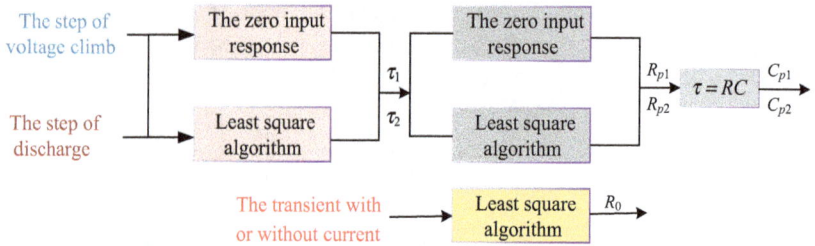

Figure 3.25   Working steps for identifying model parameters

The step expression of the voltage rise is the zero-input response. The discharge step is expressed by the least square method. Finally, a combination of concentration polarization resistance and capacitance is proposed. The curve-fitting approach uses a system in which the whole pulse discharge curve is balanced by a standard function of parameters to obtain model parameters. The selection of the equivalent circuit model can reasonably reflect the characteristics of the battery and the CPU processing capacity of the BMS. To investigate the dynamic behavior, the following terms are the superiority of high-order models for EVs.

This model does not require matrix calculation, and the amount of calculation is small. An extra RC network is added in the improvement process of the first-order RC model. Six parameters of the high-order modeling are ensured to reliably investigate the insight state of the battery. The model exhibits high precision for determining the dynamic voltage characteristic of the battery. The presented approach has a higher estimation accuracy and faster convergence speed comprehensively independent to understand and trusted for simulation approach.

3.7   Model parameter description

The battery model is the key to determining a precise mathematical relationship that can produce significant inputs and outputs with the development of an actual battery model, analyzing the feasibility and reliability. The model can be obtained in a variety of ways. Ultimately, the various characteristics of the battery function can be contrasted with the actual model.

3.7.1   Ampere-hour counting

The Ampere-hour (Ah) integration method, also known as the Coulomb counting method, measures the accumulative charge–discharge within a period when the SOC basis for a given time is known in advance. The actual state of the battery charge can be obtained by superimposing the two techniques. The error is obtained by adding the expected value to the experimental value. The error grows with the period and does not ensure the stability of the test results. This procedure relies on the initial value, so the correct solution to the self-discharge problem is not taken.

As the most frequently used method of SOC calculation, it aims to concentrate on the external characteristics of the battery device without taking into consideration the battery electrochemical reaction and the dynamic interaction between the parameters [147]. The principle of the Ah integral method is shown as

$$SOC_t = SOC_0 - \frac{1}{C} \int_0^t \eta I dt \qquad (3.36)$$

where SOC_0 is the original battery electrical quantity, SOC_t is the batteries electrical quantity at time t, C is the rated battery power, I is the charge–discharge current, and the discharging direction is the positive direction η is the Coulomb performance coefficient, which reflects the internal electrical quantity dissipation throughout the charge–discharge process. The working principle of the improved Ah integral is shown as follows:

$$SOC_t = \alpha \times SOC_0 - \frac{1}{\delta C} \int_0^t \eta I dt \qquad (3.37)$$

where α is the aging factor and the self-discharge correction factor δ is the correction factor of battery capacity C. The reason for adding the correction factor is that the battery will age after long-term use. The functional relationship between the correction factor δ of the total battery capacity C and the number of cycles for the lithium-ion battery is determined through experiments to improve the accuracy of SOC estimation [172]. However, the approximation Ah integral method proves that the accuracy of SOC_0 is very important to the real-time estimation. The commonly used method is to use the OCV method to measure the initial power of the battery.

By defining a correction factor and defining a re-calibration point, Ah counting errors can be reduced. The Ah Coulomb counting method has better precision than other SOC estimation methods [14]. It is simple and effective if the current measurement is correct, and the re-calibration point is usable.

3.7.2 Exponential curve fitting

The curve-fitting method is a technique in which the formal equation with the parameters matches the whole pulse discharge curve to obtain the model parameters. The data acceptance rate is much higher than the point-to-point data acquisition method. In the process, we just need to give a suitable equation to the function containing the symbols of the parameter to be obtained for curve fitting. Aiming at the relatively complex calculation parameters R_{p1}, R_{p2}, C_{p1}, and C_{p2} in the high-end battery modeling system, a curve-fitting method is designed. However, the high-order model expression can be obtained as

$$U_L = U_{OC} - IR_O - IR_p\left(1 - e^{-t/\tau}\right) \qquad (3.38)$$

The mathematical expression is abstracted to obtain a parameterized expression, as shown below:

$$y = a - b\left(1 - \exp\left(-\frac{x}{d}\right)\right) - c\left(1 - \exp\left(-\frac{x}{e}\right)\right) \tag{3.39}$$

In (3.39), y represents the coordinate vector such as terminal voltage U_L, x represents the abscess variable at time t, and a, b, c, d, and e are the coefficient parameters. Depending on the conditions of the OCV, current, and ohmic resistance, the fitted curve helps to express the experimental shift in voltage and the fitting curve of the phase pulse zero state reaction with an SOC of 1.0%. It appears a good view of the process which does not require internal ohmic resistance (see Figure 3.26).

The identification of parameters has made by each step of the SOC point using a curve-fitting tool. The polarization resistance and capacitance are done by the method, while the internal ohmic resistance is measured by the experimental data, which is between 0.1% and 1%. Considering the experimental effects of the hybrid pulse-power test, the curve fitting is done by using the least square method.

3.7.3 Recursive least square

The recursive least-square (RLS) algorithm is an online parameter identification method that can estimate the current parameters of the model in real time. It can reduce the error of the discharge rate change in the offline parameter identification. The following takes the Thevenin model as an example to introduce online parameter identification.

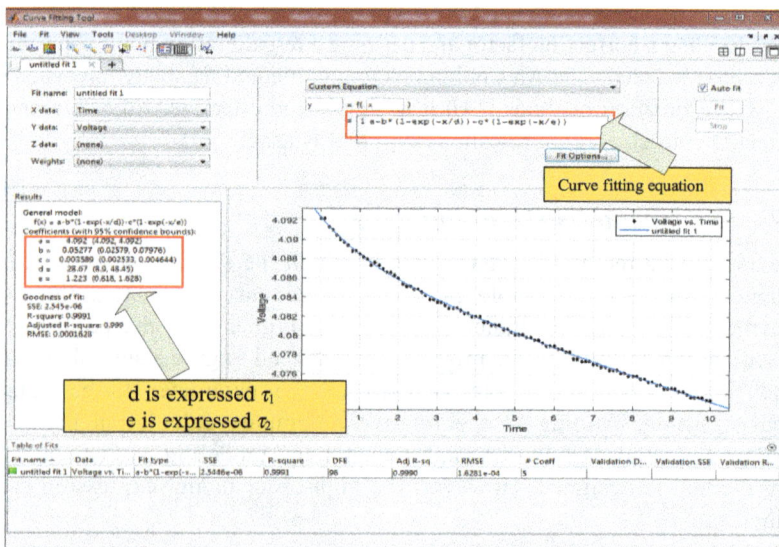

Figure 3.26   Curve fitting of identification parameters

In the improved Thevenin model, the output voltages and currents can be measured, so it is necessary to identify the four parameters of open-circuit voltage U_{OC}, ohmic resistance R_0, polarization resistance R_P, and polarization capacitance C_P. Regarding OCV, it can be obtained by the relationship between OCV and SOC by the hybrid pulse-power characterization test. For OCV, three remaining parameters need to be identified.

The identification methods include offline and online algorithms. The offline algorithm can identify the accuracy of the parameter, but the results are only applicable to specific working conditions and environments. The generality of this algorithm is not good. Quite the opposite, the online algorithm can estimate the parameters with the changing of working conditions. Therefore, in this chapter, the RLS algorithm is used to calculate the parameters. RLS algorithm can not only adapt to different working conditions but also has high parameter identification accuracy.

The principle of RLS is based on the least-square algorithm, and its implementation is mainly based on the principle of minimum mean square error. With the voltage of the lithium-ion battery as the input and current as the output, it can be regarded as a single-input-single-output model. Therefore, (3.40) can be obtained

$$\begin{cases} U_a(s) = U_{OC}(s) - U_L(s) \\ U_a(s) = R_0I(s) + \dfrac{R_PI(s)}{1 + \tau s} + v(s) \end{cases} \tag{3.40}$$

In (3.40), the equations can be obtained by Laplace transformation, where v is the measurement noise. Therefore, the discrete system as shown in the following equation can be obtained with bilinear transformation

$$y(k) = -ay(k - 1) + bu(k) + cu(k - 1) + v(k) \tag{3.41}$$

where a, b, and c are the parameters that need to be identified. To realize the least-square principle for parameter identification, the discrete system equation can be expressed as the least square form

$$\begin{cases} y(k) = x(k)^T \cdot \theta(k) + v(k) \\ x(k) = [y(k - 1) \quad u(k) \quad u(k - 1)]^T \\ \theta(k) = [-a \quad b \quad c]^T \end{cases} \tag{3.42}$$

Equation (3.42) is the expression of least square. The equations to calculate the parameters of a lithium-ion battery are

$$\begin{cases} \theta(k) = \theta(k - 1) + \gamma \cdot P(k - 1)x(k)[y(k) - x^T(k)\theta(k - 1)] \\ \gamma = [x^T(k)P(k - 1)x(k) + 1]^{-1} \\ P(k) = [I - \gamma \cdot P(k - 1)x(k)x^T(k)]P(k - 1) \end{cases} \tag{3.43}$$

After the parameter identification results are obtained, the ohmic resistance R_0, polarization capacitance C_P, and polarization resistance R_P can be derived from the expression

$$\begin{cases} a = \dfrac{T - 2\tau}{T + 2\tau} \\[2mm] b = \dfrac{TR_P + TR_0 + 2R_0\tau}{T + 2\tau} \\[2mm] c = \dfrac{TR_P + TR_0 - 2R_0\tau}{T + 2\tau} \end{cases} \tag{3.44}$$

where τ represents the time constant of the RC parallel circuit and T is the sampling time point. According to the above equation and the obtained parameters by the RLS algorithm, the parameters of the Thevenin model can be calculated. The parameter identification under Beijing bus dynamic stress test (BBDST) working conditions is shown in Figure 3.27, where R_0 is the internal resistance of the Thevenin, R_p and C_p are the polarization resistance and capacitance, respectively, U_1 is the measured voltages that are obtained by the experiment, U_2 is the simulation results of the parameters that are identified by using the RLS method, and E is the error of simulation voltage. It can be seen from the figure that this method can identify the model parameters well.

3.7.4   Full model parameter identification

For the parameter identification of the lithium-ion battery model, a full identification method based on the RLS algorithm is introduced. According to the bilinear

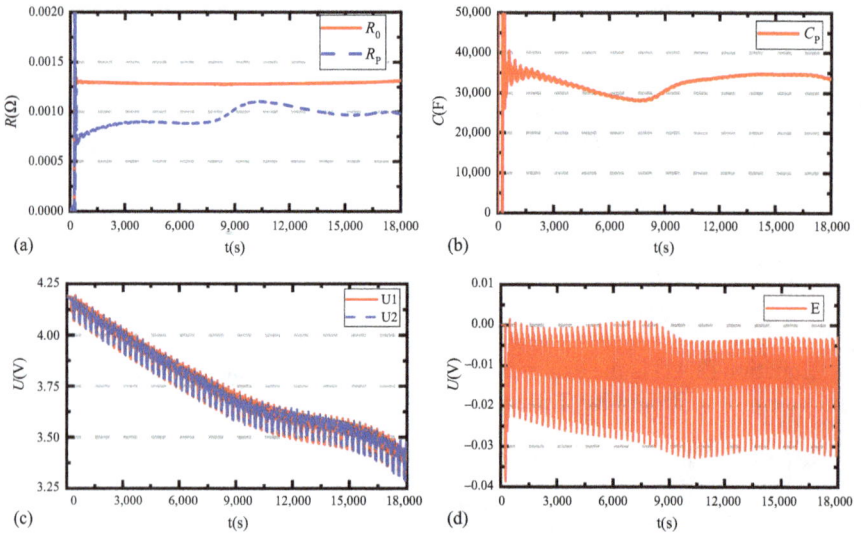

Figure 3.27   Parameter identification results: (a) internal and polarization resistance, (b) polarization capacitance, (c) simulation voltage, and (d) simulation voltage error

transformation, the mathematical expression can be obtained after the discretization of the Thevenin equivalent circuit model, which is shown as

$$\begin{cases} y(k) = y_{OCV}(k) - y_{UL}(k) \\ y(k) = -ay(k-1) + bu(k) + cu(k-1) + n_1 v(k) + n_2 v(k-1) \\ y_{UL}(k) = y_{OCV}(k) + ay_{OCV}(k-1) - ay_{UL}(k-1) - bI(k) - cI(k-1) - n_1 v(k) - n_2 v(k-1) \end{cases}$$

$$(3.45)$$

In (3.45), $y_{UL}(k)$ represents the terminal voltage, $y_{OCV}(k)$ is the OCV in the EEC. The coefficients of the discrete equation are described by a, b, and c. According to the recursive calculation equations, the identified parameters and verification results under BBDST conditions are shown in Figure 3.28.

Figure 3.28(a) shows the comparison between the experimental open-circuit voltage curve and the estimated open-circuit voltage curve which is predicted by an

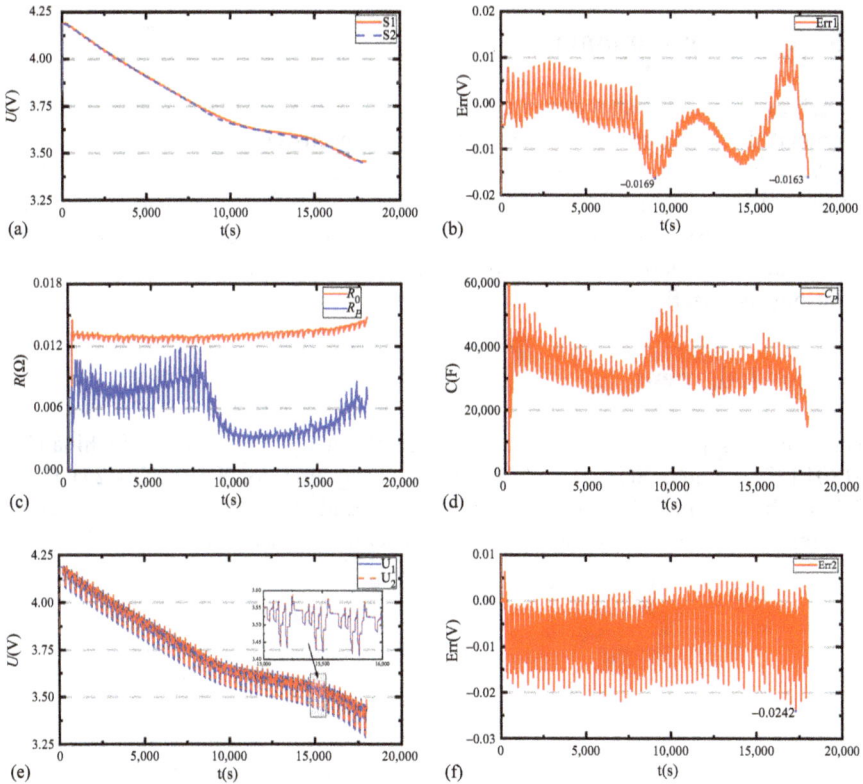

Figure 3.28   *The full parameter identification results: (a) open-circuit voltage, (b) error of open-circuit voltage, (c) internal and polarization resistances, (d) polarization capacitance, (e) terminal voltage, and (f) error of terminal voltage*

online full parameter identification algorithm, where S_1 is the true value and S_2 is the estimation curve.

Figure 3.28(b) shows the online OCV estimation error. The maximum value of the error is 16.9 mV, which illustrates the prediction result is reliable.

Figure 3.28(c) and (d) presents the remaining parameters in the proposed EEC. Due to internal chemical reactions, the internal resistance, polarization resistance, and polarization capacitance of the battery will change with the charge–discharge rate. Therefore, the fluctuation of the parameters is mainly caused by the charge–discharge rates.

Figure 3.28(e) shows the simulation voltage by the identified parameters and the measured terminal voltage by the sensor where U_1 is the simulation value and the U_2 is the measured value. The error of the simulation voltage compared with the actual value is shown in Figure 3.28(f) where the maximum error is 24.2 mV. As can be seen from the experimental results, this EEC has high precision and the parameter identification result is efficient.

3.8   Chapter summary

To achieve accurate battery state estimation, an effective battery modeling method is crucial. Different battery modeling types are discussed elaborately in this chapter. Various battery modeling strategies are presented in this section, such as Rint, PNGV, and Thevenin model. Significantly, four types of standard battery modeling are studied. Also, their advantages and limitations are provided under different connections. Furthermore, the high-order modeling with its mathematical analysis is introduced briefly.

Acknowledgment

The work is supported by the National Natural Science Foundation of China (No. 61801407), Sichuan Science and Technology Program (No. 2019YFG0427), China Scholarship Council (No. 201908515099), and Fund of Robot Technology Used for Special Environment Key Laboratory of Sichuan Province (No. 18kftk03).

Chapter 4
Extended Kalman filtering and its extension

Abstract

The Kalman filtering extension strategies for accurate battery state estimation are analyzed, especially the extended Kalman filtering and fractional-order extended Kalman filtering algorithms. The battery equivalent circuit modeling methods include second-order Thevenin modeling and other modelings. The procedure design and verification are introduced in this chapter. The model parameters are identified by the recursive least-square method, the forgetting factor least-square method, and the hybrid pulse-power characteristic (HPPC) experimental tests, in which the results are verified by pulse-current cycling experiments. The battery equivalent modeling strategy is explored to obtain its state-space expression, which is then used to realize its state of health and state of charge estimation. The fractional experiments are then conducted, including real-time platform implementation, HPPC tests, and capacity tracking experiments. The experimental procedure is designed, including the whole experiment structure and the detailed procedure flowchart. After that, the experimental platform is built, and the designed experiments are realized. Consequently, this chapter introduces the extended Kalman filtering algorithm into the state of health and state of charge estimation, which can realize the dynamic parameter estimation of the battery system and improve the state estimation accuracy for lithium-ion batteries under the HPPC, BBDST, and dynamic stress test conditions.

Keywords: Extended Kalman filtering; Equivalent circuit modeling; Parameter identification; Recursive least square; Fractional experimental test; Experimental procedure design; State of health estimation; Time-varying correction; State-space expression; State of charge estimation; Internal resistance increasing

4.1 Kalman filtering extension strategies

The Kalman filtering (KF) algorithm offers high optimization estimation accuracy in a linear system; however, as battery SOC estimation is a typical nonlinear system, it is difficult to use the traditional KF algorithm in the battery management system (BMS). The extended Kalman filter (EKF) algorithm is one of the most common SOC estimation algorithms. This algorithm not only solves the state estimation problem in the nonlinear system but also ensures fast correction and therefore achieves adaptive target tracking in the event of a relatively big error of the initial value. The EKF approach is combined with other algorithms to improve the robustness of BMS and the SOC estimation accuracy, to achieve accurate estimation of the lithium-ion battery state.

4.1.1   Kalman filtering algorithm

The KF algorithm is mainly used for the estimation of linear time-invariant systems. The observable output estimation error is used in the prediction process [173], and the recursive linear minimum variance estimation method is used to repair the unobservable state estimation error, which greatly reduces the noise interference in the data stream and improves the estimation accuracy of the new system [174]. First, a set of recursive equations is established to describe the battery characteristics system and perform recursive operations [175]. The state-space equation of the system is then obtained, including signal and noise. Finally, the mathematical methods are used to perform optimal regression data processing on the estimation result of the previous step and the existing measurement data to obtain the current optimal estimation result. The advantage of the method is that it can perform a highly accurate measurement of the dynamic state. Meanwhile, the accuracy of the battery model is required relatively high, and the operation is complicated.

There are several widely used techniques for calculating states for the dynamics. The error of the ampere-hour (Ah) integral process is immense [176]. The open-circuit voltage (OCV) approach is primarily used for offline estimation. Highly accurate estimation approaches include neural networks, fuzzy inference, and KF [177]. The EKF algorithm linearizes the nonlinear functions by the Taylor series expansion of first-order that ignore the remaining high-order terms. The transform nonlinear problems with linear problems, which can contribute to higher accuracy [178].

A modification of time and measurement is required for the process of calculation. The time shift is a means of predicting current state variables and providing a previous estimate of the next time point, which also refers to the projection [179]. The measurement-update process, also known as the adjustment process, is the feedback course on the measured values and the variances correction.

The core of the KF process is the method of Ah integration, which can track and correct the external current according to the voltage value obtained [180]. When using KF to estimate the state, it is necessary to establish an equivalent model, and the accuracy of the KF algorithm is closely related to the model. Standard KF is only applicable to linear systems. First, the nonlinear system is linearized in the EKF algorithm. Its advantage is high accuracy even in the case of currency fluctuations and noise; it also has a good correction effect. The disadvantage is that it requires an accurate battery model and a state-space description. The flowchart of the Kalman filtering algorithm is shown in Figure 4.1.

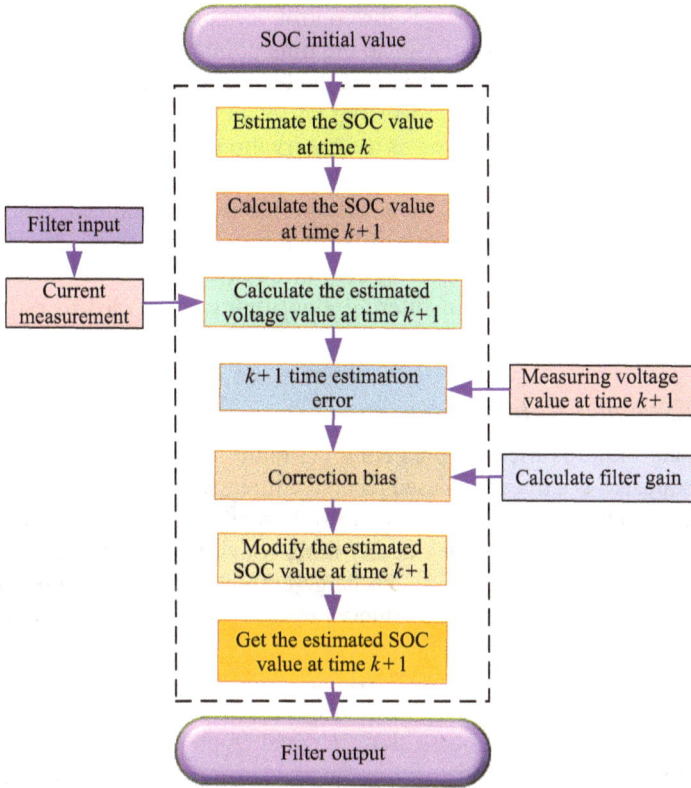

Figure 4.1   Flowchart of the Kalman filtering algorithm

When the KF algorithm is used to estimate the battery state, an effective equivalent battery model must be created. The battery is considered a power system, and SOC is the system input parameter [181,182]. The battery current charge–discharge rate is assumed to be the input of the device, and the terminal voltage may be considered the output. By comparing the observed terminal voltage value and the predicted state error, the device state is continuously modified to achieve the lowest estimated variance [183]. It is an optimal self-regressive data processing algorithm that is suitable for both stationary and non-stationary processes. It has got an accurate real-time output that is quick to be realized. However, the algorithm has a heavy dependency on the model and can only be used in the linear system [184]. The state equation and the observation equation are defined as

$$\begin{cases} X_k = A_{k-1}X_{k-1} + B_{k-1}Z_{k-1} + W_{k-1} \\ Y_k = C_{k-1}X_k + D_kZ_k + V_k \end{cases} \tag{4.1}$$

In the above expression, the state variable at the time point of k is described by the parameter X_k, and Y_k is the systematic observation variable at the time point of k. Z_k is the system input, as the control variable. A_k is the transfer matrix of state x from the time point of $k-1$ to the time point of k. B_k is the input matrix [185]. C_k is the measurement matrix. D is the feedforward matrix. W_k is the noise of the system state equation as processing noise, and its variance is Q_k. V_k is the noise of the measurement equation as the observation noise, whose variance is R_k. Its crucial step can be realized to establish a reasonable state equation and observation equation. However, its effect applied to the nonlinear system is not ideal due to the strong nonlinearity of the battery system [186]. Therefore, it is proposed to linearize the nonlinear system through the Taylor series expansion. After linearization, the Kalman filtering algorithm can be used to estimate the state variables [187].

4.1.2   Extended Kalman filtering

The EKF algorithm is used for predicting the future state of a system based on a previous state. KF is a linear unbiased recursive filter, which is constantly predicted and corrected in the calculation process [188]. Whenever new data is observed, new predicted values can be calculated at any time, which is very convenient for real-time processing. Due to the charge–discharge current rate, temperature, and complex internal chemical reaction, the battery presents a nonlinear state. Based on the Kalman filter algorithm, the Jacobian matrix is obtained by using the Taylor equation for linearization, and the EKF algorithm is obtained. The algorithm iteration process is described as shown in Figure 4.2.

In the EKF calculation process, SOC, U_1, and U_2 are used in the state equation as the state variables. The current I is the system input, and U_{oc} is the system output, according to which the expression of the state equation and the observation equation is constructed, as shown below:

$$\begin{cases} x_k = A_{k-1}x_{k-1} + B_{k-1}x_{k-1} + w_{k-1} \\ y_k = C_k x_k + D_{k-1}u_k + v_k \end{cases} \tag{4.2}$$

where x is the system state variable; u is the system input and y is the system output; w_k and v_k are the system noise, with the covariance matrices being Q and R, respectively. The A_k, B_k, C_k, and D_k matrix of the above-mentioned equation is expressed as

$$\begin{cases} A_k = \begin{bmatrix} 1 & 0 & 0 \\ 0 & exp\left(-\dfrac{T}{\tau_1}\right) & 0 \\ 0 & 0 & exp\left(-\dfrac{T}{\tau_2}\right) \end{bmatrix}, C_k = \left[\dfrac{dU_{OC}(\text{SOC})}{d(\text{SOC})}, 1, 1\right]^T \\ B_k = \left[-\dfrac{\eta T}{Q_N}, R_1\left[1 - \exp\left(-\dfrac{T}{\tau_1}\right)\right], R_2\left[1 - \exp\left(-\dfrac{T}{\tau_2}\right)\right]\right], D_k = R_0 \end{cases}$$

$$\tag{4.3}$$

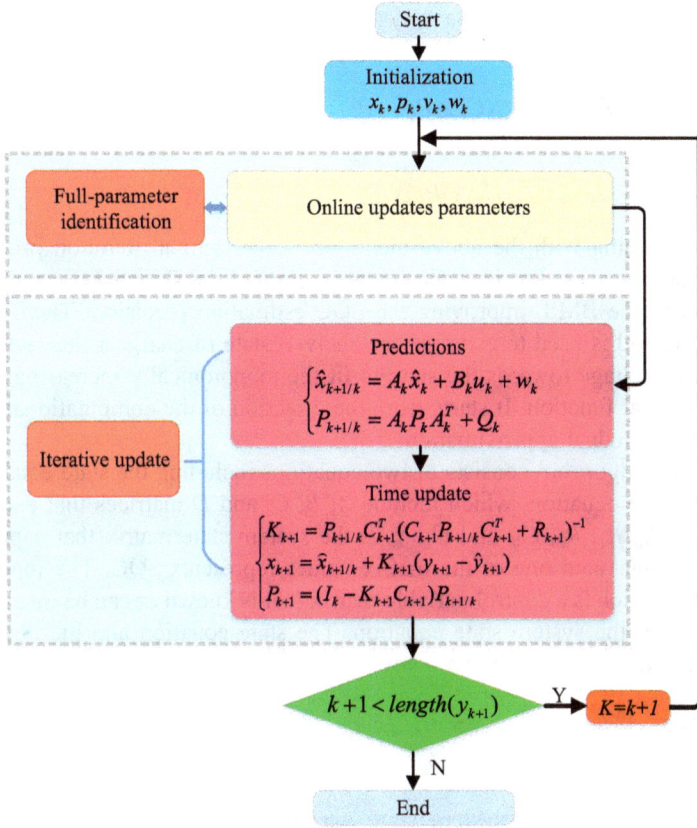

Figure 4.2   The EKF algorithm iteration flowchart

The EKF algorithm uses the first-order Taylor expansion of the OCV equation to transform the nonlinear system into a linear system, and its calculation is shown as follows:

$$\frac{dU_{OC}(\text{SOC})}{d(\text{SOC})} = \frac{K_1}{\text{SOC}} - \frac{K_2}{1 - \text{SOC}} \tag{4.4}$$

The parameter values estimated in real time by the multi-innovation least squares (MILS) algorithm and the EKF algorithm are combined to realize the iterative calculation of SOC. The calculation of the state prediction equation and state covariance prediction equation is shown below:

$$\begin{cases} \widehat{x}_{k+1/k} = A_k \widehat{x}_k + B_k u_k + w_k \\ P_{k+1/k} = A_k P_k A_k^T + Q_k \end{cases} \tag{4.5}$$

where P_k is the covariance matrix. To obtain the optimal filter gain matrix K_{k+1}, the optimal state matrix x_{k+1}, and the optimal covariance matrix value P_{k+1}, the above prediction equation needs to be updated over time, and the updated equation is shown as follows:

$$\begin{cases} K_{k+1} = P_{k+1/k}C_{k+1}^T(C_{k+1}P_{k+1/k}C_{k+1}^T + R_{k+1})^{-1} \\ x_{k+1} = \hat{x}_{k+1/k} + K_{k+1}(y_{k+1} - \hat{y}_{k+1}) \\ P_{k+1} = (I_k - K_{k+1}C_{k+1})P_{k+1/k} \end{cases} \tag{4.6}$$

By combining with the above-mentioned EKF optimal iteration process, the MILS algorithm is used to identify the parameters of the E model, to achieve state estimation of the BMS, improving the SOC estimation precision. Then, a binary iterative method is used to estimate the shelved state of charge value, which uses open-circuit voltage towards the state to fit the monotonically increasing function of a polynomial function. It starts from the direction of the computational thought to realize the gradual approximation.

The EKF algorithm consists of two equations including the state equation and the observation equation, which include A, B, C, and D matrices that can be realized using R_0, R_1, R_2, C_1, and C_2. x_k is the system state matrix that captures the system dynamics and one of the matrix values represents SOC. The input of the system is u_k which is a control variable matrix that is known or can be measured. w_k is the noise of the system state equation. The state equation and the observation equation are described as shown below:

$$\begin{cases} X_k = AX_{k-1} + BI_{L,k-1} + w_k \\ U_{L,k} = CX_k + DI_{L,k-1} + v_k \end{cases} \tag{4.7}$$

where X_k represents the system state variable at time k, $U_{L,k}$ is the system observed variable at time k, $I_{L,k-1}$ is the system input that is used as the control variable, A is the transfer matrix of state x from $k-1$ to k, B is the input matrix, C is the measurement matrix, D is the feed-forward matrix, and w_k is the noise of the system state equation. The KF algorithm is used for state prediction and estimation. It consists mainly of five equations, which can be divided into the stage of prediction and the stage of correction. The recursive relationship between the estimated value of state and covariance in the prediction stage (time update) is shown as follows:

$$\begin{cases} \hat{X}_k^- = \hat{X}_{k-1} + BI_{L,k-1} + w_k \\ \hat{P}_{k|k-1}^- = A\hat{P}_{k-1}A^T + Q_w \end{cases} \tag{4.8}$$

According to the model, the last moment of the state estimate of $k-1$, X_{k-1}, and its covariance matrix P_{k-1} directly calculates the forecast of this moment, X_k^- and its covariance matrix $P_{k|k-1}$. Q_w is the covariance matrix of process noise w_k. The estimation of Kalman gain is shown as follows:

$$K_k = \hat{P}_{k|k-1}^- C^T \left(C\hat{P}_{k|k-1}^- C^T + v_k \right) \tag{4.9}$$

The state correction stage is then performed for further computations to arrive at an appropriate equation that can be used to effectively make sure the appropriate parameters are identified. This can be achieved by the mathematical analysis as

$$\begin{cases} \hat{X}_k = \hat{X}_k^- + K_t\left(U_{L,k} - C\hat{X}_k^-\right) \\ \hat{P}_{k|k-1} = \hat{P}_{k|k-1}^- - K_t C\hat{P}_{k|k-1}^- \end{cases} \tag{4.10}$$

The moment of state estimation x_k and $P_{k/k-1}$ is realized after this and the KF algorithm is completed in one iteration, and an iterative estimation is carried out for each observation, with good real-time performance.

The EKF algorithm includes a linearization process of the nonlinear system. It estimates the value of the next time point according to the previous time point. The state variables are continuously updated with the input and output observations of the system to achieve the optimal estimation [189,190]. When it is used to estimate the battery state, the processing and observation noises are required to be white noise with approximate Gaussian distribution. It is also a limitation of all Kalman filtering methods.

In this case, the covariance of processing and observation noises can be easily controlled. In the process of estimation, the Taylor series expansion algorithm is implemented to expand the system model of the battery. Then, a first-order linear model is left after removing the high-order terms. The classical algorithm is designed for linear systems, while the battery system is nonlinear. It uses the Taylor series expansion to get the approximate linear space equation and then uses the KF algorithm to estimate the present state, which is suitable for the discrete nonlinear system. The expression and observation equations of the discrete nonlinear system are expressed as

$$\begin{cases} X_{k+1} = f(X_k,k) + W_k \\ Z_k = h(X_k,k) + V_k \end{cases} \tag{4.11}$$

The first part of Equation (4.11) represents the state equation and the second part represents the observation equation, k is a discrete-time point, X_{k+1} is the n-dimensional state vector, Z_k is the m-dimensional observation vector, and W_k and V_k are independent Gaussian white noise. To apply the Kalman filter, the first-order Taylor series expansion of nonlinear functions $f(*)$ and $h(*)$ is carried out around the estimated value. To apply the Kalman filter, the nonlinear functions are extended to the first-order Taylor series, and the results are shown as

$$\begin{cases} f(X_k,k) \approx f\left(\hat{X}_k,k\right) + \dfrac{\partial f(X_k,k)}{\partial X_k}\Big|X_k = \hat{x}_k\left(X_k - \hat{X}_k\right) \\ h(X_k,k) \approx f\left(\hat{X}_k,k\right) + \dfrac{\partial h(X_k,k)}{\partial X_k}\Big|X_k = \hat{x}_k\left(X_k - \hat{X}_k\right) \end{cases} \tag{4.12}$$

The new state transformation matrix and observation driving matrix can be accomplished by using the Taylor series expansion and ignoring the high-order term to update the linear space equation. The values of A_k, B_k, C_k, and D_k can be calculated as

$$\begin{cases} A_k = \dfrac{\partial f(X_k, k)}{\partial X_k} \Big| X_k = \hat{X}_k, B_x = f\left(\hat{X}_k, k\right) - A_k, \hat{X}_k \\[4mm] C_k = \dfrac{\partial h(X_k, k)}{\partial X_k} \Big| X_k = \hat{X}_k, D_x = h\left(\hat{X}_k, k\right) - C_k, X_k \end{cases} \qquad (4.13)$$

This approach is based on the state of the system and measurement equation. The prediction state equation involves the Ah integral measurement process, and the observation equation describes the analogous model of the lithium-ion batteries. As for (4.11), the coefficient can be linearized as

$$\begin{cases} X_{k+1} = A_k X_k + Z_k B_k + \omega_k \\[2mm] Y_k = C_k X_k + D_k + u_k \end{cases} \qquad (4.14)$$

The recursive method is accomplished by applying the basic equation to the discreet model, obtaining the basic equations for the linearized model. However, the recursive process of the EKF algorithm is obtained by applying equations to the linearized equation (4.14), which is described as

$$\begin{cases} \hat{X}_{\bar{k}+1} = f\left(\hat{X}_k\right) \\[3mm] \hat{P}_{\bar{k}+1} = A_k \hat{P}_k A_k^T + Q_{k+1} \\[3mm] K_{k+1} = \hat{P}_{\bar{k}+1} C_{k+1}^T \left(C_{k+1} \hat{P}_{\bar{k}+1} C_{k+1}^T R_{k+1}\right)^{-1} \\[3mm] \hat{X}_{k+1} = \hat{X}_{\bar{k}+1} + K_{k+1}\left[Z_{k+1} - h(X_{\bar{k}+1})\right] \\[3mm] \hat{P}_{k+1} = [I - K_{k+1} C_{k+1}] P_{\bar{k}+1} \end{cases} \qquad (4.15)$$

wherein P is the mean square error and K is the Kalman gain. The parameter I should be initialed as $n \times m$ unit matrix. Q and R are the variances of w and v, respectively, which generally do not change along with the system. The initial state value is $X(0) = E[X(0)]$, and its variance is $P(0) = \text{Var}[X(0)]$. The calculation steps for the time point of $k+1$ can be described as follows. First, the state and mean square error of the present time can be estimated by the last time point to obtain the initial state and its mean square error. Then, the Kalman gain of this time point is calculated. Finally, K_{k+1} is used to modify the initial state to get the present state. The prior mean square error is modified to get the current mean square error. The working process of the EKF algorithm is shown in Figure 4.3.

The calculated value is equal since there is an absolute error between the statistical linearization error plus the original expected value and the estimated measured value [191,192]. There is an individual error between all the calculated values derived by the statistical model and the measured values displayed in Figure 4.4.

The EKF approach is improved by using the standard measurement technique and linearization technology. The approximate parameters are linearized, and a

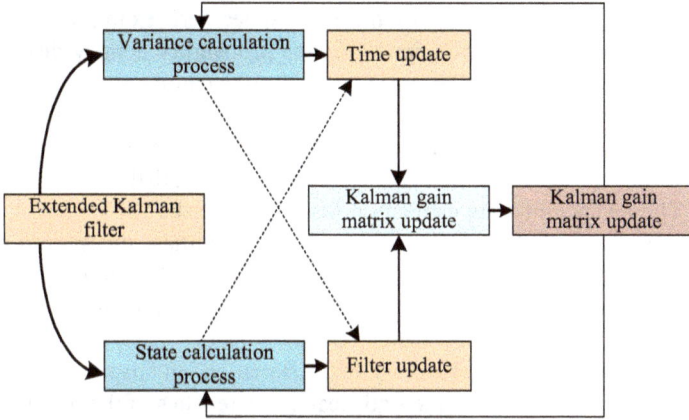

Figure 4.3   The processing and updating processes of the EKF algorithm

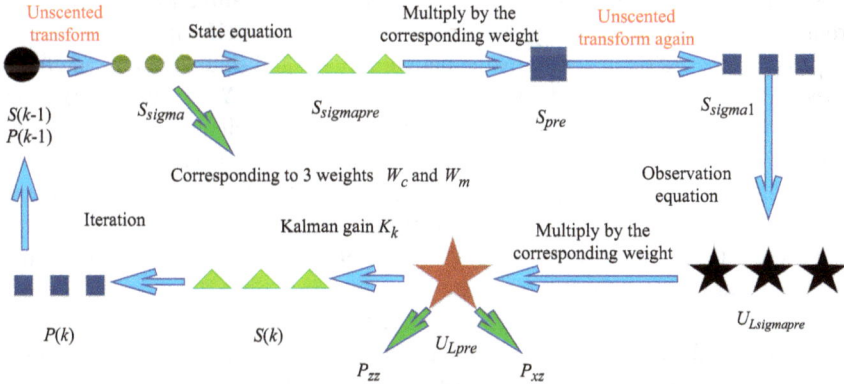

Figure 4.4   Double unscented transform extended Kalman filtering process

linear approximation is carried out [193]. Thus, the battery state can be measured under which the device and transmission sounds are usually approximated as white noise, consistent with the Gaussian distribution.

This method has a small amount of calculation and a good filtering effect and has been widely used. It utilizes linear Kalman filtering instead of the conventional nonlinear function linearization process. For one-step predictive equations, the non-tracking transformation is used to process the nonlinear transfer of mean and covariance. The algorithm uses a linearization method that converts nonlinearity into the probability distribution of state variables.

It does not ignore higher-order term errors, and it does not need to compute the Jacobian matrix continuously. As a result, the computational complexity is reduced. The transformation is the key to the algorithm, as it has a significant impact on the estimation outcome. Symmetric sampling, reduced skew simplex sampling, and simplex sampling of the hypersphere are examples of sampling techniques. The UKF algorithm approaches the probability density distribution of nonlinear functions by using a stronger sampling procedure.

It uses a series of samples to approximate the posterior probability density of the state, rather than approximate a nonlinear function. It does not ignore higher-order terms. It has high calculation accuracy for the statistics of nonlinear distribution. The essence of state estimation is to combine the OCV method with the Ah integration method. The Ah integral is used to estimate the system state and feedback with the voltage. The idea of the dual Kalman filtering algorithm is to alternately estimate the model and system with two Kalman filter lines and then adjust the feedback to get the estimated result.

To improve the accuracy of battery state estimation, an adaptive Kalman filtering algorithm is used to estimate the statistical characteristics of the battery based on the improved model. The simulation results show that the state estimation accuracy of the algorithm is significantly higher than that of the EKF algorithm. It effectively reduces the noise interference in the state estimation process and has certain reliability and practicability, although the UKF has higher estimation accuracy and stronger robustness and stability than the classical KF and EKF algorithms.

It is also based on accurate mathematical models and statistical characteristics of the system to deal with noise and observation noise. When the surrounding environment of the carrier changes or the motion state changes drastically, the statistical characteristics of the system processing and observation noise will change greatly. At this time, the accuracy and stability of the standard UKF estimation of the battery state need to be greatly improved. When the working current changes drastically over time, the negative determination problem of covariance may be encountered in the later operation.

In the calculation process, the covariance of the state variable is negative, and the Cholesky decomposition requires the matrix to be positive semi-definite. Otherwise, the algorithm that invalidates the filter will not continue due to rounding errors in the numerical calculations. To solve this problem, the square root method derived from the UKF is applied. The square root covariance of the state variable can be used to participate in iterative operation instead of the covariance. It can ensure the positive semi-definiteness and numerical stability of the state variable covariance matrix and overcome the discreteness caused by filtering.

In this algorithm, the most expensive operation is to recalculate a new set of the Sigma points at each update. The difference between the square root and UKF algorithms is that the first method is to replace the error covariance of the state variable with the square root of the error covariance of the state variable. The square root of the covariance can be passed directly, avoiding re-decomposition at each step. Although the square root of the covariance should be a component of the UKF algorithm, it is still a recursively updated covariance.

When S is the square root of the covariance matrix P, $SS^T = P$, as long as $S \neq 0$, P can be guaranteed to be non-negative. It uses three powerful linear algebra techniques, including QR decomposition, Cholesky factor update, and effective least square.

4.1.3   Fractional-order adaptive correction

When $t = 0$, a constant voltage u is applied for the capacitor to generate current i $(t) = V/(ht^n)$ where $0 < n < 1$, $t > 0$. h is a constant value related to the capacitance, which means that the capacitance is fractional. Fractional derivative is a generalization of integer derivative. Common fractional derivatives include Riemann–Liouville (R–L), Caputo, and Grunwald–Letnikov (G–L) fractional derivatives. Adaptive fractional-order extended Kalman filter (F-EKF) designs an algorithm to estimate SOC based on the G–L definition, which is defined as

$$
\begin{cases}
f^{(n)}(x) = \lim_{h \to 0} \dfrac{\Delta_h^n f(x)}{h^n} = \lim_{h \to 0} h^{-n} \displaystyle\sum_{k=0}^{n} \binom{n}{k}(-1)^k f(x - kh) \\[2mm]
\dbinom{n}{k}(-1)^k = (-1)^k \dfrac{n(n-1)\cdots(n-k+1)}{k!} = \dfrac{\Gamma(k-n)}{\Gamma(k+1)\Gamma(-n)}
\end{cases}
$$

$$(4.16)$$

In Equation (4.16), h is the displacement distance and n is the derivative order, which can be extended to any fractional derivative. Fractional derivatives have a better global correlation than integer derivatives. The G–L fractional derivative definition is used, and the fractional derivative calculation is conducted for EKF state variables and taking sampling time as

$$
\begin{cases}
\Delta^n x_{k+1} = \displaystyle\sum_{j=0}^{n} \binom{n}{j}(-1)^j x_{k+1-j} = x_{k+1} + \sum_{j=1}^{n} \binom{n}{j}(-1)^j x_{k+1-j} \\[2mm]
x_{k+1} = \Delta^n x_{k+1} - \displaystyle\sum_{j=1}^{n} \binom{n}{j}(-1)^j x_{k+1-j}
\end{cases}
$$

$$(4.17)$$

Equation (4.17) can describe the state variables of stochastic linear systems where Δ is the difference operator and n is the different order for x_{k+1}. Combining with the state equation, the linear fractional-order stochastic discrete state-space equation can be obtained as

$$
\Delta^n x_{k+1} = A_d x_k + B_k u_k + \omega_k \tag{4.18}
$$

where $A_d = A - I$. I can be used to describe the identity matrix. All of them are fractional-order representations of one-dimensional state variables. Since the traditional EKF state uses two-dimensional state variables, it is extended to the high-dimensional fractional state-space equation as

$$
\begin{cases}
\Delta^\gamma x_{k+1} = A_d x_k + B_k u_k + \omega_k \\[2mm]
x_{k+1} = \Delta^\gamma x_{k+1} - \displaystyle\sum_{j=1}^{k+1}(-1)^j \gamma_j x_{k+1-j}
\end{cases}
$$

$$(4.19)$$

$$\begin{cases} \gamma_j = \mathrm{diag}\left[\begin{pmatrix} n_1 \\ j \end{pmatrix} \cdots \begin{pmatrix} n_N \\ j \end{pmatrix} \right] \\ \\ \Delta^\gamma x_{k+1} = \begin{bmatrix} \Delta^{n_1} x_{1,k+1} \\ \vdots \\ \Delta^{n_N} x_{N,k+1} \end{bmatrix} \end{cases} \tag{4.20}$$

Equations (4.19) and (4.20) can be applied to the calculation of multi-dimensional state variables, $n_1, \ldots,$ and n_k. It is the order of the fractional-order, and N is the dimension of the state variable. The one-step optimal linear prediction of the state variable can be obtained by defining the traditional Kalman filter, as shown in the following equation:

$$\begin{aligned} \hat{x}_{k|k-1} &= \hat{E}[x_k | Y^{(k-1)}] \\ \\ &= \hat{E}\left[A_d x_{k-1} + B u_{k-1} + \omega_{k-1} - \sum_{j=k-M+1}^{k} (-1)^j \gamma_j x_{k-j} | Y^{(k-1)} \right] \\ \\ &= A_d \hat{x}_{k-1} + B u_{k-1} - \sum_{j=k-M+1}^{k} (-1)^j \gamma_j \hat{x}_{k-j} \end{aligned} \tag{4.21}$$

The fractional-order EKF (F-EKF) algorithm has a strong dependence on historical data. Although the global nature of the iterative algorithm is guaranteed, too much historical data will cause data saturation. Besides, the battery state is only related to the state of the past period. As the two aspects, when the data exceeds 20, the memory length M ($M = 20$) of the historical data is used in this study, the "outdated" old data is forgotten, and the 20 data closest to the current time are retained. At the same time, the corresponding adaptive process noise and observation noise are considered, which are defined mathematically as

$$\begin{cases} Q_k = (1 - dk)Q_{k-1} + dk(KR_{e_k}K^T + P_k - AP_{k-1}A^T) \\ R_k = (1 - dk)R_{k-1} + dk\left(R_{e_k}^2 - CP_{k-1}C^T\right) \end{cases} \tag{4.22}$$

where d_k is the weight factor in the process of noise adaptation. The sliding window method is adopted to calculate the mean error for the output voltage error of fixed length. When the new data is obtained and updated, the old data with the longest window residence time is replaced. According to the memory length $M = 20$, the window length is also 20, and the output terminal voltage error calculation equation is

$$\begin{cases} e_k = U_L(k) - (U_O(k) - U_P(k) - i(k)R_o(k)) \\ R_{e_k} = \dfrac{1}{M} \sum_{k-M+1}^{k} e_j, dk = \dfrac{1-b}{1-b^{k+1}} \end{cases} \tag{4.23}$$

where e_k is the innovation sequence of terminal voltage, and the average error corresponding to fractional differential historical data is calculated through historical data of fixed length. Then, the prediction covariance matrix $P_{k/k-1}$ and the Kalman gain K_k are calculated according to the principle of minimum mean square error as

$$\begin{cases} P_{k|k-1} = E\left[\left(x_k - \widehat{x}_{k|k-1}\right)\left(x_k - \widehat{x}_{k|k-1}\right)^T\right] \\ = (A_d - \gamma_1)P_k(A_d - \gamma_1)^T - \sum_{j=2}^{n}\gamma_j P_{k-j}\gamma_j^T + Q_{k-1} \quad (4.24) \\ K_k = P_{k|k-1}C_k^T\left(C_k P_{k|k-1}C_k^T + R_k\right)^{-1} \end{cases}$$

According to the state variable x_k and the one-step optimal linear prediction, the state variable is updated to obtain the optimal state variable value after obtaining $P_{k/k-1}$, and K_k. The covariance matrix P_k is updated, as shown below:

$$\begin{cases} \widehat{x}_k = x(k|k-1) + K_k[U_L(k) - C_k * x(k|k-1)] \\ P_k = (I - K_k C)P_{k|k-1}\left(I - K_k C\right)^T + C R_k C^T \end{cases} \quad (4.25)$$

The adaptive noise factor is added to Kalman gain and the covariance matrix can correct the errors caused by the fractional-order process. Based on the F-EKF, the memory length and adaptive factors are added by using the sliding window model to improve the estimation accuracy and make the result to be smooth. The algorithm flowchart is shown in Figure 4.5.

First, $x(k/k-1)$ is predicted according to the fractional state variable equation, and then the terminal voltage error e_k of the model is calculated through the obtained $x(k/k-1)$. Then, the adaptive noise factor is calculated and brought into the Kalman gain K and covariance matrix $P_{k/k-1}$ to correct the noise error.

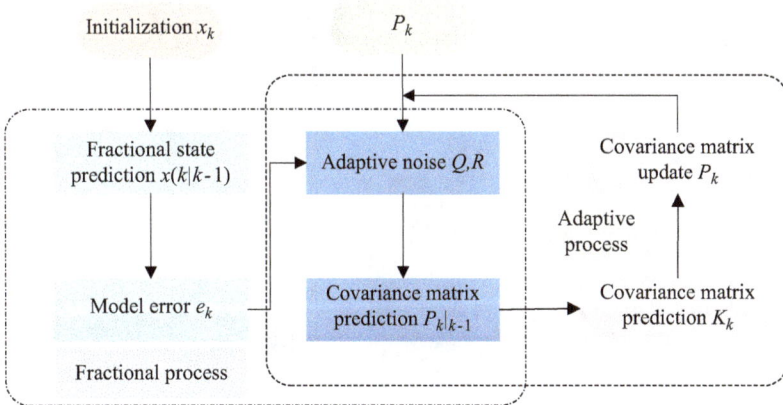

Figure 4.5   A-F-EKF flowchart

4.2   Equivalent circuit modeling

To better reflect the dynamic and static characteristics of the battery, it is proposed to use multiple RC parallel circuits connected with the series to simulate the polarization reduction process of lithium-ion batteries, that is, based on the Thevenin circuit model, multiple RC parallel circuits are connected in series to form a multi-stage Thevenin circuit model [194].

4.2.1   Second-order Thevenin modeling

Through the identification of the internal parameters of the Thevenin equivalent circuit model of different orders, combined with the experimental data, the accuracy of the battery model is verified, and it is concluded that the second-order Thevenin circuit model can reflect the dynamic response to the change process between the two poles of the battery. The dynamic performance characteristics of the battery model are very strong. The improved Thevenin equivalent circuit model is shown in Figure 4.6.

In Figure 4.6, U_{OC} represents the open-circuit voltage of the lithium-ion battery pack. R_0 represents the ohmic resistance. R_0 characterizes the ohmic effect of the battery. R_1 and R_2 represent the polarization resistance of the lithium-ion battery. C_1 and C_2 represent the polarization capacitance, and the double RC parallel circuit characterizes the polarization effect of the battery. The load current can be indicated by parameter I, where the discharge direction is positive, and U_L indicates the closed-circuit voltage when the battery pack is externally connected. According to the Kirchhoff law, the equations can be established as

$$\begin{cases} U_L = U_{OC} - IR_0 - U_{p1,t} - U_{p2,t} \\[2mm] I_{t1} = \dfrac{U_{p1,t}}{R_{p1}} + C_{p1}\dfrac{dU_{p1,t}}{dt} \\[2mm] I_{t2} = \dfrac{U_{p2,t}}{R_{p2}} + C_{p2}\dfrac{dU_{p2,t}}{dt} \end{cases} \qquad (4.26)$$

For the selected second-order equivalent model, [SOC U_{P1} U_{P2}] are selected as the state variables, combining the equation and the definition of SOC after

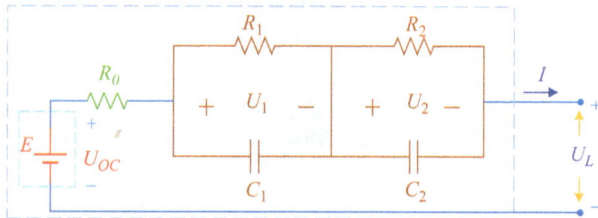

Figure 4.6   Improved Thevenin equivalent circuit model

discretization, the state-space equation can be listed as

$$
\begin{cases}
\begin{bmatrix} \text{SOC}_{k+1} \\ U_{p1,k+1} \\ U_{p2,k+1} \end{bmatrix} = \begin{bmatrix} 1 & 0 & 0 \\ 0 & 1 - \dfrac{t}{\tau_1} & 0 \\ 0 & 0 & 1 - \dfrac{t}{\tau_2} \end{bmatrix} \\[3em]
U_{L,k+1} = U_{oc}(\text{SOC}, k+1) - U_{p1} - U_{p2} - IR_0
\end{cases} \tag{4.27}
$$

In the above equations, t is the sampling time. After the above iterative update and correction, accurate SOC value estimation is inseparable from the internal dynamic characteristics of the battery; hence, it is particularly important to create an effective equivalent model and accurately identify the parameters of the selected model offline [195].

In this research work, the utilized battery model for validating the proposed battery aging model is a second-order Thevenin equivalent circuit battery model. It is a stand-alone Thevenin model based on the battery modeling technique that has been updated from the previous version. The primary version of the second-order equivalent circuit battery model is known as the first-order equivalent circuit model. The first-order Thevenin equivalent circuit model has relatively higher accuracy than other electrical-based battery models; also, the circuit element in the first-order Thevenin model is not complex and has a distinct physical explanation.

A fundamental first-order Thevenin model is already presented, in which U_{oc} (V) denotes a nonlinear voltage source which is the function of SOC and the first-order RC network consists of a capacitor C_1 (F) and a resistor R_1 (Ω). The capacitor and resistor are the polarization capacitance and resistance. R_0 ($\mu\Omega$) denotes the battery ohmic resistance, I (A) denotes the changing current, and U_t (V) denotes the battery terminal voltage. If considering the voltage of the RC network as U_1, according to Kirchhoff's law, the following equations are obtained for the first-order Thevenin equivalent circuit, where τ_1 is the time constant. The calculation procedures are

$$
U_t = U_{oc} - U_1 - IR_0 \tag{4.28}
$$

$$
U_1 = IR_1 \left(1 - e^{-t/\tau_1} \right) \tag{4.29}
$$

$$
\begin{cases}
I_{U_1} = I_{R_1} + I_{C_1} \\
I_{U_1} = \dfrac{U_1}{R_1} + C_1 \dfrac{dU_1}{dt}
\end{cases} \tag{4.30}
$$

Although the first-order Thevenin equivalent circuit battery model seems relatively accurate as to the experimental battery model, it is not efficient. The shortcoming of this model is the fixed open-circuit voltage U_{oc}. Therefore as a solution to this challenge, an advanced Thevenin model with an additional RC network in series with the voltage source is suggested by Thanagasundaram *et al.* [196] for the very first time and named it a second-order Thevenin model. As a

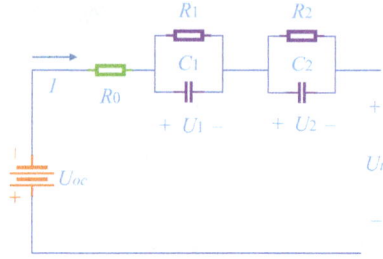

Figure 4.7   Second-order Thevenin model

result, it can increase the efficiency of the battery model and against the influential battery response, dynamic open-circuit voltage, and electrochemical polarization. The suggested dual-polarization model or second-order Thevenin equivalent circuit battery model is presented in Figure 4.7.

Here, the new RC network also consists of a polarization capacitor C_2 (F) and a polarization resistor R_2 (Ω). If considering the voltage U_2 in the second RC network, the equation obtained from the circuit model will be derived by (4.31), (4.32), and (4.33), where τ_1 and τ_2 represent the time constant for U_1 and U_2, respectively. Then, the current of the circuit will be derived by

$$U_t = U_{oc} - U_1 - U_2 - IR_0 \tag{4.31}$$

$$U_1 = IR_1\left(1 - e^{-t/\tau_1}\right) \tag{4.32}$$

$$U_2 = IR_2\left(1 - e^{-t/\tau_2}\right) \tag{4.33}$$

$$I = \frac{U_1}{R_1} + C_1\frac{dU_1}{dt} = \frac{U_2}{R_2} + C_2\frac{dU_2}{dt} \tag{4.34}$$

The battery model presented in Figure 4.7 is the main battery model that is used in this research work to validate the proposed battery aging model. Equations (4.31)–(4.34) are the fundamental equations for this battery model. The next chapter will describe the mathematical analysis and the implementation of this battery model to validate the proposed battery aging model.

4.2.2   Identification procedure design

The hybrid pulse-power characterization (HPPC) test is an experiment to obtain the performance of lithium-ion batteries through pulse charge–discharge treatment. The constant temperature is set to be 25 °C, and the HPPC experimental test flowchart is shown in Figure 4.8.

The HPPC experiment is to perform pulse charge–discharge experiments on lithium-ion batteries and obtain the current–voltage curve of the battery under the premise of fully considering safety. According to the experimental data and curves, the parameters of the model are identified to obtain the corresponding relationship

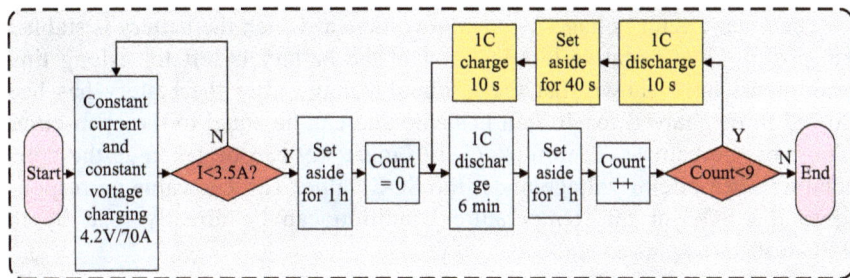

Figure 4.8   HPPC experimental test flowchart

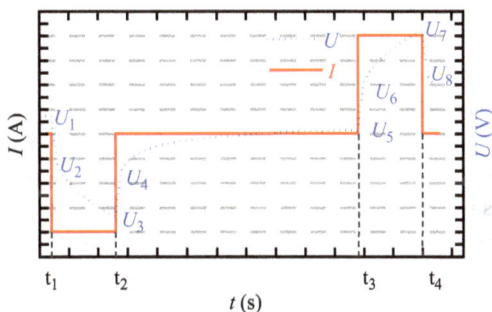

Figure 4.9   Single HPPC test voltage and current curve

between them and the SOC. The current and voltage curve of the HPPC experiment at 25 °C is shown in Figure 4.9.

By analyzing the HPPC experimental voltage variation curve, the variation characteristics can be extracted to obtain the model parameters. The features in Figure 4.9 can be extracted by analysis. The discharge starts at time t_1. The lithium-ion battery terminal voltage changes from U_1 to U_2, mainly due to the ohmic resistance of the lithium-ion battery. From t_3 to t_4, the lithium-ion battery terminal voltage is changed from U_3 to U_4, which is also due to the voltage change caused by the ohmic resistance of the lithium-ion battery. Therefore, the two voltage differences can be selected to divide the voltage by the current for the average value to obtain the ohmic resistance. From t_2 to t_3, the lithium-ion battery terminal voltage drops slowly from U_2 to U_3. This is due to the battery polarization effect. The discharge current charges the polarization capacitor, which is the zero-state response of the dual RC series circuit. The time-domain analysis is performed on the circuit and selects the data from t_2 to t_3 to obtain the terminal voltage U_L as a function of time t, as shown below:

$$U_L(t) = U_2 - I\left[R_{P1}\left(1 - e^{-t/\tau_1}\right) + R_{P2}\left(1 - e^{-t/\tau_2}\right)\right] \tag{4.35}$$

The open-circuit voltage U_{OC} is the voltage at which the battery is stable at both positive and negative terminals when the battery is left for a long time. Experiments have shown that the terminal voltage after the battery has been allowed to be shelved for 40 min is stable and can be equal to the open-circuit voltage of the battery. Therefore, U_1 in Figure 4.9 can be taken as the corresponding open-circuit voltage under the SOC value. The U_{oc} value corresponding to the SOC at this temperature condition can be directly read by the HPPC test.

4.2.3   Identification effect verification

Based on previous research, a polynomial fitting equation for each parameter of the second-order model is obtained through actual testing of various performance parameters of lithium-ion batteries. The mathematical expression is introduced into Simulink to obtain the equivalent model verification output voltage curve, as shown in Figure 4.10.

The high accuracy of the equivalent model can be testified from the error curve in Figure 4.10. The verification deviation is toggled within the effective accuracy range, and the maximum error is 3.96%. The model has high experimental accuracy, which verifies the feasible value of the model. A new equivalent circuit model is established based on the improved Thevenin model, and the HPPC test curve is used to perform parameter identification and verification in Simulink to realize an accurate description of the working process of lithium-ion batteries. Considering the nonlinear operating characteristics of lithium ions, a state of charge estimation method based on the EKF is established to estimate the SOC value of lithium-ion batteries.

The simulation platform proves that the adaptive correction capability of the EKF algorithm can well realize the online state estimation. As an approached model-based battery system, it is necessary to establish a modeling-based simulation direction. The simulation model can compare experimental data and identified parameters. To build the simulation model, a variety of working processes has been done.

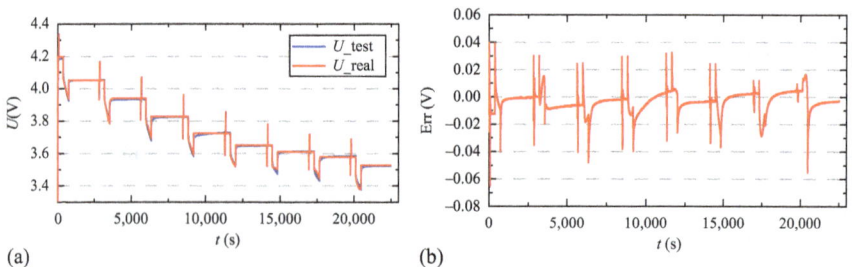

Figure 4.10   Model validation output curve: (a) comparison of real and test voltages and (b) output voltage error curve

4.2.4   Corroboration of model parameters

To verify the validity of the high-order equivalent circuit model, under the condition of cyclic discharge maintenance, the characterization of the actual battery voltage and current data is imported into the developed high-order equivalent circuit model. The model is verified by integrating the effects of the identification of the identified parameters. The present state of SOC values of the battery is measured and combined with the nonlinear function relationship between SOC and U_{oc} using the Ampere-time necessary procedure. The corresponding OCV value of the battery can also be accessed. The battery terminal voltage measured in the equivalent model is substituted by U_{oc}. The calculated value is compared to the experimental terminal voltage value. The result of the comparison and the resulting error is seen in Figure 4.11.

Figure 4.11 indicates a relation between the approximate value of the battery terminal voltage and the experimental voltage of the battery discharge under the experimental conditions of the cyclic discharge. The solid blue line X_1 is the simulation voltage that can be compared with the actual voltage. The solid red line X_2 is the experimental terminal voltage value. However, the simulation error by the comparison of two simulation results is presented in Figure 4.12.

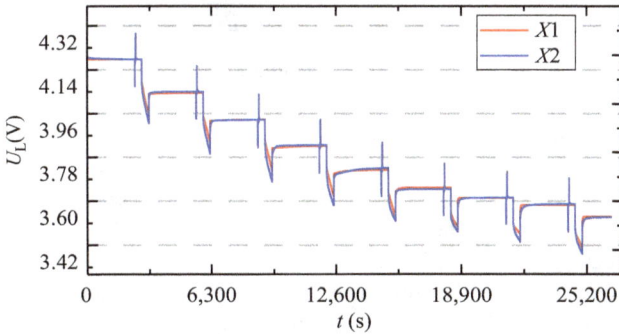

Figure 4.11   Simulation indication of a waveform

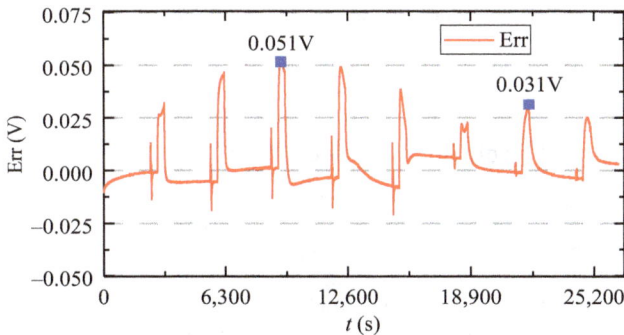

Figure 4.12   Error indication of simulation result

Figure 4.12 shows the variation of the simulation error. The figure indicates that the projected value has a strong monitoring impact on the experimental value. The average estimated variance is roughly 0.051 V, which may describe the battery voltage value at the end of the experiment. A study of the voltage contrast error between the voltage calculation deviation increases at the end of the battery discharge. In this respect, the battery voltage varies significantly before the final discharge. A calculation error accompanies the simulation result. The model accuracy is 99.14%, while the highest lithium-ion battery voltage is 4.2 V.

4.3   Model parameter identification

The equivalent models should be used to illustrate the problematic electrochemical response function of the battery physical circuit. In short, the equivalent refers to the same current in the battery system. The developed equivalent model will effectively track the voltage. The voltage error is minimal, with a highly accurate effect. Identification of the model parameter shall be carried out with the state of charge and other factors concerning the battery power, the current rate of charge–discharge, the degree of cell aging, and the ambient temperature.

4.3.1   Recursive least-square calculation

As the battery system has strong nonlinear characteristics, it is a difficult problem in the process of battery modeling to identify its parameters effectively. At present, the classical identification methods include the least-square algorithm, particle swarm optimization (PSO) algorithm, genetic algorithm, and so on. These algorithms can effectively identify the equivalent model parameters online or offline.

To make the model reflect the actual running state of the battery more accurately, it is necessary to identify the model parameters. The recursive least-square (RLS) algorithm is one of the most widely used estimation methods in the field of parameter and system identification. Consequently, the RLS algorithm is used to identify the modeling parameters.

For a single-input single-output (SISO) system, only the input and output characteristics of the system need to be understood, not its internal structure. Suppose an SISO stochastic system, the discrete equation of the model to be identified is shown as

$$G(z) = \frac{y(z)}{u(z)} = \frac{b_1 z^{-1} + b_2 z^{-2} + \cdots + b_n z^{-n}}{1 + a_1 z^{-1} + a_2 z^{-2} + \cdots + a_n z^{-n}} \qquad (4.36)$$

The corresponding difference equation is shown as

$$z(k) = -\sum_{i=1}^{n} a_i y(k - i) + \sum_{i=1}^{n} b_i u(k - i) + v(k) \qquad (4.37)$$

where $z(k)$ is the k-th observation value of the system output, $u(k)$ is the k-th value of the system input, $y(k)$ is the k-th true value of the system output, and $v(k)$ is the random noise with the mean value of 0.

The parameters are initialed as $h(k) = [-y(k-1), \ldots, y(k-n), u(k-1), \ldots, u(k-n)]$, $\theta = [a_1, a_2, \ldots, a_n, b_1, b_2, \ldots, b_n]^T$, $V_m = [v(1)\ v(2)\ \ldots\ v(m)]^T$, where θ is the parameter to be identified. Then, the matrix form is

$$Z_m = H_m\theta + V_m \tag{4.38}$$

For the above equation, the idea of the least-square method is to find the estimated value of θ, so that the sum of squares of the difference between the measured value and the estimated value is obtained by estimating the minimum, as shown below:

$$J\left(\hat{\theta}\right) = \left(\hat{Z}_m - H_m\hat{\theta}\right)^T\left(Z_m - H_m\hat{\theta}\right) \tag{4.39}$$

According to the extremum theorem, finding the minimum value of the above equation is equivalent to finding its derivative and then solving it, that is, the least-square estimation of θ is shown below:

$$\hat{\theta} = \left[H_m^T H_m\right]^{-1} H_m^T Z_m \tag{4.40}$$

The idea of the RLS algorithm is to use the new measurement value to correct the last estimation based on the last estimation result until satisfactory accuracy is achieved. Considering that the measurement data may be obtained under different conditions, the measurement accuracy is affected by many factors. Therefore, the data obtained each time may have the problem of confidence. As a result, the weighted method is used to treat each measurement value. Finally, the recursive equation of the least-square method is obtained as

$$\begin{cases} \hat{\theta}_{m+1} = \hat{\theta}_m + K_{m+1}\left[z(m+1) - h(m+1)\hat{\theta}_m\right] \\ P_{m+1} = P_m - P_m h^T(m+1)[W^{-1}(m+1) + h(m+1)P_m h^T(m+1)]^{-1}h(m+1)P_m \\ K_{m+1} = P_m h^T(m+1)[W^{-1}(m+1) + h(m+1)P_m h^T(m+1)]^{-1} \end{cases} \tag{4.41}$$

where W_m is a weighted matrix, which is an asymmetric positive-definite matrix, usually diagonal matrix, $P_m = [H_m T W_m H_m]^{-1}$, and gain matrix $K_{m+1} = P_{m+1}hT$ $(m+1)W(m+1)$. Taking the Thevenin circuit model as an example, Laplace transform was carried out on its expressions of voltage and current, as shown below:

$$G(s) = \frac{U_{rc}(s)}{I(s)} = \frac{R_0 + R_p + R_0 R_p C_p s}{1 + R_p C_p s} \tag{4.42}$$

The Euler method is used to discretize the unknown parameter s of the continuous transfer function so that the above-mentioned Thevenin equivalent circuit model can be transformed into

$$U_{rc}(k) = aI(k) + bI(k-1) - cU_{rc}(k-1) \tag{4.43}$$

Among them, *a*, *b*, and *c* are the abstract model parameters, and the identification parameter vector is $\theta = (a, b, c)$. After the identification results are obtained, the actual parameters of the battery can be obtained by parameter separation, as shown below:

$$
\begin{cases}
R_0 = a \\
R_p = \dfrac{b - ca}{c} \\
\tau_p = R_p C_p = \dfrac{T}{1 + c} \\
C_p = \dfrac{T}{b - ca}
\end{cases}
\tag{4.44}
$$

The main idea of the least-square method is to solve the unknown parameters so that the sum of squares of the difference between the theoretical value and the observed value is minimized, and the fitting object is infinitely close to the target object. The offline least-square algorithm is used to identify nonlinear polarization resistance and polarization capacitance in equivalent model parameters. The least-square method is widely used in mathematical optimization problems. According to the minimum variance criterion, the best function matching of data is found in this method, which is widely used in error estimation, system identification, and prediction. In the field of parameter identification, it solves a group of unknown parameters, so that the sum of squares of the difference between the theoretical values and the observed values of these parameters is minimized.

The RLS algorithm is based on the improvement of the least-square algorithm, and the model parameters of each sequence data are based on the model parameters at the previous moment and the "innovation" caused by the data at this moment to achieve the purpose of real-time estimation. The flowchart of RLS parameter identification based on the lithium-ion battery is shown in Figure 4.13.

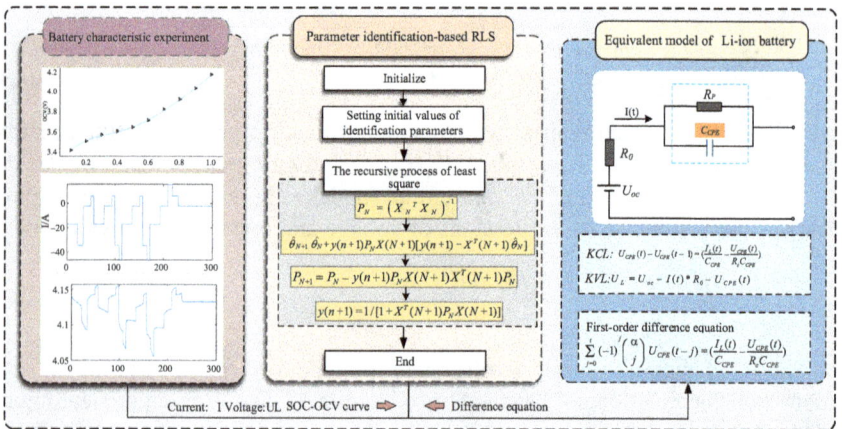

Figure 4.13   Recursive least-square identification model parameter structure diagram

4.3.2 Forgetting factor—RLS algorithm

Due to the phenomenon of the filtering saturation in the least-square method, the values of gain k and p will become smaller and smaller as the number of iterations of algorithm data increases. This makes the ability to correct the data gradually weaker and increasing the degree of data saturation, which eventually leads to larger and larger errors in parameter identification. Therefore, it is considered to add a forgetting factor based on least-square identification to improve the online estimation capability of the RLS algorithm. The forgetting factor is used to offer less weight to the long-running data while the latest observation data takes up more weight in the identification process. The forgetting factor λ $(0 < \lambda < 1)$ is introduced to weaken the influence of old data and enhance the feedback effect of new data. The improved objective function is shown as below:

$$J = \sum_{k=n+1}^{N} \lambda^{N-k} [e(k)]^2 \tag{4.45}$$

After the forgetting factor is added, the recursive formula of the least-square method is modified as

$$\begin{cases} \hat{\theta}(k+1) = \hat{\theta}(k) + K(k+1)\left[y(k+1) - \phi^T(k+1)\hat{\theta}(k)\right] \\ K(k+1) = P(k+1)\phi(k+1)[\phi^T(k+1)P(k)\phi(k+1) + \lambda]^{-1} \\ P(k+1) = \lambda^{-1}[I - K(k+1)\phi^T(k+1)]P(k) \end{cases} \tag{4.46}$$

In (4.46), the closer the forgetting factor is to 1, the better the simulation result is. Generally, the value is between 0.95 and 1. The smaller the value, the better the tracking effect of the algorithm, but it will cause algorithm fluctuations. When λ is equal to 1, it is the standard RLS algorithm.

4.3.3 Adaptive PSO

To judge whether the equivalent model of the battery can accurately characterize the working characteristics of the lithium-ion battery, the parameter identification of the equivalent circuit model should be carried out first. The curve fitting, partial least square, Kalman filtering, and other widely used methods for the detection of parameters are currently used. The lithium-ion battery is a typical nonlinear system. Considering the limitation of the least-square method and curve fitting method to a nonlinear system, an improved PSO algorithm is used to identify circuit parameters.

The standard PSO algorithm originates from the study of bird predation behavior. The PSO algorithm uses random particles in the M-dimensional solution space as the solution of the problem to be solved and uses its position to represent. Each particle judges its degree of conformity with the optimal solution through the fitness function. At the same time, the particles update the current position through speed and position transformation. The update of speed and position is not random, but the global search is performed by the globally optimal particles and the

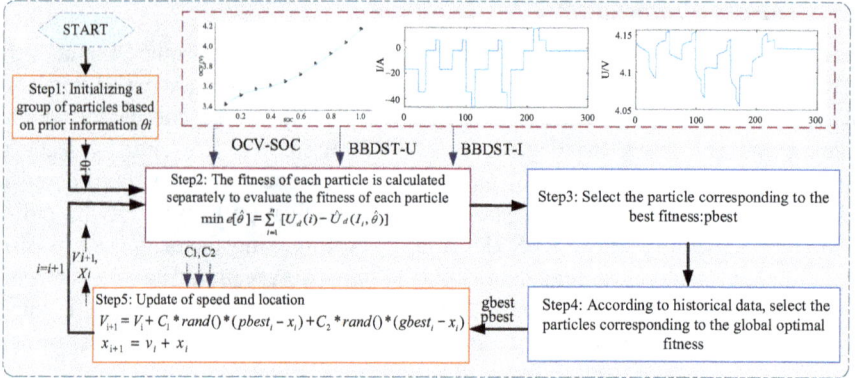

Figure 4.14   PSO algorithm flowchart

individual historical optimal to converge to the optimal solution. The iterative process of the PSO algorithm is shown in Figure 4.14.

As shown in Figure 4.14, the improved PSO algorithm is used for parameter identification, and the specific steps are as follows.

Step 1: The parameters are initialed as $i = 0$ and a group of particles is initialized with dimension m. Each particle represents a possible solution in the solution space. To have a faster convergence speed and calculation accuracy, it is necessary to set the scope of the solution space and the maximum search speed according to historical experience. Initialize a group of three-dimensional particle swarm according to the number of parameters to be identified, including the number, speed, and position setting of particle swarm.

Step 2: The fitness of particle population at the time i is calculated. The fitness of each particle is calculated. The fitness function is a standard for judging the advantages and disadvantages of the particle position. To make the identified parameters accurately characterize the space state of the lithium-ion battery, the absolute difference between the model output value corresponding to the particle position and the actual output value of the battery is selected as the fitness function J, as shown below:

$$J = \sum_{j=1}^{n} \left(\left| U_m(k) - U_m\left(\hat{k}, \theta\right) \right| \right) \tag{4.47}$$

In (4.47), $U_m(k)$ represents the actual output voltage of the lithium-ion battery, and $U_m\left(\hat{k}, \theta\right)$ represents the model output voltage under the identified parameters.

Step 3: Particles are selected accordingly. Selecting the particle with the best fitness from the particle swarm at the time $t = i$, even if j is the local optimal value. Then, from each local optimum in $t = 1$ to $t = i$, the position with the best fitness is selected. Judging whether it meets the convergence condition, if so, the global search ends, and the identification result is the corresponding position, if not, the procedure turns to the next step.

Step 4: The speed and position of the particle swarm are updated. The speed update equation and position update equation are

$$
\begin{cases}
v_{id}^{k+1} = w * v_{id}^{k} + c1 * \eta_1 * \left(P_{id}^{k} - x_{id}^{k} \right) + c2 * \eta_2 * \left(P_{gd}^{k} - x_{id}^{k} \right) \\
x_{id}^{k+1} = x_{id}^{k} + r v_{id}^{k+1}
\end{cases}
\tag{4.48}
$$

Among them, v_{id}^{k+1} is the speed at the time point of $k+1$, ω is the inertia weight coefficient, which is generally taken as a value between 0.4 and 0.9. The larger the value of w, the larger the range of particle search, but the convergence speed will also slow down correspondingly. $c1$ and $c2$ are self-cognitive weight coefficients and social cognitive weight coefficients, respectively, and their values determine the following situation of particle position with local optimum and global optimum. η_1 and η_2 are random numbers between $(-1,1)$.

Steps 2–4 are carried out cyclically and iteratively and keeps approaching the optimal value until the convergence condition is reached.

PSO algorithm is easy to fall into local optimum, so it is considered to improve PSO algorithm with an adaptive adjustment strategy to improve the convergence speed and estimation accuracy of the algorithm as a whole. The search speed of particles changes with the value of inertia weight w and a larger inertia factor is beneficial to jump out of the local minimum and facilitate global search, while a smaller inertia factor is conducive to an accurate local search of the current search area to facilitate algorithm convergence. However, if w is too large, it will easily lead to premature convergence and oscillation near the global optimal solution in the later stage of the algorithm. To make a proper trade-off between search speed and accuracy of results, a PSO algorithm based on adaptive weight is established, and an improved approach is established, whose inertia weight dynamically changes with iteration times and particle position. The equation of dynamic change of inertia factor w is

$$
\begin{cases}
w = w_{\min} + \dfrac{(w_{\max} - w_{\min}) * (J - J_{\min})}{J_{\mathrm{avg}} - J_{\min}}, J < J_{\mathrm{avg}} \\
w = w_{\max}, J > J_{\mathrm{avg}}
\end{cases}
\tag{4.49}
$$

In (4.49), J represents the current target function value of particles, and J_{avg} represents the average target value of particles. According to the iteration times of the algorithm and fitness function of particles, the inertia weight w will increase when the target values of each particle tend to be consistent or local optimum but decrease when the target values of each particle are scattered. Meanwhile, for particles whose target function value is better than the average target value, the corresponding inertia weight factor is smaller, thus retaining the particle. On the contrary, for particles whose target function value is worse than the average target value, the corresponding inertia weight factor is larger, which makes the particle orientation better.

The PSO algorithm is a process of iterative optimization, so it needs the discrete state-space equation of the lithium-ion battery. According to the current–voltage

Table 4.1   Difference equation of equivalent circuit model

Step 1: List the state equation of the model	$\begin{cases} U_{oc} = IR_0 + U_p + U_L \\ I = \dfrac{U_p}{R_p} + C_p * \dfrac{dU_p}{dt} \end{cases}$
Step 2: Discretization of the equation	$\begin{cases} \dfrac{U_p(k)U_p(k-1)}{T} = -\dfrac{1}{R_pC_p}U_p(k+1) + \dfrac{I}{C_p} \\ U_p(k) = U_{oc} - U_L(k) - I(k)*R \\ U_p(k+1) = U_{oc} - U_L(k+1) - I(k+1)*R \end{cases}$
Step 3: Organize and simplify	$U_m(k) = \frac{R_pC_p}{R_pC_p+T}U_m(k-1) - \left[\frac{TR_p}{R_pC_p+T}+R_0\right]I(k) + \frac{RR_pC_p}{R_pC_p+T}I(k-1)$

relationship of the equivalent circuit model and Kirchhoff's law, the differential equation is discretized to obtain its difference equation, and its recursion process is shown in Table 4.1.

4.3.4   Parameter extraction results

This analysis will increase the complexity of the online parameter detection to improve accuracy. For this cause, the method of offline recognition is chosen for the experiment at 25 °C. The battery is subjected to the HPPC experiment. The parameters of the battery model are obtained by analyzing the operational characteristics. Related practical measures have been taken below in the HPPC experiments.

The battery is subjected to continuous current constant voltage charging at a current rate of 1 C. The cut-off voltage charge is set to 4.2 V and then transformed to a constant voltage charge, with a current cut-off charge of 0.5 C to guarantee that the battery is fully charged.

To reach a stable battery voltage, the battery must be fully assembled after charging, and the preferred standing time must be 30 min.

The lithium-ion battery has steadily been discharged at a rate of 1 C for a period of 10 s. The battery should be permitted to stand for 40 s after the discharge has ceased.

At a time 10 s, the lithium ion is subject to continuous current charging at a rate of 1 C. The battery should be allowed to stand for 40 s after the discharge has ceased.

The practical operation (3) and (4) are a complete HPPC trial. To investigate the characteristics of the reaction of the battery under various SOC values. The battery has been discharged for 6 min under the existing discharge conditions with a capacity rate of 1 C to reduce the SOC battery by 0.1 and obtain a new SOC experimental design. After 40 min, the HPPC experiment continued until the battery power is 0, and the experiment is over.

The fitted parameter is then shown with a polynomial of the sixth order, and the following results are obtained. The fitting results of the curve are R_0, R_{P1}, R_{P2},

C_{P1}, and C_{P2}, respectively, reflecting the parameters for the equivalent circuit of high-order modeling. The fitted results are

$$
\begin{cases}
\begin{cases}
R_0(S) = .00377 * u^6 - 0.01272 * u^5 + 0.01776 * u^4 \\
-0.01371 * u^3 + 0.006676 * u^2 - 0.002097 * u + 0.001634
\end{cases} \\[4pt]
\begin{cases}
R_{p1}(S) = 0.2705 * u^6 - 0.9849 * u^5 + 1.417 * u^4 \\
-1.018 * u^3 + 0.3783 * u^2 - 0.06759 * u + 0.004558
\end{cases} \\[4pt]
\begin{cases}
R_{p1}(S) = 0.2705 * u^6 - 0.9849 * u^5 + 1.417 * u^4 \\
-1.018 * u^3 + 0.3783 * u^2 - 0.06759 * u + 0.004558
\end{cases} \\[4pt]
\begin{cases}
C_{p1}(S) = 1.136 * 10^7 * u^6 - 3.991 * 10^7 * u^5 + 5.509 * 10^7 * u^4 \\
-3.754 * 10^7 * u^3 + 1.294 * 10^7 * u^2 - 2.031 * 10^6 * u + 1.241 * 10^5
\end{cases} \\[4pt]
\begin{cases}
R_{p2}(S) = -0.1694 * u^6 + 0.6867 * u^5 - 1.08 * u^4 + 0.8263 * u^3 \\
-0.3155 * u^2 + 0.05511 * u - 0.002991
\end{cases} \\[4pt]
\begin{cases}
C_{p2}(S) = -1.09 * 10^6 * u^6 + 4.437 * 10^6 * u^5 - 7.464 * 10^6 * u^4 \\
+6.696 * 10^6 * u^3 - 3.353 * 10^6 * u^2 + 8.561 * 10^5 * u - 5.348 * 10^4
\end{cases}
\end{cases}
$$

$$(4.50)$$

According to the HPPC terminal voltage curve as shown in Figure 4.9, some experiments have been carried out on HPPC. The battery that is shelved after 6 min of discharge is selected for double exponential fitting. The parameters identified results are shown in Table 4.2 according to the HPPC experimental data at each point. The values of the high-order equivalent circuit model parameters corresponding to the SOC values can be obtained.

The parameters are identified by the HPPC test at the temperature condition of 25 °C. Each parameter reflects each condition searchable of lithium-ion batteries. However, the effect of electrochemical polarization and concentration resistance is carried out by curve fitting results in Figure 4.15.

Table 4.2 *Discharge-profile criteria under specific SOC levels*

S1 (%)	OCV (V)	R_0 (Ω)	R_{p1} (Ω)	R_{p2} (Ω)	C_{p1} (μF)	C_{p2} (μF)	I_L
10	3.4545	0.00148	0.00069	0.0001	17,529	4,495	70
20	3.5367	0.0014	0.00003	0.00045	13,469	27,034	70
30	3.5900	0.00135	0.00003	0.00038	17,552	31,568	70
40	3.6163	0.00132	0.00034	0.00003	33,144	29,758	70
50	3.6511	0.00131	0.00003	0.00034	20,421	31,815	70
60	3.7366	0.00131	0.00004	0.00057	20,176	23,233	70
70	3.8309	0.0013	0.00004	0.00064	18,860	25,931	70
80	3.9360	0.0013	0.00004	0.00056	23,504	25,388	70
90	4.0513	0.0013	0.00003	0.00048	19,710	25,491	70
100	4.1840	0.00131	0.00003	0.00045	26,053	28,427	70

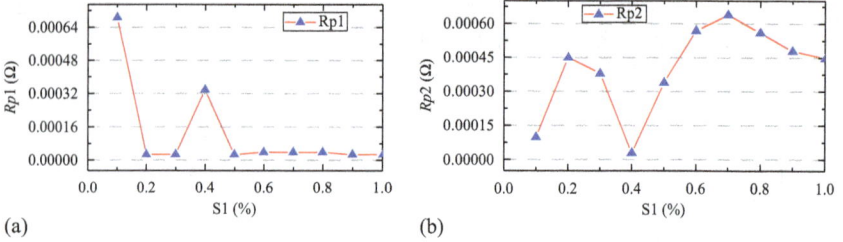

(a)   (b)

Figure 4.15   High-order resistance fitting waveform: (a) polarization resistance fitting and (b) concentration resistance fitting

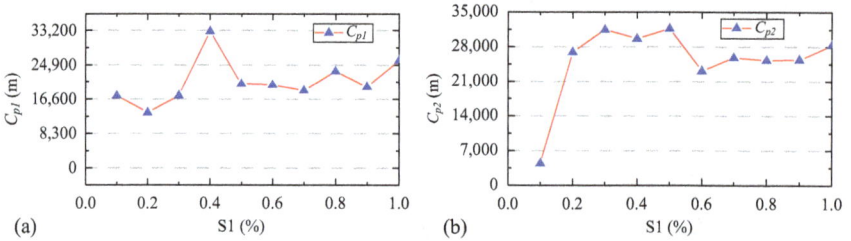

(a)   (b)

Figure 4.16   High-order resistance fitting waveform: (a) polarization capacitance fitted waveform and (b) concentration capacitance fitted waveform

The hybrid pulse-power curve of the typical experimental voltage variance is obtained. The curve fitting results showed that it is very anastomotic to the experimental result. The variation characteristics can be derived to obtain the model parameters. The polarization and concentration capacitance are fitted by the experiment data in Figure 4.16.

The trends of R_0, R_{p1}, C_{p1}, and C_{p2} are coupled with state combinations, respectively. As can be shown from the diagram, the internal parameters fluctuate within the particular spectrum of the SOC. These parameters may then be replaced by their average values or by a look-up table if the precision effect is not good. One way to use model parameters is to provide a limited amount of processing.

4.4   Fractional experimental test

The traditional hybrid pulse-power experimental research was performed on lithium-ion batteries, according to the freedom car battery experiment manual. By using the same process already carried out, the battery power has been extracted.

4.4.1   Real-time platform implementation

The battery test platform arrangement is introduced by the secondary leading test equipment "CT-4616-5V100A-NTFA" produced by Shenzhen Xinwei New Energy

Technology Corporation. The platform capable of charge–discharge is exposed in a ternary battery module with a maximum voltage of 380 V at the frequency of 50/60 Hz and the maximum current is 100 A. Major parameters such as voltage, current, and temperature have been calculated by equipment to be promising. The host computer is installed with the BTS-7.6 software of the experimental program, which is used for real-time processing of data collection. The measuring data has been transmitted via Transmission Control Protocol/Internet Protocol (TCP/IP) ports to the host computer. The incubator has three independent layers manufactured by Dongguan Bell Experimental limited company qualified to monitor the high- or low-temperature chamber (DGBELL-BTT-331C). This experiment is placed at a constant temperature of 25 °C, as shown in Figure 4.17.

The battery ages due to recycling and other reasons and the actual battery capacity will be significantly different from the calibration capacity. The actual discharge capacity of the battery is essential for the SOC estimation of the lithium-ion battery. Therefore, the calibration capacity of the lithium-ion batteries must be performed first. In this study, the identification of the online parameter will increase the complexity of the algorithm, and the accuracy is not improved. Therefore, an offline identification model is constructed.

The battery is subjected to the pulse discharge experiment, and the parameters of the battery model are obtained by analyzing the operating characteristics of the battery during operation. The battery has been subjected to the pulse discharge experiment, and the parameters of the battery model have been obtained by analyzing the operating characteristics of the battery during operation. The lithium-ion

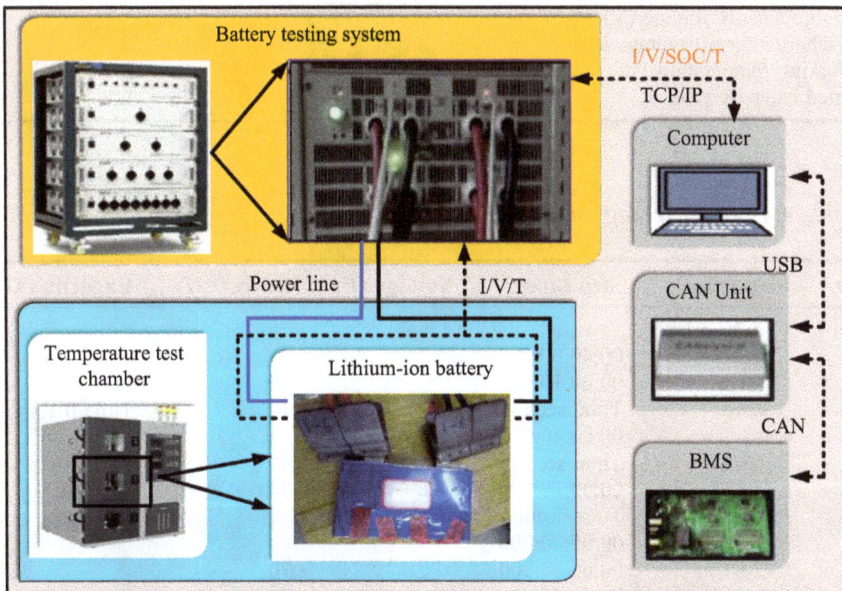

Figure 4.17   The schematics of the battery test platform

ternary battery is picked as the experimental object in this experiment. The primary technical parameters of the battery are shown in Table 4.3.

The experiment has been carried out using specified requirements with corresponding parameters. The experiment is done in the laboratory in a day according to these parameters. Therefore, an offline method of identification is selected. The battery has been subjected to the pulse discharge experiment, and the parameters of the battery model have been obtained by analyzing the operating characteristics of the battery during operation.

4.4.2   Test step procedure

The HPPC test is intended to determine dynamic power capability over the useable voltage range using a test profile that integrates both discharge and regen pulses. In this analysis, the battery at ambient temperature is subjected to HPPC and the parameters of the battery model are obtained by examining the battery operating characteristics during operation. The following steps describe the basic experimental procedure of the HPPC experimental test, as shown in Table 4.4.

Table 4.3   Key functions of battery parameters

Criteria	Parameters
Size: length * width * height (mm)	200 * 80 * 180
Rated voltage (V)	3.75
Maximum load current rate (C)	1.50
Charge cut-off voltage (V)	4.20
Discharge cut-off voltage (V)	3.00
Working temperature (°C)	25.00
Rated capacity (Ah)	4.00

Table 4.4   Steps setting for HPPC test

ID	Step name	Step time	Voltage (V)	Current (I)	Capacity (Ah)
1	CCCV_ Chg		4.2	70	
2	Rest	00:40:00:000			
3	CC_DChg	00:00:10:000		70	
	Record	Time section	0 s	10 s	Time 0.1 s
4	Rest	00:00:40:000			
	Record	Time section	0 s	40 s	Time 0.1 s
5	CC_DChg	00:00:10:000		70	
	Record	Time section	0 s		
6	Rest	00:05:00:000			
7	CC_DChg	00:06:00:000		70	
8	Cycle	Begin ID: 2	Times 9		
9	End				

The battery is completely charged first and then left for 30 min to rest. A discharge pulse of 10 s is implemented and left to rest for 40 min, and then a charge pulse of 10 s is implemented and left to be rest for 30 min. This is replicated 9 times, and the data is registered at various SOC stages to define the parameters in the circuit. The measurement is performed from 0.9 to 0.1 at 0.1 SOC intervals. When a current is loaded, there is a pulse charge or discharge, a voltage increase or decrease occurs and this can be used to measure the parameters.

4.4.3 HPPC test

The HPPC test is a battery performance test method that is described in the freedom car battery test manual for power-assist hybrid electric vehicles. The model parameters can be sufficiently identified via HPPC tests. The HPPC experimental process has been described. The lithium-ion battery is discharged at 1 C for 10 s, which is then charged with 0.75 C for 10 s after shelved for 40 s. In the cyclic test, the battery is subjected to a pulse-power characteristic test of an equally spaced state point. The SOC varies from 0% to 1%.

To return the battery to electrochemical and thermal equilibrium at evenly spaced stages, the battery must be left for a longer time between adjacent pulse checks. According to the HPPC experiment protocol, the entire experiment consists primarily of a single repeated charge–discharge pulse test, as shown in Figure 4.18.

All the model parameters can be obtained through the hybrid pulse-power experimental test. However, it takes a long time of about 12 h. It generates excellent consumption of the experimental equipment and operators. Most time of the HPPC test is spent on the repeated shelved stage. A hybrid pulse test is conducted consisting of a series of HPPC profiles, constant current discharge pulses, and rests. The HPPC test will also be used to drive parameters of the circuit module equivalent to the battery.

Figure 4.19(a) and (c) presents a single cycle HPPC test voltage and current HPPC curves, respectively. Figure 4.19(b) and (d) shows the full cycle current and voltage. The batteries of similar capacity are selected to conduct the HPPC test simultaneously, making batteries of similar capacity for the same batch. It is used to ensure the consistency of the quantities that affect the experimental results except for the experimental quantities.

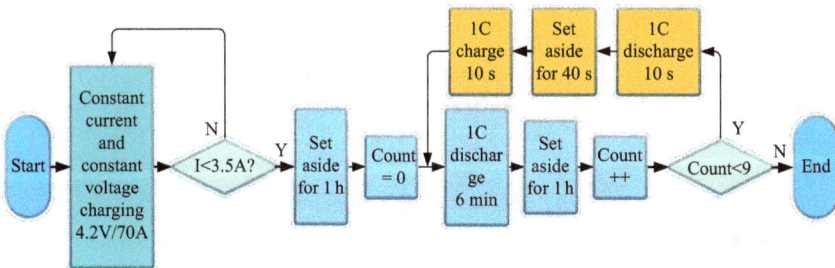

Figure 4.18   HPPC test procedure

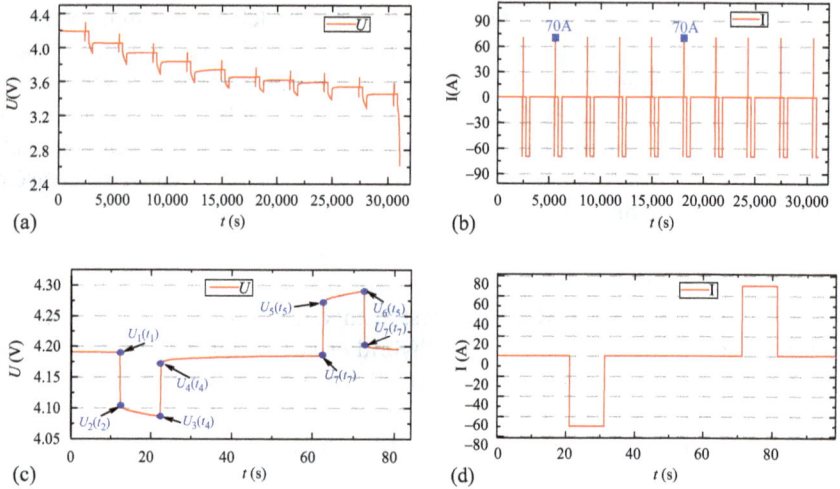

Figure 4.19   Complete current and voltage discharge profile during the HPPC
test: (a) closed cycle voltage response curve, (b) closed cycle current
response curve, (c) single-cycle voltage tracking curve, and (d)
single-cycle voltage response curve

Figure 4.19(c) reflects the transitory and steady-state characteristics of lithium-ion batteries. Each voltage point should satisfy the following criteria.

The discharge starts on time t_1. The lithium-ion terminal voltage drops abruptly from U_1 to U_2. It is mainly because of the voltage changes caused by the ohmic resistance.

During t_2 and t_3, due to the influence of battery polarization, the lithium-ion terminal voltage gradually drops from U_2 to U_3. The current discharge charges the polarizing capacitor, which is the zero-state response of the two RC network circuits.

The lithium-ion batteries' terminal voltage abruptly increases from U_3 to U_4 during the period from t_3 to t_4, which is also due to the voltage shift caused by ohmic resistance.

The terminal voltage rises slowly from U_4 to U_5 during the period from t_4 to t_5, which is the process that discharges the polarization capacitance to the polarization resistance and the two RC networks circuit zero-state response.

Overall, the HPPC program includes 10 repetitions of the voltage–current curve, as shown in Figure 4.19, except for the constant current discharge at 10% depth of discharge (DOD) with 1 C current rate. A continuous current discharge follows an hour rest time. The rest period should allow the battery to achieve thermal and electrochemical equilibrium. The procedure continues until DOD reaches 90%, after which another 1 C rate discharge is performed until the battery is 100% DOD.

4.4.4   Capacity tracking experiments

The first step in charging the lithium-ion battery is to release all the remaining lithium-ion battery power that reaches the $S = 0$ set at the start. Generally, as the lithium-ion battery voltage decreases to 2.50 V, the lithium-ion battery discharge is assumed to be

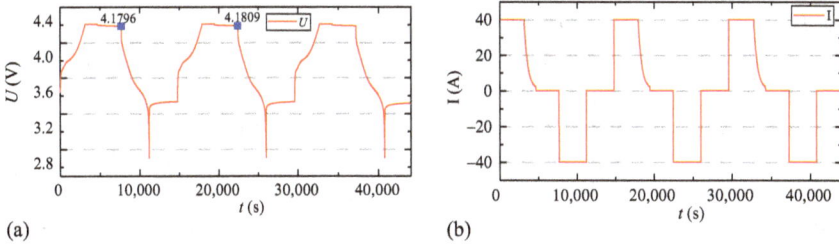

Figure 4.20   Voltage and current transition waveform

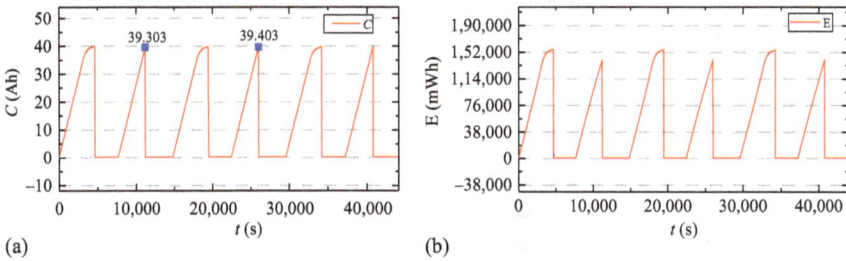

Figure 4.21   Capacity and energy transition waveform

completed. After the continuous current method of charging the various currents for a certain amount of time, the lithium-ion battery is steadily charged up to around 1 h of charging as the device cannot identify the lithium-ion battery charge status at any time. The fixed stop state is adjusted to stop the power supply until the battery voltage exceeds 3.65 V. The waveform voltage is obtained, as shown in Figure 4.20.

As its primary function is to charge the lithium-ion battery, the waveform of power and energy fluctuates in the process. The constant current charge–discharge cycles are carried out, in which the capacity determination takes into account the strain of material reactions. However, energy and power are different since the voltage component is still used. The higher the voltage platform, the higher the output. The change in time and the changing direction of the two variables over time are approximately the same as seen in Figure 4.21.

The error waveform analysis reveals that the modified algorithm is adequate. The error is significantly decreased relative to the error. Since the given initial value varies significantly from the initial value, the internal polarization effect is improved at the initial discharge point. The error in the state calculation of two algorithms at the beginning of the discharge is greater than the latter level.

4.5   Extended Kalman filtering-based state of charge estimation

To verify the model accuracy, robustness, and reliability in the SOC estimation, a model-based method is proposed to create a relationship between experimental accuracy and model accuracy.

4.5.1   State of charge determination

SOC refers to the level of charge for an electric battery relative to its capacity. Its measurement unit is percentage points ranging from 0% when empty to 100% as full. The SOC is the amount of power left in the battery and so it is the ratio of the amount of battery remaining to the rated capacity of the battery under a certain discharge condition. It is the residual capacity of the battery at that specific moment. The SOC cannot be measured directly, it can only be measured indirectly through other external parameters of the battery, which plays an important role in battery health evaluation. Its value is defined as the ratio between the current residual capacity and the capacity in the fully charged state when the battery is fully static at a certain time, as shown below:

$$SOC_t = \frac{Q_t}{Q_0} \times 100\% \tag{4.51}$$

where Q_t is the residual capacity of the battery at time t and Q_0 is the rated capacity. The discharge capacity can be defined as the actual time integral. It is considered that the lithium-ion battery is affected by various factors in the charge–discharge process, such as ambient temperature, experimental current, and cycle number. The SOC of batteries cannot be estimated accurately. Therefore, a more precise approximation is shown as follows:

$$SOC = \left(1 - \frac{\xi Z_I}{Z_n}\right) \tag{4.52}$$

where Z_I means the discharged battery power for the parameter of I and ζ is the battery efficiency parameter. It is also known as the Coulombic efficiency, which refers to the ratio of battery discharge capacity to charge capacity during the same cycle. Consequently, it is also the percentage of discharge capacity to charge capacity, which can be obtained through hybrid pulse experimental data. Z_n is the rated capacity of the lithium-ion battery. The method of measuring the remaining battery power has always been the focus and sophistication of the battery control system. To increase the lithium-ion battery service life and ensure optimum discharge performance [197], it is crucial to calculate the batteries remaining energy with high precision. Consequently, the Coulombic efficiency is used to describe the discharging characteristic, which refers to the ratio of the discharge capacity and the charging capacity for the same cycle process.

4.5.2   Application requirements

At present, power application and safety management of the lithium-ion battery packs have entered an important period of overcoming difficulties. Key technologies such as lithium-ion battery equivalent modeling, accurate estimation of battery state of charge are in urgent need of breakthroughs. In the state estimation of the lithium-ion battery packs, equivalent modeling is the cornerstone of obtaining battery characteristics and mathematical expression. The estimation of the battery state of charge can provide a reliable basis for other important battery states [198]. Therefore, studying the equivalent modeling and state estimation methods of lithium-ion batteries is a key means to eliminate their safety threats, and it is of great significance to improve their energy conversion efficiency and extend their life.

The state of charge of a lithium-ion battery is an index describing the remaining power of the battery, and it is also one of the most important parameters in the battery usage process. In recent years, relevant scientific researchers have done a lot of research in this field. Lithium-ion battery state of charge estimation methods mainly includes the discharge test method, Ah integration method, OCV method [11], artificial NN method, PF and KF-based [199] algorithm, and so on. With the deepening of research and the further improvement of the accuracy requirements for the estimation of the state of charge of lithium-ion batteries, researchers are also constantly exploring some improved methods to achieve accurate estimation of the state of charge of the battery.

4.5.3   Time-varying correction

The KF method is an optimal autoregressive data processing algorithm. The basic principle of the EKF algorithm is as follows: Taking the minimum mean square error as the best estimation criterion, using the state-space model of signal and noise, deriving the relationship between the state variables and the observation variables by establishing the state equation and observation equation of the model, and updating the estimation of the state variable by using the estimated value of the previous time and the observation value of the current time. The essence of the KF algorithm to estimate the SOC of the lithium-ion battery is to calculate SOC by the Ah integration method, which is then used to correct the SOC value obtained by using the measured voltage value. This approach reduces noise interference effectively in the measurement process by increasing the precision degree. However, due to the strong nonlinearity of the battery system, the effect of the KF algorithm for the nonlinear system is not ideal. EKF can achieve the purpose of linearization by making a real-time linear Taylor approximation to use for the battery system.

Temperature, SOC value, current, and other factors may affect battery parameters. If the battery parameters are assumed to be constant during the iterative EKF calculation process, the estimation error of the experimental results will steadily increase over time. Therefore, SOC estimation algorithms are proposed by using the time-varying correction EKF methods [169,170,190,194,199,200,214]. In this proposed algorithm, the battery parameters in the EKF equations are updated according to the SOC value and discharge current. The basic iteration calculation process for SOC is shown in Figure 4.22.

When the EKF algorithm estimates the SOC of the lithium-ion battery, the battery is regarded as a power system. SOC is the state variable and the real-time terminal voltage is the observed variable. The algorithm predicts and updates the battery SOC in each iteration process and adjusts the estimated value continuously

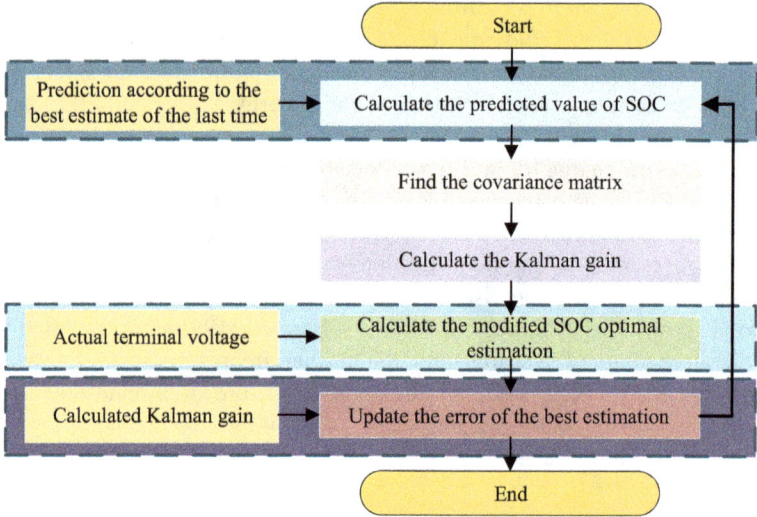

Figure 4.22   Time-varying correction extended Kalman filtering process

by using the deviation between the estimated value of the observed variable and the experimental value.

4.5.4   Simulation interfacing process

The simulation system model is implemented to verify the SOC estimation effect of the algorithm. According to the above extended Kalman iteration method, a simulation structure of SOC estimation for lithium-ion batteries using the EKF algorithm is built-in Simulink. The current and voltage data obtained from the HPPC experiment are imported into the model. The crucial simulation part is to write the S-function (System function) module of the EKF program. The input module is the current, voltage, and six polynomial functions created by the model. The parameters are including R_0(SOC), R_{p1}(SOC), C_{p1}(SOC), R_{p2}(SOC), C_{p2} (SOC), and OCV–SOC, and the output is the SOC value. The simulation model is shown in Figure 4.23.

In Figure 4.23, the actual current I_L, the available capacity Z, and the original SOC of the device at a given working condition can be obtained employing an Ampere-time integration process. The exact value of the SOC has been taken as the standard for the effects of the EKF calculation. The U_L voltage calculation data in response to the I_L state is used to determine the residual. The short sample time of the experimental instruments is 0.01 s. The SOC value obtained by amp-time integration of the current data by setting the exact initial SOC value has a high precision, which can be assumed to be the experimental value according to the improved EKF algorithm. The value obtained by the improved EKF algorithm is compared to the actual value, and the observation algorithm is used to

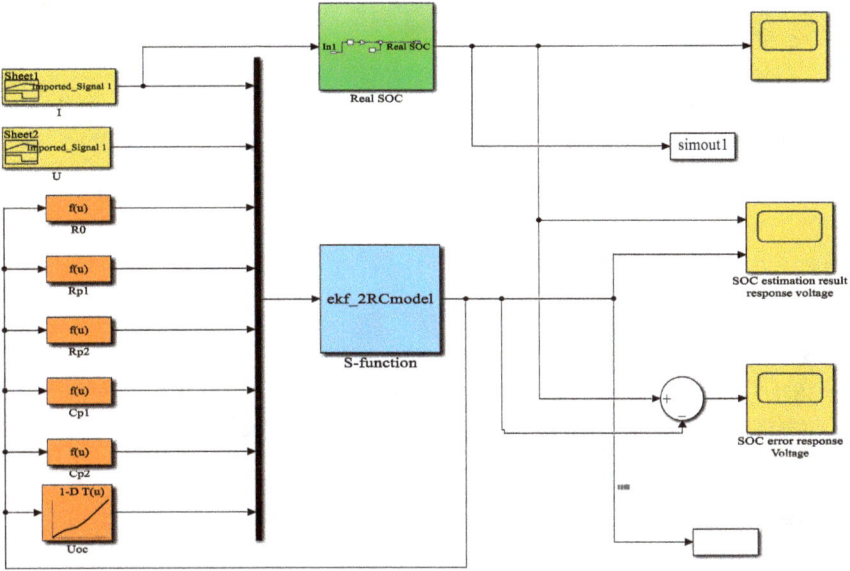

Figure 4.23   EKF-based high-order modeling structure

estimate the effect. However, the estimation of the EKF simulation code is shown below.

```
Function [sys,x0,str,ts]=ekf_2RCmodel(t,x,u,flag)
switch flag
case 0
[sys,x0,str,ts]=mdlInitializeSizes(t,x,u);
case 1
sys=mdlDerivatives(t,x,u);
case 2
sys=mdlUpdate(t,x,u);
case 3
sys=mdlOutputs(t,x,u);
case 4
sys=mdlGetTimeOfNextVarHit(t,x,u);
case 9
sys=mdlTerminate(t,x,u);
otherwise
DAStudio.error('Simulink:blocks:unhandledFlag', num2str(flag));
end
function [sys,x0,str,ts]=mdlInitializeSizes(-,-,-)
sizes=simsizes;
sizes.NumContStates=0;
sizes.NumDiscStates=12;
```

```
sizes.NumOutputs=1;
sizes.NumInputs=8;
sizes.DirFeedthrough=0;
sizes.NumSampleTimes=1;
sys=simsizes(sizes);
x0=[0.5;0;0;0.01;0;0;0;0.01;0;0;0;0.01];
str=[];
ts=[0.1 0];
function sys=mdlDerivatives
sys=[];
function sys=mdlUpdate(-,x,u)
X1=[x(1);x(2);x(3)];
P=[x(4),x(5),x(6);x(7),x(8),x(9);x(10),x(11),x(12)];
q=0.000001;Q=[q,0,0;0,q,0;0,0,q];
R=0.00002;
A=[1,0,0;0,exp(-0.1/(u(4)*u(5))),0;0,0,exp(-0.1/(u(6)*u(7)))];
B=[-0.1/250580;u(4)*(1-exp(-0.1/(u(4)*u(5))));u(6)*(1-exp(-0.1/(u(6)*u(7))))];
C=[-12.032*X1(1)^3+21.978*X1(1)^2-10.82*X1(1)+1.966,1,1];
I=[1 0 0;0 1 0;0 0 1];
X1=A*X1+B*u(1);
P=A*P*A'+Q;
K=P*C'*(inv(C*P*C'+R));
X1=X1+K*(u(2)-[0,1,1]*X1-u(1)*u(3)-u(8));
P=(I-K*C)*P;
x(4)=P(1,1);x(5)=P(1,2);x(6)=P(1,3);x(7)=P(2,1);x(8)=P(2,2);x(9)=P(2,3);
x(10)=P(3,1);x(11)=P(3,2);x(12)=P(3,3);
x=[X1;x(4);x(5);x(6);x(7);x(8);x(9);x(10);x(11);x(12)];
sys=x;
function sys=mdlOutputs(-,x,-)
sys=x(1);
function sys=mdlGetTimeOfNextVarHit
sampleTime=1;
sys=t + sampleTime;
function sys=mdlTerminate(-,-,-)
sys=[];
```

According to the EKF iteration program and simulation model, an online reliable SOC estimation model of the ternary lithium-ion battery is constructed in the S-function module. According to the extended Kalman iteration process mentioned above, the simulation framework of the SOC calculation of the lithium-ion battery using the EKF algorithm has developed into Simulink.

4.5.5   *Pulse-current estimation effect verification*

The high-order equivalent circuit model shall be calculated by parameter identification with the experimental hybrid pulse. Modeling data and experimental data are

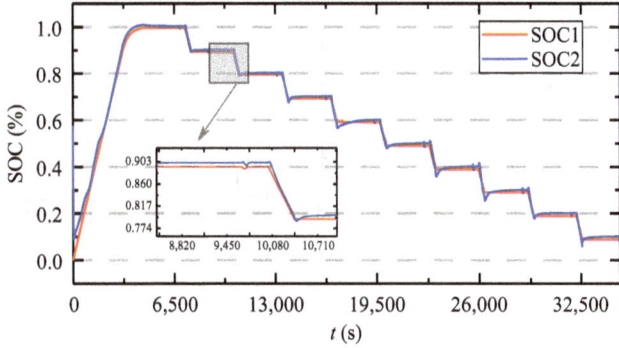

Figure 4.24   EKF estimation results with SOC

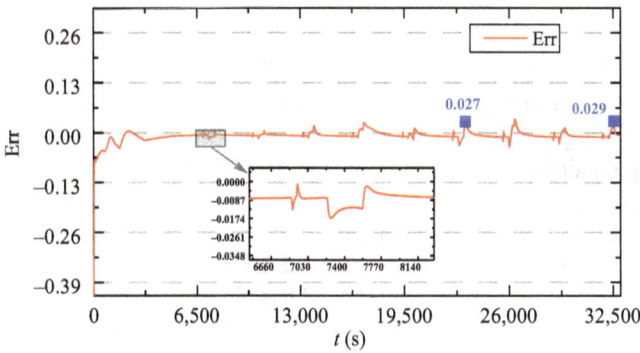

Figure 4.25   EKF-based SOC estimation error

compared and evaluated with other battery state data to check the model validity. The continuous power discharge at a given time is used to model the batteries under different operating conditions. The current value in the experimental data collected from the test instruments shall be taken as the input state. Simulation terminal voltage is obtained from a simulation model. The experimental terminal voltage is compared to the effects as seen in Figure 4.24.

In Figure 4.24, SOC1 is marked by the red curve of experimental terminal voltage data obtained by the test equipment. The SOC2 is marked by the solid blue curve output terminal voltage curve obtained by the simulation model under the input current condition. The deviation of the simulation curve is close to that of the experimental test curve, which may help to stimulate the process of discharge. The initial value calculates the error. The actual value is arbitrarily seen in Figure 4.25, which is the error curve produced by subtracting two SOC value curves.

In Figure 4.25, the solid red curve is the difference between the EKF estimation of the EKF modeling estimated value and the actual SOC value. The experimental findings suggest that the initial SOC error is minimal. This algorithm can be

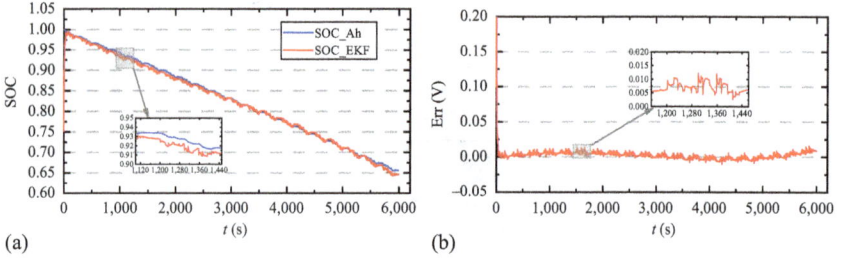

Figure 4.26   EKF-based lithium-ion battery SOC estimation

quickly addressed and has strong robustness. Under complex conditions, the SOC prediction error is less than 0.029%, which has a good effect on the battery state estimation.

4.5.6   Estimation for BBDST conditions

For the verification of the SOC estimation method, the current and voltage data of BBDST operating conditions are used for testing. The current and voltage data of the working condition is input into the simulation system and then the parameter identification result is combined to estimate the SOC value accurately. The verification result is shown in Figure 4.26.

It can be found from the figure that the EKF algorithm can well reflect the working state of the lithium-ion battery and has high accuracy. In the early stage of the algorithm, because the initial value of the SOC of the lithium-ion battery is not known, there will be a large error, but as time accumulates, the algorithm will continue to iterate to the true value. The maximum error of SOC estimation results based on the EKF algorithm is less than 2.14%.

4.6   EKF-based state of health estimation

This section presented a dual estimation method including two directions that internal resistance increasing and capacity fading. The state of health (SOH) variation process of the lithium-ion batteries is evaluated in terms of both short-term and longer-term use for the first time. It is straightforward to implement in a low-cost approach while enabling an improvement of the electronic modeling accuracy used in the BMS.

4.6.1   Estimation model establishment

The SOH expression based on the increasing internal resistance is built, in which the resistance detector is used to detect the variation of the internal resistance of the battery in a short-term operating condition. Then, the SOH estimation of the battery is realized accordingly in a short period. The internal resistance of the battery is calculated by the improved EKF algorithm and the SOH value is estimated accurately in the longer-term and different operating conditions. Temperature is used as an argument, the battery capacity fading at different temperatures is analyzed and the mathematical contraction between the state of health and the fading of the capacity is established. The innovation is that the variation of internal resistance of the lithium-ion battery is divided into longer-term and short-term conditions, the short-term conditions are further divided into 0–1 s and 1–10 s, and the prediction of variation in internal resistance in various longer-term operating conditions is realized by improving the EKF algorithm and had the advantages of the simple measurement method and quick parameter verification.

The state of health of a lithium-ion battery can be defined from a variety of perspectives, including internal resistance, capacity, self-discharge rate, cyclic life, and so on. Therefore, according to the conditions of laboratory equipment, it is easy to measure and implement SOH from the perspective of internal resistance increase and capacity attenuation. For HEV, the SOH is usually characterized by internal resistance increases. It is known that the battery work will lead to an increase in internal resistance. If the internal resistance increases to a certain extent and the service power of the battery have been limited, then the battery will reach the end of life (EOL) state.

Furthermore, in the process of battery operation, the anode material has structural damage, including microstructure cracks on the particle surface and intergranular fracture, among others, giving rise to active material structure damage. With the charge–discharge cycles under different conditions, lithium deposition is easy to occur in negative graphite materials, which consume active lithium ions. At the same time, the cathode material reacts with the electrolyte and forms a passivation film on its surface, which is called the solid electrolyte interface, constantly consuming lithium ions and solvents in the dielectric. These three reasons all lead to irreversible attenuation of the lithium-ion battery capacity. Hence, the degree of the battery capacity fading should also exist as a measure for the battery reaching the state for the EOL. To sum up, SOH can be defined as

$$\begin{cases} \mathrm{SOH}_1 = \dfrac{R_{EOL} - R}{R_{EOL} - R_{BOL}} * 100\% \\[4mm] \mathrm{SOH}_2 = \dfrac{C - C_{EOL}}{C_{BOL} - C_{EOL}} * 100\% \end{cases} \tag{4.53}$$

where R_{EOL} is an internal resistance for the battery has reached a state for the EOL, which is generally ten times for the internal resistance of the new battery used in HEV [201], R_{BOL} is the internal resistance of the new battery, and R is the actual internal resistance measured by the experiment.

C_{BOL} is the capacity of the new battery that is also the initial capacity. C is the actual capacity, which is based on the amount of electric quantity discharged from the complete discharge experiment. C_{EOL} is the capacity for the battery that has reached a state that the EOL is generally 80% of the capacity for a new battery [202]. These two methods are mainly used to comprehensively estimate the SOH, and the second-order RC model is used as the research object to realize the simulation of the operating conditions of the battery.

4.6.2   Model parameter verification

First, an equation is listed according to Kirchhoff's law to calculate the zero-state response during the HPPC experiment. To reduce input variables and improve the accuracy of parameter verification, simplify the equation, the time constants are made as τ_1 and τ_2, and parameters of a, b, and c are adopted to replace the part of original elements, as shown below:

$$\begin{cases} U_L = U_{OC} - I_L R_0 - I_L R_1 \left(1 - e^{-t/R_1 C_1}\right) - I_L R_2 \left(1 - e^{-t/R_2 C_2}\right) \\ y = a - b\left(1 - e^{-t/\tau_1}\right) - c\left(1 - e^{-t/\tau_2}\right) \end{cases} \tag{4.54}$$

The terminal voltage U_L under the zero-state response in the time domain is obtained according to Kirchhoff's law of voltage and current, and the parameter verification expression is simplified by partial parameter substitution. The corresponding relationship is shown in Equation (4.55).

$$\begin{cases} a = U_{OC} - I_L R_0, b = I_L R_1, c = I_L R_1 \\ \tau_1 = R_1 C_1, \tau_2 = R_2 C_2 \end{cases} \tag{4.55}$$

The discharging voltage mutation in the HPPC test is caused by ohmic resistance R_0, which can be acquired directly from the Ohm law. U_1 is the voltage before the mutation at the beginning of discharge. U_2 is the mutation voltage at the beginning of discharge. U_3 is the voltage before the mutation at the end of discharge. U_4 is the mutation voltage at the end of discharge. I_L is the discharging current. To improve the calculation accuracy, which of the two processes is calculated by using the voltage variation of discharge and charge in one cycle, and the average value of ohmic resistance in the two processes is calculated as

$$R_0 = \frac{|U_1 - U_2| + |U_3 - U_4|}{2 I_L} \tag{4.56}$$

Parameter verification is mainly aimed at the polarization resistance and capacitance of the battery. Importing the terminal voltage U_L and the time t under the zero-state response, the simplified equation can be obtained. As the internal resistance of the battery is too low, the internal resistance will increase rapidly, and the estimated SOH result will decrease rapidly. However, the experimental performance of the battery will not be affected too much at this time. The internal resistance of the battery may be too small when it is nearly fully charged. In this state, the estimated state of health may be lower than the actual situation. As for SOC, when the estimated internal resistance is meaningless, the state of health is lower than 30% or higher than 80%. The data from 80% to 30% of SOC is mainly selected and identified with each 10% decrease. Since the battery has been completely standstill after each HPPC test cycle, the results are stable at different initial SOC levels each time. The results are shown in Table 4.5.

In Table 4.5, the discharging current is denoted by I_L, and a, b, c, τ_1, and τ_2 are the results identified directly. According to the mathematical relationship between them and the parameters, the results are shown in Table 4.6.

Table 4.5   The results of parameter verification

SOC	a	b	c	$I_L(A)$	τ_1	τ_2
0.8	3.732	0.06380	0.02238	68.4771	27.12	0.4055
0.7	3.631	0.07757	0.02208	68.4740	32.34	0.4073
0.6	3.528	0.05834	0.02185	68.4678	25.65	0.4103
0.5	3.456	0.05124	0.02161	68.4709	30.33	0.4025
0.4	3.422	0.01984	0.04562	68.4709	0.4245	26.8500
0.3	3.396	0.05502	0.02052	68.4740	32.02	0.4251

Table 4.6   The results of the calculation

SOC	U_{OC} (V)	R_0 (Ω)	R_1 (Ω)	C_1 (F)	R_2 (Ω)	C_2 (F)
0.8	3.9368	0.00215142	0.00093170	29,108.1340	0.00032683	1,240.7267
0.7	3.8323	0.00210102	0.00113284	28,547.7525	0.00032246	1,263.1096
0.6	3.7252	0.00204085	0.00085208	30,102.8295	0.00031913	1,285.6905
0.5	3.6527	0.00200420	0.00074835	40,529.3208	0.00031561	1,275.3141
0.4	3.6185	0.00198594	0.00028976	1,465.0150	0.00066627	40,299.0720
0.3	3.5907	0.00197412	0.00080352	39,849.8270	0.00029968	1,418.5330

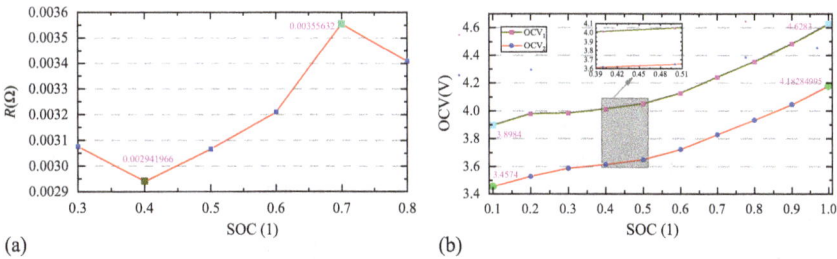

Figure 4.27   The average internal resistance for (a) different SOC and (b) the OCV–SOC curve of the battery

In Table 4.6, the ohmic resistance of the lithium-ion battery is proportional to the initial SOC. Since the time of each HPPC cycle is short and the battery is completely shelved before each test, the polarization reaction is relatively weak, which makes the polarization resistance much smaller than the ohmic resistance that constitutes the main part of the internal resistance. Also, to further analyze the influence of SOC on battery characteristics, which is fully charged and discharged, and the OCV–SOC curves of the tested battery are obtained by measuring the open-circuit voltage of the battery after fully resting after each test. The average internal resistance in each HPPC test and the OCV–SOC curves of the battery are shown in Figure 4.27(a) and (b).

As can be seen from Figure 4.27, the average internal resistance of the lithium-ion battery is greatly affected by SOC. Hence, to research variation in internal resistance during the short term, it is necessary to keep the SOC constant and detect the internal resistance, obtain the functional relationship between the two, and then evaluate the state of health of the lithium-ion battery.

OCV_1 and OCV_2 are the open-circuit voltage in the charging and discharging processes after each HPPC cycle. As can be seen from Figure 4.27, in the case of the same SOC, the difference between the two is roughly equal. Attributing to the OCV is composed of the supply voltage and the voltage at both ends of the load during charging. While the supply voltage is composed of the open-circuit voltage and the voltage at both ends of the load during discharging, the current rates in the charging and discharging stages are 1 C. The overall trend is proportional to the SOC, the same battery has a similar degree of aging, and the open-circuit voltage will be higher with the more remaining electric quantity. When SOC ranges from 30% to 80%, the change rate of OCV is relatively slow, and the battery is in the plateau phase. However, when SOC $< 30\%$ or $> 80\%$, the change rate of OCV is faster, and the battery is in the polarization region.

The variation of the internal resistance at a short time of the battery in the process of 10 s C–C discharging (the charging stage is the same) with 1 C current of HPPC experiment under different SOCs, a resistance measuring instrument is used to record the battery internal resistance change in real time. Curve fitting is conducted again, and it is found that the variation process of internal resistance in a short time is roughly proportional to time. Besides, the two temporal intervals include 0–1 s and 1–10 s, the variation process of internal resistance is affected by the initial SOC and presents different trends. The internal resistance R and time t are imported into the command window, and the functional relationship can be set as shown below:

$$R = at + b \qquad (4.57)$$

where R is the internal resistance, t is the measurement duration, and a, b are parameters. As the experiment goes on, the polarization reaction of the battery will be intensified; hence, R is no longer linear with t. It can reflect the ohmic resistance of the battery better when the test time is shorter, and the influence of polarization resistance is weakened. The fitting relationships between internal resistance and time in two different time intervals in the discharging stage of the HPPC cycle are shown in Table 4.7.

Table 4.7   Relationship between internal resistance and time in discharging stage

SOC	Parameters of the fitting curve within 0–1 s			Parameters of the fitting curve within 1–10 s		
	a	*b*	R^2	*a*	*b*	R^2
30	6.6671	25.075	0.9950	1.4455	32.811	0.8845
40	6.5852	24.723	0.9919	1.7110	31.854	0.8963
50	6.6593	24.995	0.9892	2.1564	31.015	0.9292
60	6.5847	24.698	0.9938	2.7394	29.518	0.9687
70	6.6157	24.854	0.9934	3.5778	27.510	0.9949
80	6.6397	24.612	0.9922	4.9394	22.249	0.9872

Here, R^2 is variance. According to the fitting results of the above table, on account of the slope, a is bigger than zero, the internal resistance of the battery will increase as the experiment goes on. In the first 1 s of the experiment, there is almost no difference in the growth rate of internal resistance under different SOC, indicating that the current is the main factor affecting the variation of internal resistance in the time domain. From 1 to 10 s, the larger the SOC is, the faster the growth rate of the corresponding internal resistance will be, indicating that SOC has a significant influence in this time domain. To evaluate the state of health of lithium-ion batteries in the short-term operating process, the internal resistance changes in different intervals under different SOC conditions can be analyzed. According to the established mathematical calculation equation, the variation process of the state of health in the period of 0–1 s and 1–10 s can be obtained.

4.6.3   *State of health estimation for the HPPC test*

Under the HPPC condition, the ternary lithium-ion battery with a rated capacity of 70 Ah is filled with energy to the voltage of 4.20 V through the CC–CV charging method, and then fully rest. A charge–discharge cycle of the battery includes that charging in 1 C current for 10 s, rest for 40 s, discharging at 1 C current rate for 10 s, rest for 3 min, discharging in 1 C current for 6 min, again rest for 40 min. This process is repeated 10 times until the lithium-ion battery is discharged to a cut-off voltage of 2.75 V. Due to the inevitable occurrence of heat release in the battery operating process, and the temperature of the battery can greatly affect the reaction rate inside the battery, to facilitate the analysis of the changes in the state of health of the battery during the whole HPPC experiment, the surface temperature of the battery is detected and recorded. At the same time, the electric quantity released in the experiment is also been exhibited. The operating characteristics of the lithium-ion battery in HPPC condition are shown in Figure 4.28(a)–(d).

As shown in Figure 4.28, the temperature of the battery in the HPPC test experienced a violent fluctuation process. At the beginning of the reaction, the temperature rises immediately when the battery discharges, and then the battery enters the resting state, and the temperature drops rapidly. Therefore, there are ten cycles in the HPPC experiment, and the temperature of the battery has ten peak time points.

For the SOH estimation based on the increase of internal resistance, under different initial SOC conditions, two ranges of 0–1 s and 0–10 s are analyzed. These two intervals are conducted in the first cycle of grounding discharge. The scrap internal resistance is 20 mΩ, and the new battery internal resistance is 2 mΩ. The variations of SOH during two different intervals are shown in Figure 4.29(a) and (b).

From Figure 4.29, in the discharging process within 10 s, SOH changes in a roughly linear manner. This is because, in the short-term operating process, the internal polarization reaction is relatively weak, the polarization resistance is relatively small, the main part of the internal resistance is composed of the ohmic resistance of the battery. Within 0–1 s, the initial SOC value has little effect on the variation of SOH during a short period of discharging, there is little difference in

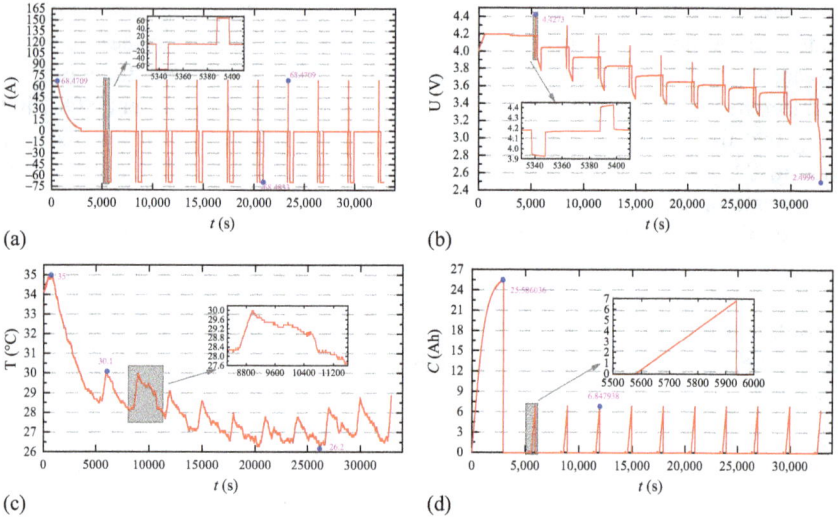

(a)   (b)   (c)   (d)

Figure 4.28   The battery operating characteristics in HPPC conditions: (a) variation of current, (b) variation of voltage, (c) variation of the temperature, and (d) variation of the capacity

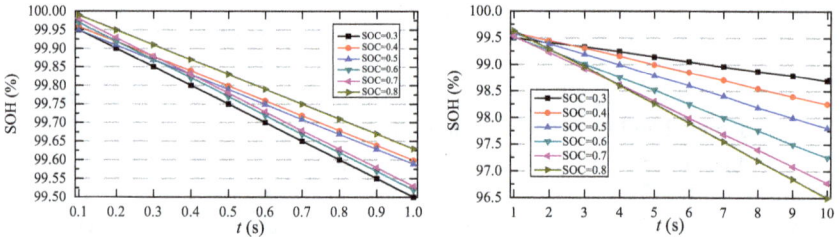

Figure 4.29   Variation of SOH in different time quantum: (a) variation of SOH in 0–1 s and (b) variation of SOH in 1–10 s

the state of health of the lithium-ion battery under the different initial state of charge. Within 1–10 s, the rate of decline in the state of health is proportional to the initial value of SOC. This indicates that the state of charge has a significant impact on SOH during this period.

It is no longer possible to measure the internal resistance and state of health of a lithium-ion battery using linear methods due to the rapid polarization reaction in the battery after a long period of use. The improved KF method can be used to estimate the internal resistance in the whole process of the HPPC test from SOC = 100% to 0%, and the variation of SOH can be obtained by taking advantage of the established mathematical equation. The internal resistance and SOH variation process in HPPC conditions obtained by using the improved Kalman filtering method are shown in Figure 4.30(a) and (b).

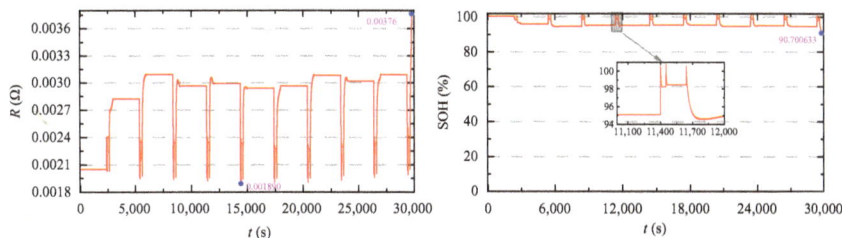

Figure 4.30   Variation of internal resistance and SOH in HPPC conditions: (a) variation of internal resistance and (b) variation of SOH

It should be pointed out at first that the research of the state health of the lithium-ion battery in the early and late stages of the experiment is meaningless, which is only for reference and comparison. From the above figures, the improved KF algorithm estimated the internal resistance changes throughout the entire process of the HPPC experiment, in the early stage of the experiment, the internal resistance of the battery almost did not change, attribute to the polarization reaction in the battery is relatively weak at this stage, and the polarization resistance is far less than the ohmic resistance which almost did not experience variation.

With the progress of the experiment, the polarization reaction rapidly intensified, and the polarization resistance increased quickly and became the main component of the internal resistance; hence, the fluctuation is large. Near to cut-off voltage, the internal resistance increased rapidly, which is because the battery is about to put out the electric quantity entirely, the polarization reaction is rapidly intensified, polarization resistance increased quickly become the main part of internal resistance. However, the state of health of the battery does not change much in the whole process and only drops quickly when the battery is near the end of discharge. This indicates that the performance of the battery has not decreased significantly throughout the entire process.

4.6.4   State of health variation for BBDST

Based on the BBDST, the operating condition of the ternary lithium-ion battery is divided into morning peak, evening peak, and smooth period. Constant power charging and discharging method are adopted to simulate the actual operating condition of the battery, and the battery is finally discharged to the cut-off voltage of 2.75 V. One cycle includes 20 operating conditions, involving C–P discharging and C–P charging. This experiment is started at room temperature. The operating characteristics of the lithium-ion battery in BBDST conditions are shown in Figure 4.31(a)–(d).

In the process of the BBDST experiment, the current and voltage are all experienced a complex variation. In each cycle, the product of current and voltage is always constant, which is determined by the fact that the simulated electric bus is always running at constant power. Through a complex process, the BBDST experiment can effectively evaluate the electric characteristics of the battery. Then,

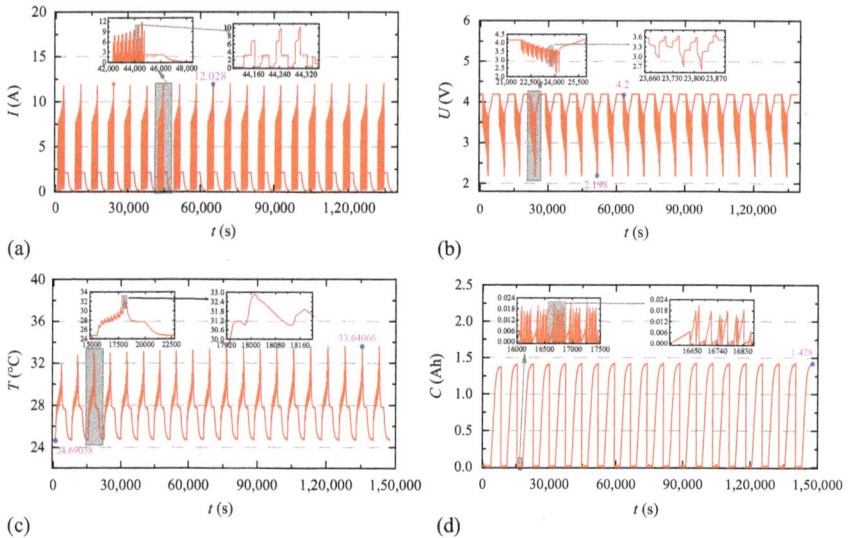

Figure 4.31   The operating characteristics of the lithium-ion battery in BBDST conditions: (a) variation of current, (b) variation of voltage, (c) variation of the temperature, and (d) variation of the capacity

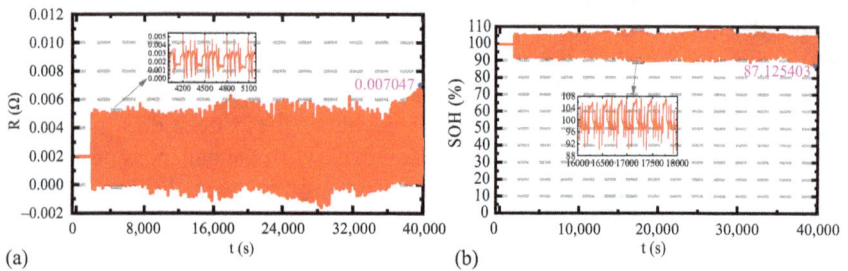

Figure 4.32   Variation of internal resistance and SOH in BBDST conditions: (a) variation of internal resistance and (b) variation of SOH

by importing the experimental current and voltage, the improved KF algorithm is used to estimate the internal resistance and SOH variation under this operating condition, as shown in Figure 4.32(a) and (b).

From Figure 4.32, under the BBDST operating condition, the internal resistance fluctuation of the battery is relatively obvious. It is an attribute that the battery is charged and discharged repeatedly, and the heat in the battery accumulates, which continuously intensifies the polarization reaction. The state of health of the battery did not change significantly during the whole process, because although the battery is charged and discharged continuously, the duration is relatively short; hence, the performance of the battery is not significantly affected.

4.6.5   State of health estimation of dynamic stress test

Dynamic stress test (DST) is performed on the ternary lithium-ion battery for a cyclic constant current charge–discharge test, with each discharge time of 4 min and a current of 1 C. The charging time is 2 min, the current is 0.5 C, and the battery is finally discharged to the cut-off voltage of 2.75 V. A total of 20 cycles are carried out, and the current and voltage changes of the battery under DST conditions are finally obtained. At the same time, the battery temperature and capacity are detected during the experiment. The operating characteristics of the lithium-ion battery in HPPC conditions are shown in Figure 4.33(a)–(d).

The experimental process includes the CC–CV process in the early stage and the process of the battery is charged to the rated voltage. In the process of the experiment, except for the CC–CV charging in the early stage, the current in the BBDST experiment presented periodic changes, and the voltage generally showed a downward trend, which would suddenly fluctuate violently when the electricity of the battery is about to be completely discharged. Then the EKF algorithm is used to import the experimental current and voltage to estimate the resistance variation of the battery under this operating condition, as shown in Figure 4.34(a) and (b).

From Figure 4.34, different from other operating conditions, under DST operating conditions, the internal resistance of the lithium-ion battery has little overall fluctuation, but which is increasing rapidly when it is about to discharge to the cut-off voltage. Similarly, the state of health does not decrease significantly during the earlier stage, but it does attenuate significantly near the complete discharging.

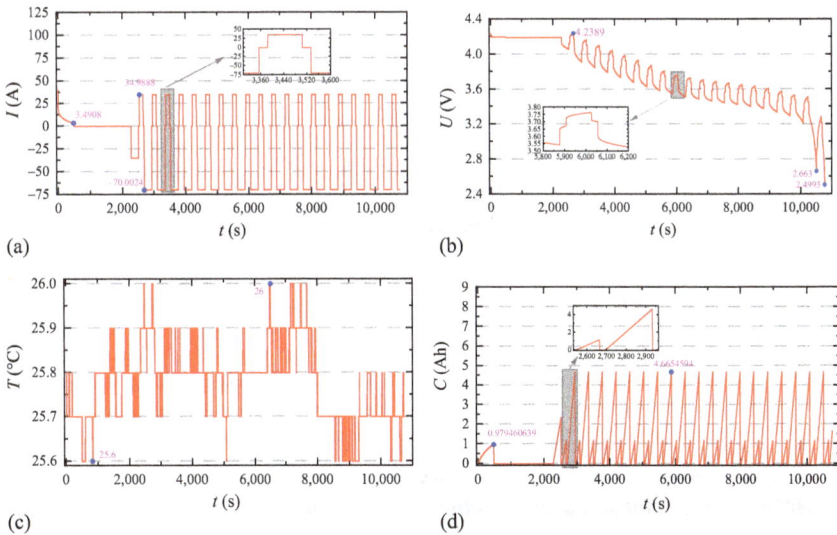

Figure 4.33   *The operating characteristics in DST conditions: (a) variation of current, (b) variation of the voltage, (c) variation of the temperature, and (d) variation of the capacity*

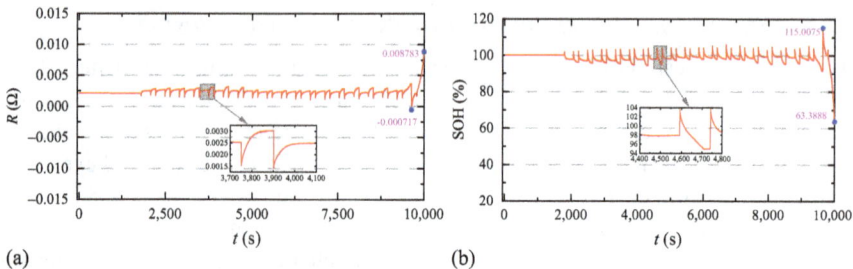

(a)   (b)

Figure 4.34   Variation of internal resistance and SOH in DST conditions: (a)
               variation of internal resistance and (b) variation of SOH

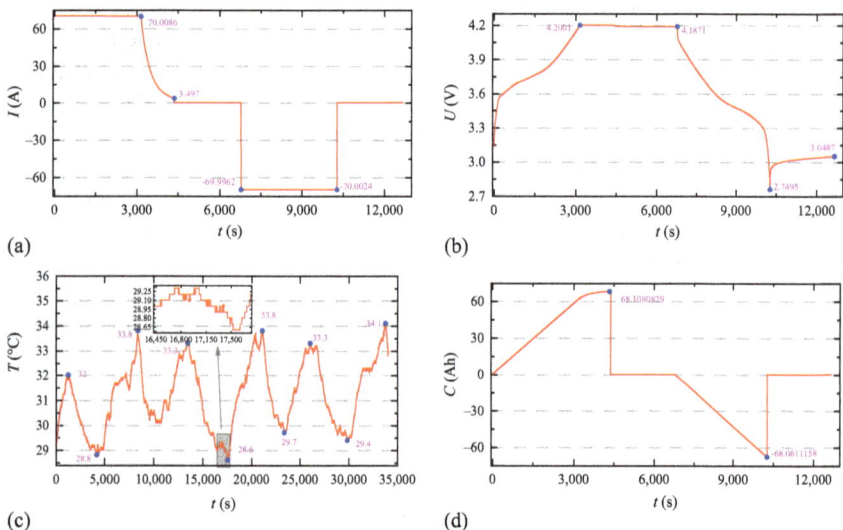

(a)   (b)

(c)   (d)

Figure 4.35   The operating characteristics in the capacity calibration experiment:
               (a) variation of current, (b) variation of the voltage, (c) variation of
               the temperature, and (d) variation of the capacity

4.6.6   State of health estimation with capacity fade

The complete charging and discharging tests are carried out at different temperatures
to represent the actual capacity of the battery. First, CC–CV charging with 1 C
current rate and 4.2 V are carried out until the battery is charged fully, then resting for
40 min, and the fully charged battery is carried out under CC discharging conditions
with 1 C current rate. Besides, various exothermic reactions inside the battery make
the temperature fluctuate, hence affecting the duration of the experiment. Besides, at
different temperatures, there is also a difference in the electric quantity released when
the battery is discharged fully. Relevant experiments are carried out at room temperature. The operating characteristics of the lithium-ion battery in the capacity
calibration experiment are shown in Figure 4.35(a)–(d).

As can be seen from Figure 4.35, in the process of CC–CV charging, the lithium-ion battery experienced the C–C charging firstly until the voltage raised to about 4.2 V, then the C–V charging began until the current decreased to about 0.05 C (3.5 A). In the C–C charging stage, the voltage raised relatively fast at the beginning, and then the rise rate decreases, but the change is opposite in the C–C discharging stage. Also, the voltage picked up in the resting stage. The actual capacity of the new battery is less than its rated capacity. The capacity went through a linear change, which reflected the resistance characteristic of the battery, and then went through a nonlinear process, which responded to the capacitance characteristic. Also, when the external temperature is constant, the overall temperature fluctuation of the battery is still very violent. At the beginning of the experiment, due to the large charging current, the battery temperature raised rapidly. The high crest value of the battery indicated that the reaction had reached its peak, and then it entered the resting state. The low crest of the battery revealed that the resting stage had ended and the battery had begun to recharge or discharge.

To research the difference in the discharging capacity of the battery at different temperatures, for the same type of lithium-ion battery, the process is repeated at different temperatures to obtain the detailed change process of the battery capacity. The rated capacity of the experimental battery is 70 Ah, and the scrap capacity is 56 Ah from the conclusion. It should be clear that the discharging test capacity at each temperature is based on the initial capacity of the new battery. The results of the discharging test measuring the battery capacity and SOH of lithium-ion batteries are shown in Figure 4.36(a) and (b).

As can be known from the above test results, under 39 °C, as temperatures rising, the released electric quantity of the battery gradually increased. After more than 39 °C, the released electric quantity filled again; hence, 39 °C is the peak that the battery releases most electric quantity. When the temperature is lower than 10 °C, the capacity is meaningless to research, on account the discharged electric quantity is too low; hence, the SOH at this condition will not be analyzed. According to the state of health estimation method given, the optimum operating temperature of the battery is about 25 °C, and the battery had the best performance.

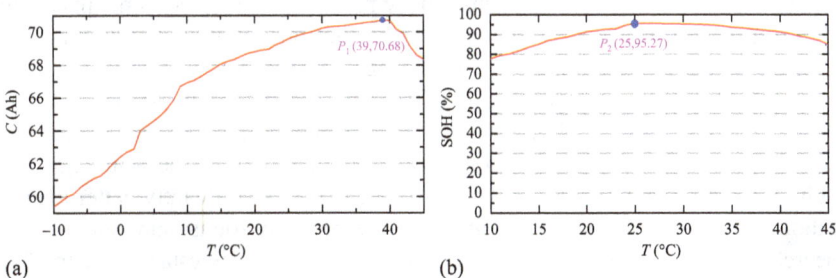

Figure 4.36 Capacity and SOH tests results: (a) capacity test results and (b) SOH test results

Besides, the tested battery is not resistant to low temperatures, and its performance deteriorates rapidly at temperatures below 10 °C.

Accurate SOC estimation approaches are presented with numerous experiments and methodologies. However, more research needs to be conducted to examine the subjects of the BMS. The focus of this study is to make the best BMS for the reliable and effective outcome for SOC estimation for EVs. However, understanding lithium-ion behavior and realistic validation simulation is a beneficial way to understand the behavior of dynamic systems under several conditions.

It presents the notion of the need for this analysis. The mathematical analysis of the EKF algorithm is presented briefly, which is used in the research study. It introduces a comparison of battery modeling with different battery modeling with the state-space explanation. Also, high-order modeling is established, which is the major part of the equivalent circuit modeling. It gives the comprehensive results of all the experiments, and the parameter extraction of the high-order model is briefly discussed. Furthermore, the simulation platform is conducted concerning high-order modeling and verification the reliability. Also, coding helps to estimate the SOC value accurately by combining the EKF algorithm briefly.

To solve the difficulties in real-time estimation difficulty and low precision under various operating conditions in a variety of working conditions in battery SOC and SOH estimation, this chapter improves the existing Kalman filter algorithm and obtains the EKF, the improved KF, and fractional-order EKF algorithm. The model parameters are identified by the HPPC test and the RLS method. The equivalent modeling strategy of the battery is studied, the state-space expression of the battery is obtained, and the SOC and SOH of the battery are estimated by the expression. To solve the problem that the error of SOC estimation increases with time, a time-varying correction EKF calculation is proposed. The battery parameters in the EKF equation are updated according to the SOC and discharge current, and the SOC estimation under different operating conditions is realized. The variation of SOH is dissected, respectively, from the two approaches of internal resistance increases and capacity fading. The effect of the SOC on internal resistance in the short-term is considered, which is set under the discharge condition, the variation of internal resistance is analyzed in two intervals of 0–1 s and 1–10 s, respectively. The improved Kalman filter algorithm is applied to accurately predict the internal resistance in different operating conditions. The temperature is adopted as the parameter, the measurement interval is extended to reflect the capacity fading at high and low temperatures by fully charging and discharging in current 1 C.

The high-order model is also verified, and the maximum simulation error comparatively is 0.051 V. Furthermore, the Ah integration and EKF method are developed for SOC estimation, and the algorithm is written by the S-function. The results from the simulation show that EKF precision is significantly better than the method of Ah integration. The result is comprehensively robust and benefits other researchers to take a reference. Finally, a modeling code is established for EKF with high-order modeling to satisfy the accurate state of charge estimation. The computational formulas based on internal resistance increasing and capacity fading are established and the estimating result of SOH is obtained.

4.7 Chapter summary

In this chapter, the experiment verification derived parameters, validated graphs, and algorithm results are outlined. A variety of tests are carried out to determine the SOC reliably, including the capacity test, and HPPC test. The model parameter is identified using the curve fitting method and the relationship between SOC and OCV is presented briefly. After that, the simulation verification using the parameters is established to compare the experimental data versus model data. A thermal dependency electrical high-order equivalent circuit of a rechargeable lithium-ion battery cell and experimental identification of the SOC estimation is carried out.

Acknowledgment

The work is supported by the National Natural Science Foundation of China (No. 61801407), Sichuan Science and Technology Program (No. 2019YFG0427), China Scholarship Council (No. 201908515099), and Fund of Robot Technology Used for Special Environment Key Laboratory of Sichuan Province (No. 18kftk03).

Chapter 5

Adaptive extended Kalman filtering for multiple battery state estimation

Abstract

Based on the electrical equivalent modeling, this chapter introduces an adaptive extended Kalman filtering (AEKF) method to estimate the battery state of charge and state of power based on the electrical equivalent modeling procedure design. Accurate online estimation of lithium-ion battery status can effectively extend battery life and improve battery safety, which is essential for the battery management system from the control perspective. Lithium-ion batteries are widely used in control fields such as electric vehicles and drones. The state estimation of the battery has a great influence on its working state, and the highly accurate lithium-ion battery state estimation is more conducive to the controller and the implementation of efficient energy management by the controller. Aiming at the shortcomings of traditional extended Kalman filtering design with insufficient prior information and dynamic environment, a carrier tracking algorithm based on AEKF is proposed. By monitoring the filter innovation or the residual dynamic change in real-time, the algorithm corrects the variance of the state noise and the variance of the observed noise by then adjusting the filter increment. The lithium-ion battery state estimation is carried out through online parameter identification and real-time correction of model parameters. Combined with the iterative calculation of the AEKF algorithm, the mean value and variance of the algorithm error are judged. Based on the second-order Thevenin equivalent circuit model, the battery state results are judged and estimated effectively. The experimental results show that the algorithm can significantly improve the convergence speed without reducing the estimation accuracy. It is practical in online state estimation applications.

Keywords: Lithium-ion battery; State-space modeling construction; State of charge; State of power; Second-order Thevenin equivalent circuit model; Extended Kalman filtering; Adaptive extended Kalman filtering; Iterative prediction; Computing framework; Parameter identification; State monitoring; Voltage traction; Pulse-current charge–discharge TEST

5.1   Introduction

The state of charge (SOC) can be directly calculated by the conversion of model equations, and the deployment of a simple Thevenin electrical equivalent circuit (EEC) model is used to obtain the open-circuit voltage (OCV) value [203]. Then, a linear fitting of a portion of the OCV–SOC curve is used to obtain the SOC value. A similar process is used to obtain the SOC with a simple EEC model, which only considers the voltage source and internal resistance [204]. The Kalman filter (KF) algorithm is also introduced to extract the OCV value in the noisy environment and then to estimate SOC based on the OCV–SOC mapping, and as it has the observability of the system as a requirement, a virtual-measurement-based method can be implemented to account for a local loss of observability [29].

The iterative process of the KF algorithm can estimate battery state parameters. For instance, both SOC and state of health (SOH) can be estimated [205]. For the model, there are different types of closed-loop methods for the SOC estimation models, such as direct feedback [206], extended Kalman filters (EKF) [207], unscented Kalman filters [208], and neural network (NN) [209]. These methods are used further to improve the SOC estimation accuracy since they can estimate the uncertainty of the existing system state recursively, which can also adjust the Kalman gain to achieve optimal estimation in the next time step of the iteration process. Electrochemical and thermal models of the lithium-ion battery are highly nonlinear [210–212], and therefore, EKF is used to estimate battery state parameters.

5.2   Iterative calculation strategies

5.2.1   Iterative predicting-updating calculation

KF is an effective method in the SOC estimation process [213]. It can continuously modify the current estimated by the Kalman gain, quickly track the real value of the SOC in the process of continuous iteration, and obtain the best estimation in the sense of minimum mean square error. The algorithm has a fast convergence speed and high estimation accuracy [214]. It can also correct the initial error in the battery estimation and have a certain inhibitory effect on interference noise. The basic principle is briefly described as follows. The state equation and observation equation of the system are described as

$$\begin{cases} x_{k+1} = f(x_k, u_k) + \omega_k \\ y_k = h(x_k, u_k) + v_k \end{cases} \tag{5.1}$$

Among them, x_k is the state variable and u_k is the system control parameter. y_k is the observed variable, w_k and v_k are the state noise and the observation noise, respectively. The iterative calculation of the KF algorithm can be divided into two stages, including prediction and estimation. The prediction stage is based on the system state equation to perform a time-based prediction update to the state variables, providing a priori estimated value for the next moment. In the measurement phase, the predicted value is corrected by the system observation value, and the deviation is corrected to update the estimated value. According to modern control theory, the state equation and the observation equation are discretized and further rewritten, as shown in the following equation:

$$\begin{cases} x_{k+1} = A_k x_k + B_K u_k + \omega_k \\ y_k = C_k x_k + D_k u_k + v_k \end{cases} \tag{5.2}$$

where x_k and x_{k+1} represent the state variable values of the system at time k and $k + 1$, respectively, u_k is the input variable of the system, y_k is the observed value of the system at time k, A_k is the state transition matrix, which predicts the system variables, B_k is the system control input matrix, and C_k is the systematic observation matrix, which drives the prediction system observations. From the previous description, it is easy to know that each iteration consists of two phases, prediction and update. Based on the analysis of the main work content of the two stages, the realization process of the KF algorithm is explained. The system state variable x_k at the time point of k and its corresponding error variance matrix P_k is supposed to be known.

5.2.1.1 Prediction stage
Substituting the state variable x_k and the system input variable u_k at time k into the system state equation, a preliminary estimate of the state variable at the time point of $k + 1$ is obtained, which is called the time update of the state variable, as shown in the following equation:

$$\hat{x}_{k+1} = A_k x_k + B_k u_k \tag{5.3}$$

At the same time, the $k + 1$ error variance matrix is estimated according to the error variance matrix P_k at the time point of k, as shown in the following equation:

$$\widehat{P}_{k+1} = A_k P_k A_k^T + Q_k \tag{5.4}$$

Calculating the Kalman gain K_k is conducted to obtain the correction coefficient as shown in the following equation:

$$K_{k+1} = \widehat{P}_{k+1} C_{k+1}^T (C_{k+1} \widehat{P}_{k+1} C_{k+1}^T + R_{k+1})^{-1} \tag{5.5}$$

5.2.1.2 Update stage
According to the Kalman gain coefficient K_{k+1}, combining the difference between the real-time observation data y_{k+1} at the time $k+1$ and the system observation prediction value, the prior estimation value is corrected. If the difference is large, the correction range is also large, and if the difference is small, the correction range is small. This measurement update called the state variable is shown in the following equation:

$$x_{k+1} = \hat{x}_{k+1} + K_{k+1}(y_{k+1} - C_{k+1}\hat{x}_{k+1} - D_{k+1}u_{k+1}) \tag{5.6}$$

At the same time, the error covariance is measured and updated as shown in the following equation:

$$P_{k+1} = (E - K_{k+1}C_{k+1})\widehat{P}_{k+1} \tag{5.7}$$

Among them, E is the identity matrix, and Q_k and R_{k+1} are the variance values of system process noise and observation noise, respectively. The core idea of the KF is to use the minimum variance to estimate the state of the dynamic system optimally, which is an autoregressive data processing algorithm. The battery is regarded as the power system, and the SOC is the system state. When the KF method is used to estimate the SOC of the battery, the charging and discharging current of the battery is used as the input of the system, and the terminal voltage is used as the output. The error between the observed value and the estimated value of the terminal voltage is used to continuously update the system state to obtain the estimated SOC value of the smallest variance. It aims to use the minimum mean square error criterion to obtain the dynamic estimation of the target for the case of linear Gaussian.

5.2.2   Nonlinear state-space extension

To deal with linear random system problems, the classic KF algorithm is mostly used, whereas the actual system is often a nonlinear random system, so it is recommended to using the EKF algorithm [215]. This algorithm is a local linearization method, which performs the Taylor expansion on the state and measurement functions, ignoring the influence of higher-order impact on the system. Then, operations are performed by using the classic KF algorithm, which is realized according to the nonlinear system state and observation equations [216]. At each moment, the functions of $f(x_k, u_k)$ and $h(x_k, u_k)$ can be expanded by the Taylor series. Ignoring the influence of higher-order terms on the function, two linearization processes are completed to obtain the following equations:

$$\begin{cases} f(x_k, u_k) \approx f(\hat{x}_k, u_k) + \dfrac{\partial f(x_k, u_k)}{\partial x_k}\bigg|_{x_k=\hat{x}_k} (x_k - \hat{x}_k) \\ h(x_k, u_k) \approx h(\hat{x}_k, u_k) + \dfrac{\partial h(x_k, u_k)}{\partial x_k}\bigg|_{x_k=\hat{x}_k} (x_k - \hat{x}_k) \end{cases} \tag{5.8}$$

$$\begin{cases} \widehat{F}_k = \dfrac{\partial f(x_k, u_k)}{\partial x_k}\bigg|_{x_k=\hat{x}_k} \\ \widehat{H}_k = \dfrac{\partial h(x_k, u_k)}{\partial x_k}\bigg|_{x_k=\hat{x}_k} \end{cases} \tag{5.9}$$

The nonlinear system is transformed into a linear system, and the transformed linear system is only related to state variables, as shown in the following equation:

$$\begin{cases} x_{k+1} \approx \widehat{F}_k x_k + \left[f(\hat{x}_k, u_k) - \widehat{F}_k x_k \right] + w_k \\ z_k \approx \widehat{H}_k x_k + \left[h(\hat{x}_k, u_k) - \widehat{H}_k x_k \right] + v_k \end{cases} \tag{5.10}$$

In (5.10), it is noted that these functions are not the linear functions of x_k. According to (5.10), the linear system is brought into the classical KF algorithm.

1. The initial conditions can be obtained for the equation state x_0 and the covariance P_0

$$x_0 = E(x_0) \tag{5.11}$$

$$P_0 = E\left[(x_0 - \bar{x}_0)(x_0 - \bar{x}_0)^T\right] \tag{5.12}$$

2. State prior estimation is conducted

$$\hat{x}_{k,k+1} = f(\hat{x}_{k-1}, u_{k-1}) \tag{5.13}$$

3. A priori estimation of the error covariance is calculated

$$P_{k,k+1} = F_{k,k+1} P_{k-1}\left(F_{k,k+1}\right)^T + Q_{k-1} \tag{5.14}$$

4. Kalman gain matrix can be calculated as well

$$K_k = P_k C_k^T \left(C_k P_k C_k^T + R_k\right)^{-1} \tag{5.15}$$

5. State posterior estimate is conducted

$$\hat{x}_k = \hat{x}_{k,k-1} + K_k[z_k - h(\hat{x}_{k-1}, u_k)] \tag{5.16}$$

6. The posterior estimate of the error covariance is updated accordingly

$$P_k = P_{k,k-1} - K_k H_k P_{k,k-1} \tag{5.17}$$

The main idea of the EKF algorithm is to convert the nonlinear system into a linear system, and then calculate the linear system according to the classic KF algorithm process [217]. This is an improvement of the classic KF, which makes it possible to solve the non-solvable nonlinear systems, making it more practical [218].

A simple principal introduction to EKF based on the linear system space. To make KF suitable for nonlinear systems, a first-order Taylor series expansion is performed on the nonlinear functions f and h, and x_k is the optimal estimated value at time k. The expansion result is shown in the following equation:

$$\begin{cases} f(x_k, k) \approx f(\hat{x}_k, k) + \dfrac{\partial f(x_k, k)}{\partial x_k}|x_k = \hat{x}_k(x_k - \hat{x}_k) \\[2mm] h(x_k, k) \approx h(\hat{x}_k, k) + \dfrac{\partial h(x_k, k)}{\partial x_k}|x_k = \hat{x}_k(x_k - \hat{x}_k) \end{cases} \tag{5.18}$$

The parameters are initialed in the following equation:

$$\begin{cases} A_k = \dfrac{\partial f(x_k,k)}{\partial x_k}\Big|_{x_k=\hat{x}_k} \\[2mm] B_k = f(\hat{x}_k,k) - A_k\hat{x}_k \\[2mm] C_k = \dfrac{\partial h(x_k,k)}{\partial x_k}\Big|_{x_k=\hat{x}_k} \\[2mm] D_k = h(\hat{x}_k,k) - C_k\hat{x}_k \end{cases} \tag{5.19}$$

Equation (5.18) can be discretized as shown in the following equation:

$$\begin{cases} x_{k+1} &= A_k x_k + B_k u_k + \omega_k \\ y_k &= C_k x_k + D_k + v_k \end{cases} \tag{5.20}$$

The state-space expression is in the form of an equation, containing the same matrix meaning, where D_k is also the system observation matrix. The basic equation of the KF algorithm is applied to the discretized model to obtain the recursive process of the EKF. The initial filter value and filter variance can be initialized, respectively. The system state equation is used to obtain the estimated value of the state variable and the mean square error at $k+1$. Wherein, the sum calculation is conducted for both the current state and the mean square error, as shown in the following equation:

$$\begin{cases} \hat{x}_{k+1}^- = f(\hat{x}_k) \\[2mm] \widehat{P}_{k+1}^- = A_k\widehat{P}_k A_k^T + Q_{k+1} \end{cases} \tag{5.21}$$

The Kalman gain K_{k+1} is calculated at the current moment, as shown in the following equation:

$$\begin{cases} K_{k+1} = \widehat{P}_{k+1}^- C_{k+1}^T (C_{k+1}\widehat{P}_{k+1}^- C_{k+1}^T + R_{k+1})^{-1} \\[2mm] \hat{x}_{k+1} = x_{k+1}^- + K_{k+1}\left[h_{k+1} - h(x_{k+1}^-)\right] \\[2mm] \widehat{P}_{k+1} = [E - K_{k+1}C_{k+1}]P_{k+1}^- \end{cases} \tag{5.22}$$

Among them, P is the mean square error and K is the Kalman gain. E is the $n \times m$ unit matrix. Q and R are the mean variances of w and v, respectively, and generally do not change with the system.

The classical KF algorithm requires a linear system. However, many systems are not linear in practical applications, so the EKF algorithm is proposed that expands the Taylor series and discards high-order components. Based on the classic KF method, EKF forcibly transforms the nonlinear relationship into a linear relationship. However, the forced conversion will cause Taylor truncation errors and may cause filtering to diverge. The Jacobian matrix needs to be recalculated every time EKF is used for estimation, so the computational complexity of the algorithm is relatively large. Besides, the solution obtained by applying EKF is not globally optimal but only a locally optimal solution. Only under certain conditions it can converge to the optimal solution.

5.2.3 Estimation model construction

The battery state estimation expression is shown in the first part of (5.23). In the measurement process, the current parameters are easy to be detected directly, so the calculation process of the battery state can be expressed in the second part of (5.23)

$$
\begin{cases}
S = \dfrac{Z_t}{Z_n} \times 100\% \\[3mm]
S_1 = \dfrac{Z_{It}}{Z_{In}} \times 100\%
\end{cases}
\tag{5.23}
$$

In the above expression, S_1 is the state of charge (SOC) value when the current condition is set to be I. Z_t is the remaining power, and Z_n is the rated capacity. After discretization, the technical implementation of the battery discrete-time state estimation can be expressed, as shown in Equation (5.24).

$$
\begin{cases}
S(t) = S(0) - \displaystyle\int_0^t \dfrac{\eta_i I(t)}{Q_n} \, d\tau \\[3mm]
S_{n+1} = S_n - \dfrac{\eta_i \Delta t}{Q_n} I_n
\end{cases}
\tag{5.24}
$$

where $S(0)$ is the initial value and η_i is the charge and discharge efficiency. It is the SOC calculation method for the special time point of the surface calculation characteristics. The realization process of the continuous-time estimation realization process can be expressed by the mathematical expression in the first part of (5.24). In practical application, the discrete state estimation model should be applied. The discrete state estimation model is convenient for the control of a digital system. The complete composite pulse-power experiment is conducted to test the battery model parameters at different state points and establish an accurate equivalent battery model, as shown in Figure 5.1.

The reliable battery model defines the relationship between the SOC and the different battery parameters. In the meantime, the effects of temperature, aging, and other variables should also be addressed. After acquiring the critical parameters in the model, the state-space equation is defined by the relationship between voltage and current.

5.2.4 Adaptive extended Kalman filtering

Currently, the SOC estimation method is mainly based on the equivalent model combined with the KF algorithm and its extended algorithm, as well as fuzzy logic and NN-related algorithms. The KF algorithm is one of the most widely used intelligent algorithms, and it is usually used in practical situations, such as path planning, target tracking, and the SOC estimation of lithium-ion batteries.

```
                          ┌─────────────┐
                          │    Start    │
                          └─────────────┘
                                 │
                    ┌─────────────────────────────┐
                    │ Initialization factor setting│
                    └─────────────────────────────┘
                                 │
                    ┌─────────────────────────────┐
                    │ UT transform generates S sigma point set│
                    └─────────────────────────────┘
                                 │
                    ┌─────────────────────────────┐
                    │ State equation prediction S sigmapre│
                    └─────────────────────────────┘
```

$$S(k) = S(k \mid k-1) + K(k)[U_L(k) - U_L(k \mid k-1)]$$

Calculate the state prediction matrix *Spre*

UT transformation generates new *S* sigma1

Observation equation prediction U_L sigmapre

$$U_L^{(i)} = h[S^{(i)}]$$

Calculate the observation prediction matrix $U_L pre$

Pxz Pzz

Calculate Kalman gain $K(k)$

$$K(k) = P_{S(k)U_L(k)}(P_{U_L(k)U_L(k)})^{-1}$$

Update status $S(k)$ and covariance $P(k)$

$$P(k) = P(k \mid k-1) - K(k)P_{U_L(K)U_L(K)}K^T(k)$$

End

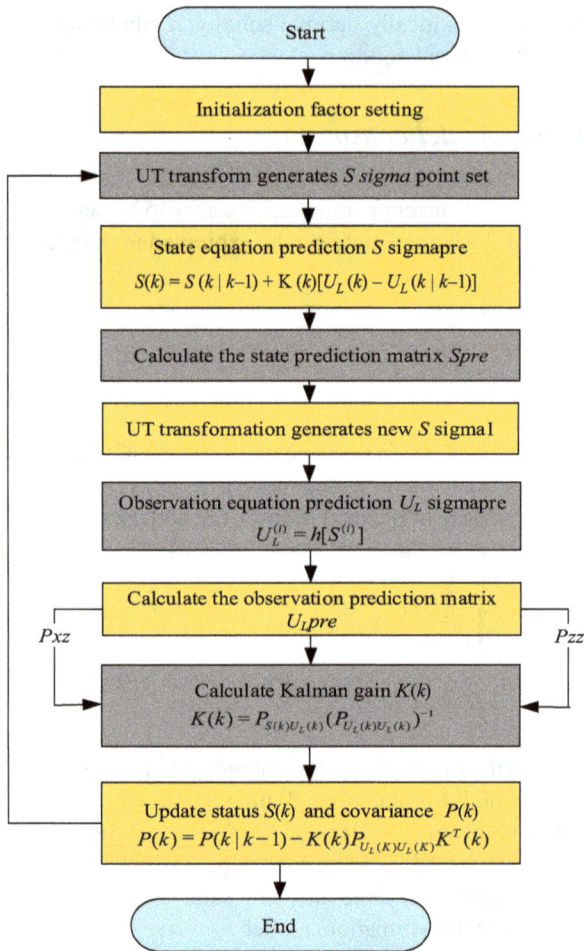

Figure 5.1   Updating iteration calculation of EKF

The basic principle of the algorithm is to take the minimum mean square error as the best estimation criterion, and by establishing a state equation and an observation equation model, a state-space model of signals and noise is used to introduce the relationship between the state variables and the observed variables.

The state variables are modified by using actual-time estimation and observation. The KF algorithm is a mathematical function that uses to estimate the state in a way that minimizes the mean squared error by calculating states in iterative steps. This technique has been providing performance efficiency in the field of parameter estimation and state transition. The improved Thevenin equivalent circuit model of the lithium-ion battery can be simplified as

$$\begin{cases} x_{k+1} = f(x_k, u_k) + w_k............A_k x_k + B_k u_k + w_k \\ y_k = h(x_k, u_k) + v_k............C_k x_k + D_k u_k + v_k \end{cases} \qquad (5.25)$$

The functions $f(*)$ and $h(*)$ are nonlinear equations and the first equation is the state equation, where x_k is the n-dimensional system state vector at time point k, and v is the n-dimensional system noise vector. The function $f(x_k, u_k)$ is a nonlinear state transition function. The second equation is an observation equation, where y is an observation vector, and v is a multi-dimensional system interference vector at time point k. The function $h(x_k, u_k)$ is a nonlinear measurement function. The above function can be explored by using the Tailor method on the prior estimation point x_k of the state x_{k+1}. The higher-order components of the process can be ignored, and linear approximations of $f(*)$ and $h(*)$ can be used as shown in the following equation:

$$\begin{cases} f(x_k, u_k) \approx f\left(\hat{x}_{k|k-1}, u_k\right) + \dfrac{\partial f(x_k, u_k)}{\partial x_k}\bigg|_{x_k = \hat{x}_{k|k-1}} \left(x_k - \hat{x}_{k|k-1}\right) \\ h(x_k, u_k) \approx h\left(\hat{x}_{k|k-1}, u_k\right) + \dfrac{\partial h(x_k, u_k)}{\partial x_k}\bigg|_{x_k = \hat{x}_{k|k-1}} \left(x_k - \hat{x}_{k|k-1}\right) \end{cases} \qquad (5.26)$$

The estimation process of the KF algorithm includes time update and measurement update. The time update process is also called the forecast process. It provides a one-step prediction of the current state variable and a preliminary estimation for the next time point. The measurement update process is the process of feeding back observation results and correcting deviations. The initial condition of the filter given by the one-step prediction means that the first state prediction has the same statistics as the initial condition of the system. The initial condition of the filter equation is given as shown in the following equation:

$$x_0 = E(x), P_0 = Var(x) \qquad (5.27)$$

When EKF is used to estimate the SOC of the lithium-ion battery, which is an integral part of the state vector. The current is used to control the quantity in the input parameter, and the output is terminal voltage calculated. The estimation time of the state vector is updated as shown in the following equation:

$$x_{k|k-1} = f(x_{k-1}, u_{k-1}) \qquad (5.28)$$

The state covariance time update process predicts the current state variable by updating the current state variable, thereby providing a prior estimate of the next time. The state covariance is updated as shown in the following equation:

$$P_{k|k-1} = FP_{k-1}F^T + Q_k \qquad (5.29)$$

The Kalman gain is the relative weight of a given measurement value and current state estimation, which can be manipulated to achieve a particular performance. The calculation of the Kalman gain coefficient is given as shown in the following equation:

$$K_k = P_{k|k-1} H^T \left(H P_{k|k-1} H^T + R_k \right) \tag{5.30}$$

The measurement updating process, also known as the correction treatment, is a feedback and deviation correction of the observed value. The state vector measurement update is given as shown in the following equation:

$$x_k = x_{k|k-1} + K_k \left(y_k - h \left(x_{k|k-1}, u_k \right) \right) \tag{5.31}$$

wherein x_k is the direct time estimate at time k. $x_{k/k-1}$ is the optimal estimated state value at the last moment. The state covariance matrix consists of the variances associated with each estimation procedure and the correlation between the errors in the state estimation. The updated state covariance matrix is given as shown in the following equation:

$$P_k = (I - K_k H) P_{k|k-1} \tag{5.32}$$

wherein P_k is the covariance update of x_k. Q_k is the covariance of process noise w. K_k is the Kalman gain. R_k is the covariance of the observed noise v. Since the covariance matrix P_k is decomposed, at least it is guaranteed that P_k is always non-negative, which can overcome the filtering divergence caused by the limited word length of the computer. The Sage–Husa algorithm adaptively updates the noise variable by comparing the final estimated value with the initial estimated value. The calculation process of the estimator-related quantity is shown in the following equation:

$$\begin{cases} \tilde{y}_k = y_k - h(x_k, u_k) - R_{k-1} \\ Q_k = \dfrac{1}{k} \sum_{i=0}^{k-1} \left(K_k \tilde{y}_k \tilde{y}_k^T K_k^T + P_k - F P_{k|k-1} F^T \right) \\ R_k = \dfrac{1}{k} \sum_{i=0}^{k-1} \left(\tilde{y}_k \tilde{y}_k^T - H P_{k/k-1} H^T \right) \end{cases} \tag{5.33}$$

To make the estimation of noise more accurate and to avoid the influence of the observed value on the estimated value, this paper considers the noise of the previous moment and the current moment at the same time. In practice, the smaller the value, the smaller the impact of the factors at the previous time point. If the value is too small, the estimated noise will oscillate, and the specific situation can be determined. Then, the calculation of the noise matrix is shown in the following equation:

$$\begin{cases} Q_k = (1 - d_{k-1}) Q_{k-1} + d_{k-1} \left(K_k \tilde{y}_k \tilde{y}_k^T K_k^T + P_k - F P_{k/k-1} F^T \right) \\ R_k = (1 - d_{k-1}) R_{k-1} + d_{k-1} \left(\tilde{y}_k \tilde{y}_k^T - H P_{k/k-1} H^T \right) \end{cases} \tag{5.34}$$

After obtaining the main parameters in the second-order Thevenin equivalent circuit model, the state-space equation is obtained using the relationship between voltage and current as shown in the following equation:

$$
\begin{cases}
E(t) = U_L + R_1 I(t) + u(t), I(t) = \dfrac{u(t)}{R_2} + C\dfrac{du}{dt} \\[2mm]
\mathrm{SOC}(t) = \mathrm{SOC}(t_0) - \dfrac{1}{Q_0}\displaystyle\int_{t_0}^{t} \eta i(t)dt
\end{cases} \tag{5.35}
$$

The three equations above are combined and discretized to obtain the state equation needed to perform further calculations leading to the conversion of non-discretized parameters into discretized ones as shown in the following equation:

$$
\begin{cases}
x(k|k-1) = A_k x(k-1) + B_k i_{k-1} + w_k \\[2mm]
A_k = \begin{pmatrix} 1 & 0 \\ 0 & e^{-t/\tau} \end{pmatrix}, B_k = \begin{pmatrix} -\dfrac{t}{Q_0} \\ R_2\left(1 - e^{-t/\tau}\right) \end{pmatrix}
\end{cases} \tag{5.36}
$$

It is important to also consider the observation equation for further successful computation and attainment of accurate results. The next step to do therefore is the observation equation which is shown in the following equation:

$$
\begin{cases}
y_k = h(x_{k-1}, i_{k-1}) + v_k \\
= U_{oc} - R_1 i_k - u_k + v_k
\end{cases} \tag{5.37}
$$

The parameters of the battery are initially nonlinear variables and have to be linearized. After linearization using the first-order Taylor series with the values of A_k, B_k, and C_k these equations are obtained as shown in the following equation:

$$
\begin{cases}
A_k = \begin{pmatrix} 1 & 0 \\ 0 & e^{-t/\tau} \end{pmatrix}, B_k = \begin{pmatrix} -\dfrac{t}{Q_0} \\ R_2\left(1 - e^{-t/\tau}\right) \end{pmatrix}, C_k = \left(\dfrac{\partial u_{oc}}{\partial s_{oc}}, -1\right)
\end{cases} \tag{5.38}
$$

When performing the mathematical calculations of the algorithm, some steps to be followed to achieve this. The first step is known as the state prediction stage. The predicted value at time k is calculated as shown in the following equation:

$$
x(k|k-1) = A_{k-1} x(k-1) + B_{k-1} i_{k-1} \tag{5.39}
$$

The second step in this calculation is the prediction of the covariance matrix. This is done by calculating the estimation error of $x(k/k-1)$, the covariance matrix of the corresponding $x(k/k-1)$ is obtained as shown in the following equation:

$$
P(k|k-1) = A_{k-1}\widehat{P}_{k-1}A_{k-1}^T + Q_k \tag{5.40}
$$

The Kalman gain is calculated in the third step to further improve the computation and arrive at the value for a specific time. The Kalman gain at time k is obtained as shown in the following equation:

$$
K_k = P_k C_k^{T}\left(C_k P_k C_k^{T} + R_k\right)^{-1} \tag{5.41}
$$

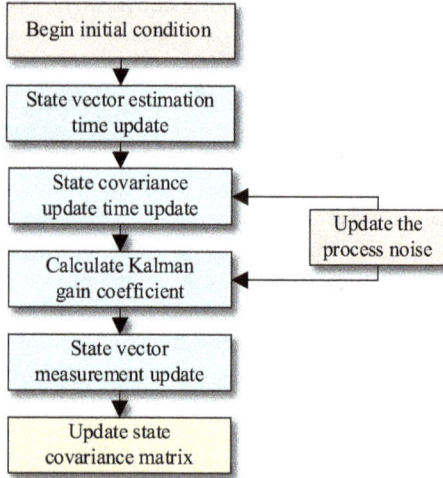

Figure 5.2   AEKF iterative process

The fourth step is the update of the status. The optimal estimated value of the existing state is estimated from the real-time measured/obtained open-circuit voltage value $U_{oc}(k)$ as shown in the following equation:

$$\hat{x}_k = x(k|k-1) + K_k[U_{oc}(k) - C_k * x(k|k-1)] \tag{5.42}$$

The fifth step is to update the noise covariance. The noise covariance is updated according to the Kalman gain. This helps to avoid errors or at least minimize them and leads to accurate state estimation. The noise covariance of the previous moment is shown in the following equation:

$$\widehat{P}_k = (1 - K_k C_k)P_k \tag{5.43}$$

In the calculation, the five steps are continually cycled in a loop, and the estimated state is continuously updated so that the estimated value is closer to the true value during the update process. The iterative calculation process for SOC using the AEKF algorithm is described as shown in Figure 5.2.

To reduce the influence of noise on state estimation and obtain more accurate results, the noise covariance matrix is introduced as an adaptive filtering method in the EKF to obtain the AEKF algorithm. When the states are estimated, the system process noise and measurement noise are estimated and corrected according to the change in the measurement data.

5.2.5   Improved adaptive extended Kalman filtering

The KF algorithm is one of the optical recursive data processing algorithms. The core idea of the algorithm is to calculate the current optimal state based on the current measured value and the prediction state and error at the last moment.

Although this algorithm is one of the best filtering algorithms, the KF algorithm is only effective for linear systems and it is difficult to meet the noise conditions when it is used. Because the lithium-ion battery is a nonlinear system, to achieve accurate state estimation, this section will introduce an improved AEKF algorithm for state estimation. The first step is the state calculation and error covariance estimation, as shown in the following equation:

$$
\begin{cases}
\hat{x}_{k+1/k} = A\hat{x}_k + Bu_{k+1} + \omega_{k+1} \\
\tilde{P}_{k+1/k} = AP_kA^T + Q_k
\end{cases}
\tag{5.44}
$$

wherein x is the state vector, A is the state transition matrix, B is the control matrix, u_k is the current at times k, P is the error covariance matrix, and Q is the noise covariance matrix, which generates by the lithium-ion battery system. Then, the second step is Kalman gain calculation, as shown in the following equation:

$$
K_{k+1} = P_{k+1}C^T\left(CP_{k+1}C^T + R_k\right)^{-1}
\tag{5.45}
$$

In the above equation, C is the measurement matrix and R_k is the observed noise covariance. The last step is state correction and the error covariance prediction for the next time point, as shown in the following equation:

$$
\begin{cases}
\tilde{y}_{k+1} = y_{k+1} - \left(C\hat{x}_{k+1/k} + Du_{k+1}\right) \\
\hat{x}_{k+1} = \hat{x}_{k+1/k} + K_{k+1}\tilde{y}_{k+1} \\
P_{k+1} = (E - K_{k+1}C)P_{k+1/k}
\end{cases}
\tag{5.46}
$$

Among them, y_k is the residual error and D is the feed-forward matrix. The state can be corrected by error. The system noise of this algorithm appears as Gaussian white noise when the KF and EKF algorithms are used, ignoring the noise characteristics. Consequently, there will be estimation errors in the application. To reduce the estimation error caused by the failure to meet the requirements of noises, the prior noise correction is leveraged to update the noise covariance matrices Q_k and R_k in the EKF algorithm. The noise covariance matrices Q_k and R_k are determined by system noise and observation noise, respectively. According to the definition of the covariance matrix, the calculation equations of Q_k and R_k are shown in the following equation:

$$
\begin{cases}
E\{\omega_k \cdot \omega_k^T\} = Q_k \\
E\{v_k \cdot v_k^T\} = R_k
\end{cases}
\tag{5.47}
$$

In (5.47), ω_k represents the system noise and v_k is the measurement noise. Therefore, the covariance matrices Q_k and R_k can be computed by the definition. After the state results are estimated, the system noise can be calculated by the prediction results, as shown in the following equation:

$$\begin{cases} \omega_{k+1} = \hat{x}_{k+1} - \hat{x}_{k+1\mathrm{o}/k} \\ \hat{x}_{k+1} = \hat{x}_{k+1\mathrm{o}/k} + K_{k+1}\widehat{y}_{k+1} \\ K_{k+1}\widehat{y}_{k+1} = \hat{x}_{k+1} - \hat{x}_{k+1\mathrm{o}/k} \\ \omega_{k+1} = K_{k+1}\widehat{y}_{k+1} \\ Q_{k+1} = E\{\omega_{k+1} \cdot \omega_{k+1}^T\} = K_{k+1}\widehat{y}_{k+1}\widehat{y}_{k+1}^T K_{k+1}^T \end{cases} \qquad (5.48)$$

To obtain the covariance matrix of the observation noise, the observation error can be obtained from the actual measured terminal voltage and the model estimated voltage. According to the definition of the covariance matrix and observation error, the covariance matrix of observation error can be obtained, as shown in the following equation:

$$\begin{cases} y_{k+1} = C\hat{x}_{k+1} + Du_{k+1} + v_{k+1} \\ v_{k+1} = \varepsilon_{k+1} = y_{k+1} - C\hat{x}_{k+1} - Du_{k+1} \\ R_{k+1} = E\{v_{k+1} \cdot v_{k+1}^T\} = \varepsilon_{k+1} \cdot \varepsilon_{k+1}^T \end{cases} \qquad (5.49)$$

wherein ε_{k+1} represents the residual energy state in the battery system, which is also the observation error of the lithium-ion battery system. To reduce the fluctuation of estimation results caused by the sudden change of noise, the forgetting factor is adopted to increase the stability of the estimation results. The adaptive correction equations are shown in the following equation:

$$\begin{cases} Q_{k+1} = (1 - d_k)Q_k + d_{k+1}\left(K_k\tilde{y}_{k+1}\tilde{y}_{k+1}^T K_k^T\right) \\ \tilde{y}_{k+1} = y_{k+1} - C\hat{x}_{k+1/k} - Du_{k+1} \\ R_{k+1} = (1 - d_k)R_k + d_{k+1}\left(\varepsilon_{k+1} \cdot \varepsilon_{k+1}^T\right) \\ \varepsilon_{k+1} = y_{k+1} - C\hat{x}_{k+1} - Du_{k+1} \end{cases} \qquad (5.50)$$

where y_{k+1} is the state observation. In the next state prediction process, the covariance matrices Q_k and R_k which are estimated by prior noise correction can be adopted into the EKF algorithm to update the original covariance matrices.

5.3   Parameter identification

5.3.1   Test platform construction

First, the experimental process is organized to include the trial context and detailed steps of each test in the form of flowcharts. After that, an experimental platform is built, and experiments are run to obtain the parameterized data and other results, such as strength, energy, current, and voltage variance curves. In addition, the open-circuit voltage change and hybrid pulse-power characteristics (HPPC) are realized. This experiment uses a GTK 3.7 V 40 Ah lithium-ion battery with a rated capacity of 40 Ah, a charge cut-off voltage of 4.20 V, and a discharge cut-off voltage of 2.75 V. The test equipment is the sub-source BTS 750-200-100-4, with a maximum charge–discharge power of 750 W, a maximum voltage of 200 V, and a maximum current of 100 A. The basic properties of the battery are shown in Table 5.1.

Table 5.1   Basic technical parameters of the battery

Factor	Specification
Size: length * width * height (mm)	$148 \times 27 \times 92$
Rated voltage (V)	3.7
Maximum load current (A)	5C
Rated capacity (Ah)	40
Charge cut-off voltage (V)	4.2
Discharge cut-off voltage (V)	2.75

To set up the experiment requires the connection of the battery to the test machine which is also connected to the computer that the data from the experiment is stored. A specific terminal is chosen from the 16 available terminals on the battery testing machine and connected to the battery for the experiment to commence. Then, the software on the computer is programmed to follow a simple logical algorithm and perform the tasks.

5.3.2   Parameter identification procedure

The number of square errors of the measured data denoted as U_{exp} and the simulated data denoted as U_{sim} for the terminal voltage at each input current sampling point is selected as the objective function to define the parameters of the lithium-ion battery model as R_0, R_1, C_1, R_2, and C_2, which is defined by L_2, as shown in the following equation:

$$L^2 : \sum_{\theta i=1}^{minN} \left[U_{exp}(t_i) - U_{sim}(\theta, t_i) \right]^2 \tag{5.51}$$

where N is the number of samples in the input current and θ is the identified parameter vector.

The battery parameter identification flowchart is based on the genetic algorithm. The optimized flowchart starts with the initialization of a random population, in which each factor represents a parameter to be identified. The output voltage of the second-order EEC is determined by the individual. The comparison of the experimental voltage versus the simulated voltage data assesses their corresponding $2L$ health. A fitness-weighted roulette game selects the most appropriate individuals in the population, and the genome of each parameter is reorganized and arbitrarily mutated to generate a population of new generations. Then, the second-order RC equivalent circuit model is used to measure the fitness values of the new population, and the optimization process runs until the optimum value of the fitness function or the number of iterations is close to the maximum, as shown in Figure 5.3.

The HPPC experiment is typically conducted to describe the model parameters and provide powerful lithium-ion battery characterization.

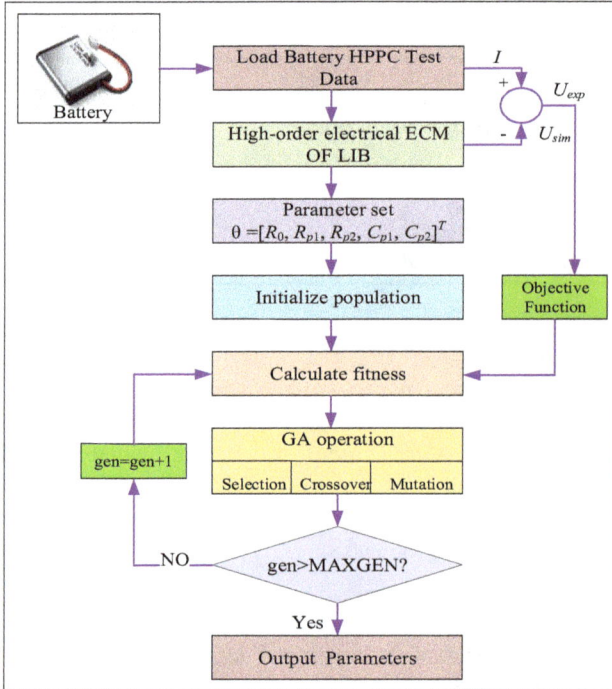

Figure 5.3   The flowchart of battery parameter identification

5.3.3   Parameter varying law extraction

The functional relationship of resistance, capacitance, voltage, and SOC is acquired by parameter recognition. Then, the circuit module that contains other submodules that assist in the processing, calculation, and optimization of the interface for better simulation is built. The EKF and the improved AEKF algorithms are incorporated in this module as well. The parameters identified in the experiment and subsequently used in SOC estimation are compared with the results from the simulated. This can be achieved by simulating the construction and completing the design of the operation function. Through direct comparison of the curves, the error in parameter identification can be observed and manipulated by changing the value of parameter input, and the optimal simulation model can be obtained by modifying the functions.

The experimental results of OCV and HPPC tests are used as the basis for parameterization, and the test data can be obtained when the SOC is 0.9. The labels U_0 to U_7 represent the values for each segment of the curve corresponding to the battery during the experiment. U_0–U_3 and U_4–U_7 show the discharge characteristics and charging characteristics of the battery, respectively, as shown in Figure 5.4.

1. The U_0–U_1 segment shows the ohmic resistance R_0, which describes the rapid drop of the voltage at the period of the battery discharge.

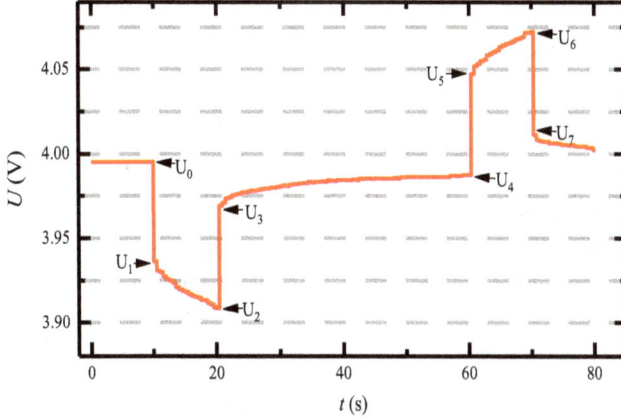

Figure 5.4   Voltage curve for a specific SOC

2. U_1–U_2 shows the steady drop in terminal voltage caused by the polarized capacitor, which is caused by the polarizing effect of the battery during charging. This polarized capacitor is the zero-state response of the second-order network.
3. U_2–U_3 shows that the battery is stationary and the disappearance of the load current, the ohmic voltage disappears, and the rapid rise of the terminal voltage.
4. U_3–U_4 shows the steady rise of the terminal voltage as the polarization capacitance discharges through the polarization resistance, forming a zero-input response of the second-order network.

5.3.3.1 R_0 calculation

The reversal of the current provided by the battery itself causes heat measured in ohms. Based on the voltage drop from U_0 to U_1 for each state of charge, the ohmic resistance R_0 can be derived. According to mathematical analysis, this can be calculated with (5.52)

$$R_0 = \frac{U_0 - U_1}{I} \tag{5.52}$$

The value of R_1, R_2, C_1, and C_2 as seen in the Thevenin equivalent circuit diagram is to make it stable and achieve possible desired outcomes. These parameters can be calculated through the following equations.

5.3.3.2 R_1 calculation

Whenever the potential of an electrode is forced away from its value at the open-circuit conditions, R_1 is calculated. This causes current to flow through the electrochemical reaction that occurs on the electrode surface. The value of R_1 can be determined by the voltage response of the battery cell to the discharging current pulse and voltage. This can be calculated by using the difference in voltage U over the current I as shown in the following equation:

$$R_1 = \frac{U_2 - U_3}{I} \qquad (5.53)$$

5.3.3.3 R_2 calculation

The accumulation of solute on the membrane surface caused by convective diffusive transport of solute in the boundary layer R_2 is calculated. The value of R_2 can be determined and calculated by the difference voltage U over the current I as shown in the following equation:

$$R_2 = \frac{U_5 - U_4}{I} \qquad (5.54)$$

5.3.3.4 C_1 calculation

A capacitor is formed when two conducting plates are separated by a non-conducting media, called the dielectric. The value of the capacitance depends on the size of the plates, the distance between the plates, and the characteristics of the dielectric. The value of C_1 can be calculated as shown in the following equation:

$$C_1 = \frac{9I}{(U_3 - U_2)\ln\left(\frac{U_3}{U_2}\right)} \qquad (5.55)$$

5.3.3.5 C_2 calculation

The part of the battery polarization resulting from changes in the electrolyte concentration is determined due to the passage of current through electrode C_2. Its value can be calculated as shown below:

$$C_2 = \frac{8I}{(U_5 - U_4)\ln\left(\frac{U_5}{U_4}\right)} \qquad (5.56)$$

With these equations, we can use the information to derive values for the parameters. These data can then be simulated in Simulink to verify the validity of the values and the modeling of the second-order Thevenin equivalent circuit. The data acquired after the calculations can be described as shown in Table 5.2.

The values of the internal ohmic resistance R_0, the electrochemical polarization resistance R_1, and concentration polarization resistance R_2 calculated by the proposed second-order Thevenin equivalent circuit are compared. According to the results in the table, the internal resistance R_0 increases steadily as the state of charge decreases.

The electrochemical polarization resistance R_1 first increases and then decreases with the decrease of the SOC. Then, it increases again when the SOC approaches the lowest point. The concentration polarization resistance R_2 first

Table 5.2 Model parameter calculation results

SOC	OCV (V)	R_0 (mΩ)	R_1 (mΩ)	R_2 (mΩ)	C_1 (kF)	C_2 (kF)
1.0	4.1917	0.0324	0.0642	0.0578	20.3215	19.2359
0.9	4.0566	0.0307	0.0532	0.0635	20.7558	19.2560
0.8	3.9422	0.0474	0.0508	0.0657	21.3673	20.2565
0.7	3.8519	0.0474	0.0635	0.0732	21.6135	20.6354
0.6	3.7496	0.0542	0.0676	0.0770	22.6417	19.6580
0.5	3.6476	0.0568	0.0730	0.0793	23.0131	22.0827
0.4	3.6191	0.0603	0.0725	0.0840	22.1528	21.8318
0.3	3.5649	0.0603	0.0832	0.0872	22.3564	20.2624
0.2	3.5274	0.0604	0.0841	0.0765	21.0251	20.4528
0.1	3.4743	0.0661	0.0762	0.0818	23.2540	20.5187

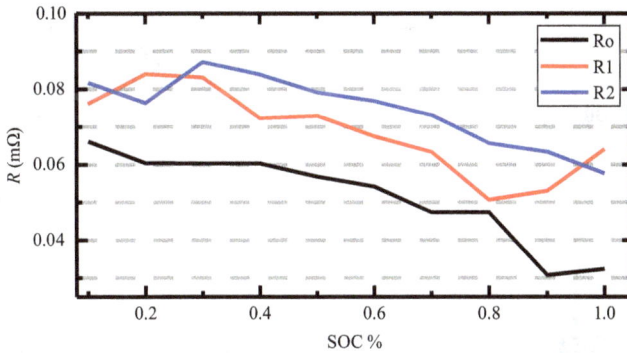

Figure 5.5 Resistance identification result

decreases then increases sharply and gently decreases as SOC decrease, as shown in Figure 5.5.

The values of the electrochemical polarization capacitance C_1 and the concentration polarization capacitance C_2 calculated from the proposed second-order Thevenin equivalent circuit are compared in Figure 5.6.

The results show the variation in the curves of the two capacitances are somewhat similar and seem to be fluctuating as SOC increases and finally decreases when SOC is above 0.7.

5.3.4 Capacity test results

The capacity experiment is conducted to calibrate the capacity, current, energy, and voltage of the battery. According to the capacity experiment, these various parameters can be derived and compared with the information provided by the manufacturer to compare and judge whether the experiment is successful. The curve of capacity variation over time can be deduced from the graph drawn in Figure 5.7.

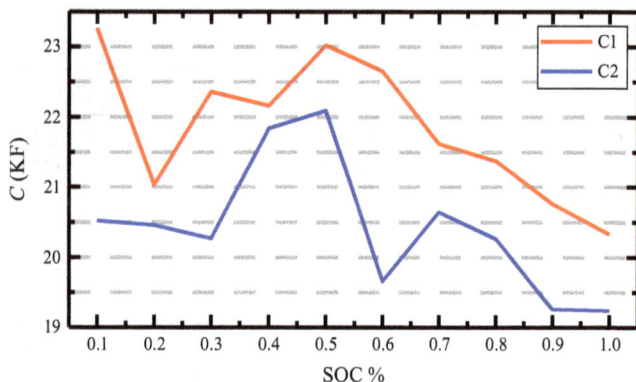

Figure 5.6   Capacitance identification result

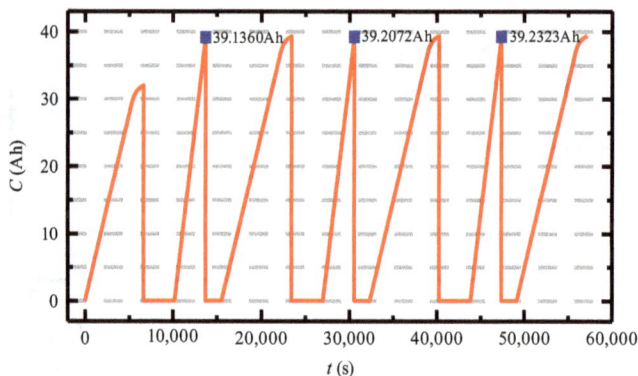

Figure 5.7   Capacity variation curve

It can be known from the results of the capacity experiment that the capacity of the battery is approximately 40 Ah. Three maximum values obtained in the experiment are 39.1360 Ah, 39.2072 Ah, and 39.2323 Ah.

Figure 5.8 shows the energy change curve and displays the three maximum energy values obtained in the experiment, which are 139,534.5534 mWh, 139,951.2741 mWh, and 139,879.0435 mWh. Therefore, the energy of the battery is approximately 140,000 mWh. The energy curve fluctuates throughout the process since its main purpose is to charge the lithium-ion battery during the constant-current charge–discharge interval.

Figure 5.9 shows the current variation curve of the capacity test and shows that the maximum and minimum current values obtained in the experiment are 20.0086 A and −40.0049 A, respectively. To ensure the safe and stable operation of the lithium-ion battery, it is necessary to measure the current in real time. This is also very important as it helps to prevent excessive current from affecting battery performance.

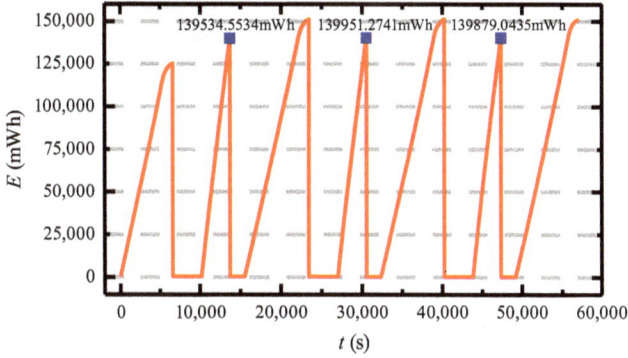

Figure 5.8   Energy variation curve

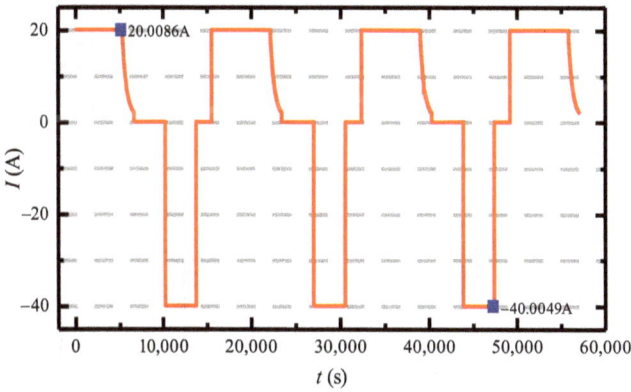

Figure 5.9   Current variation curve

Figure 5.10 shows the voltage variation curve and displays the maximum and minimum values obtained in the experiment are 4.1998 V, which is approximately the maximum voltage of the battery, as described above 4.2 V. 2.7495 V is approximately the minimum voltage of the battery, which is 2.75 V as mentioned earlier. Specifically, in the charging and discharging process of the battery, the knowledge of the minimum and maximum voltage is used to determine whether the entire battery can continue to be charged and discharged.

Figure 5.11 shows the comparison of the current and voltage curves from the experiment and how the variation is depicted to achieve the capacity of the battery. The current flow and the voltage can be obtained and analyzed.

The battery is tested for its actual capacity by extracting electrical energy from its overtime duration. The amount of energy to remove from the battery will be equal to the Ampere-hour rating of the battery. A battery is considered fit for use if its capacity is more than 80% for a one-hour discharge, which is equivalent to a full discharge for a minimum of 48 min.

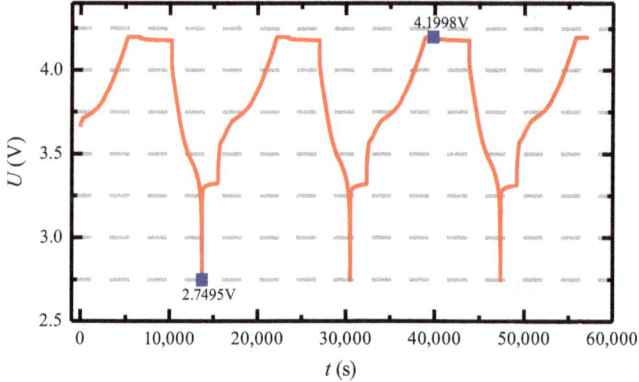

Figure 5.10   Voltage variation curve

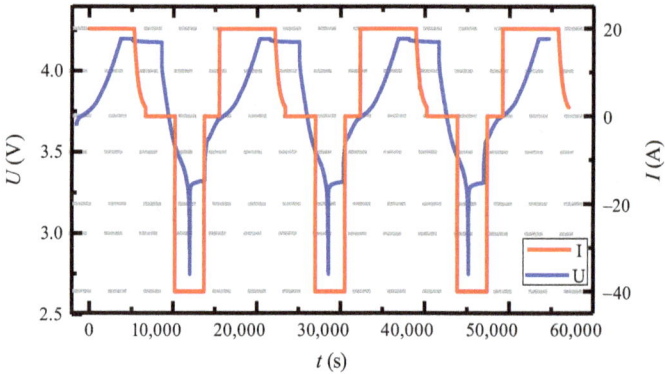

Figure 5.11   Voltage–current charge–discharge variation curve

5.3.5   HPPC test results

The results of the HPPC test are used to identify each OCV at specific SOC points. The voltage curve is also useful in the identification of battery parameters. The schematic diagram of the HPPC terminal voltage variation curve with time illustrates the experimental mechanism of the whole test process. The diagram also emphasizes the reference to a state of charge point, which shows the charge–discharge points in the experiment and helps to understand its use for parameterization as shown in Figure 5.12.

The magnitude of the voltage drop of the battery increases significantly as the discharging current rate increases. It can be seen from the above results as the number of cycles increases, the terminal voltage of the battery assumes a downward trend. The current against time figure is shown below and reveals that, as the number of cycles increases, the discharge current increases as shown in Figure 5.13.

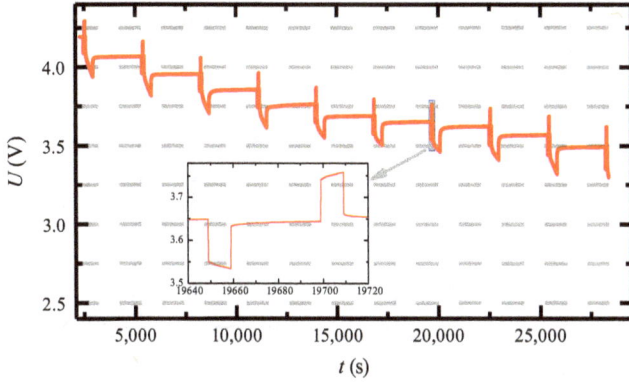

Figure 5.12   Voltage variation curve

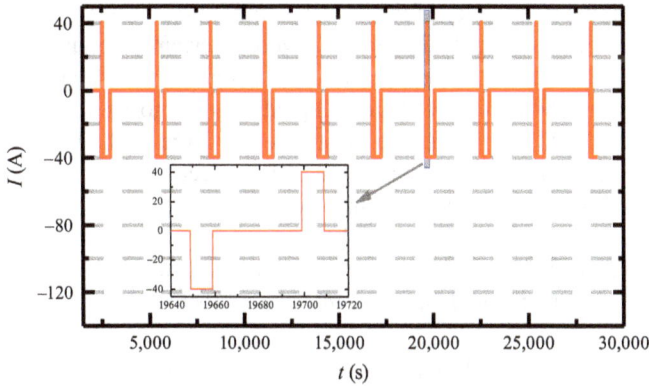

Figure 5.13   Current variation curve

Figure 5.14 shows the comparison of the voltage and current curves from the HPPC experiment and the variation with time. The overlaying curves shown in the figure depict the experimental or actual occurrence with the battery at specific times.

The battery terminal voltage drops or rises at the points in time when the battery finishes the discharge process that is connected with a load. This when is due to the influence of the internal resistance on the battery, which makes the voltage of the battery change drastically in the initial stage of discharge. The voltage decreases slowly after the first rapid drop in terminal voltage or rises slowly when the discharge ends, which is caused by the polarization effect.

5.3.6   Open-circuit voltage tests

The OCV is the stable voltage value when the battery is shelved in the open-circuit condition for a long period. The test is carried out on the lithium-ion battery to obtain data and identify the parameters of the battery. Using the Ah method, we

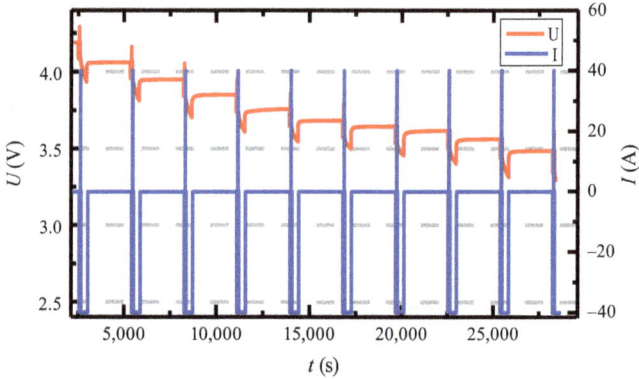

Figure 5.14   Voltage–current main-discharge variation curve

find the experimental capacity of the battery and use it for programming during the experiment. After charging the battery, the battery terminal voltage will gradually decline to a stable value when it is left in the open-circuit condition and after discharge, the battery terminal voltage will gradually rise to a stable value when the load is removed [102,219–221]. The electromotive force of the battery is equal to the open-circuit voltage of the battery. The battery electromotive force is one of the indicators used to measure the amount of energy stored in the battery. The relationship between the battery OCV and the battery SOC can be attained through this experiment. There are a few ways to obtain the open-circuit voltage of a battery and include the stationary method, also known as the direct method which is relatively more accurate.

Then, the experimental steps are set up through the flowchart. The test is composed of simple steps and follows a logical sequence and cycle to ensure consistency in the appropriate SOC points and data capture. The first step is the capacity test to calibrate the capacity of the battery and the subsequent steps capture the OCV of the battery at specific SOC points as shown in Figure 5.15.

In the test, the voltage measured at the end of each standby stage is regarded as the final open-circuit voltage. All the values of the open-circuit voltages at different SOC conditions are measured and recorded. The OCV data obtained at the various SOC levels can be obtained accordingly. These values are extracted and used in curve fitting to reveal the relationship between OCV and SOC, which is used to obtain the polynomial equation for further calculations, as shown in Table 5.3.

The OCV–SOC values are imported and using the curve fitting tool; a relationship is obtained. The variation of OCV with SOC obtained through the experimental method above is shown in Figure 5.16.

The curve is realized through the use of a polynomial fitting applied. To perform further calculation and solve mathematically for the OCV, this can be obtained to achieve the best curve fitting as

$$SOC = f^{-1}(OCV) \tag{5.57}$$

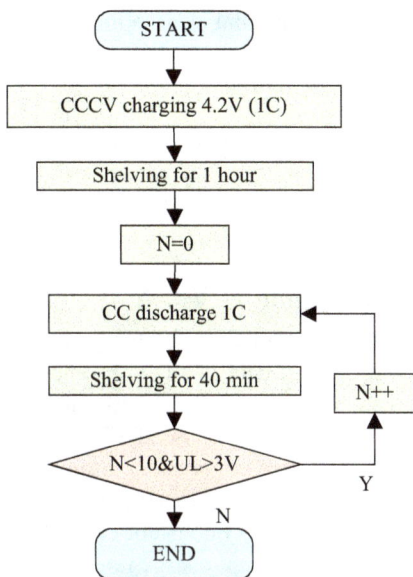

Figure 5.15 OCV experiment process

Table 5.3 Values from the SOC–OCV test

SOC	1	0.9	0.8	0.7	0.6	0.5	0.4	0.3	0.2	0.1
OCV (V)	4.1917	4.0566	3.9422	3.8519	3.7496	3.6476	3.6191	3.5649	3.5274	3.4743

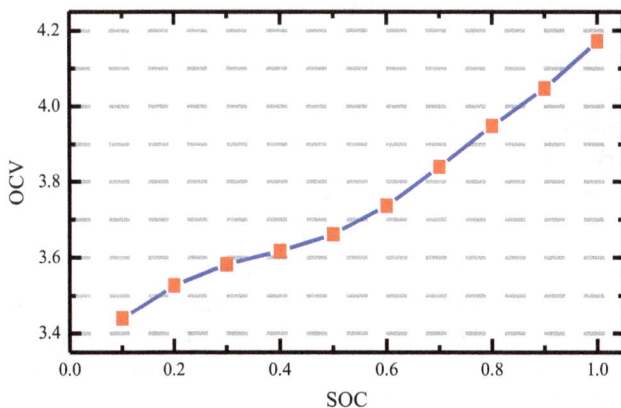

Figure 5.16 OCV–SOC curve fitting

To determine the correct polynomial term to include, the number of bends can be counted in line and add one for the model order that is needed. Mathematically, the polynomial curve reveals the following:

$$OCV = 27.57 \times SOC^6 - 86.29 \times SOC^5 + 100.7 \times SOC^4 - 52.9 \times SOC^3$$

$$+ 11.98 \times SOC^2 - 0.2793 \times SOC + 3.4743$$

(5.58)

5.3.7 Combined capacity and HPPC tests

The experiment is performed as a method for the parameter identification of the Thevenin model, and then the test data is analyzed and used in equations to calculate the parameters. Taking the GTK 3.7 V 40 Ah lithium-ion battery as the research object, the battery test equipment is BTS750-200-100-4.

The HPPC experiment is carried out with the lithium-ion battery according to the "American Freedom CAR battery experiment manual." The single HPPC working step is to take 1 C ($I = 40$ A) constant current rate for 10 s, shelve for 40 s, and 1 C ($I = 40$ A) constant current charge rate for 10 s, and then shelve. The test is conducted at 0.1 SOC intervals from 1.0 to 0.1. When the current I is loaded, a voltage rise or drop appears when there is a pulse charge or discharge and this can be used to calculate the parameters. The capacity and HPPC experiments performed on the lithium-ion battery are conducted according to the following steps.

The battery is fully charged using a constant current of 1 C followed by a constant voltage of 4.20 V. Then, the battery is discharged with a constant current to its discharge cut-off voltage of 2.75 V. The experiment is repeated three times for the difference between the discharge capacity of each measurement not to exceed 2%, and then the measured capacity is deemed to be the actual capacity of the battery.

1. The battery is left in the open-circuit condition to rest for 40 minutes to achieve electrochemical and heat equilibrium. After performing step (1) on the lithium-ion battery, SOC = 1.
2. The experimental battery is discharged at a constant current of 1 C for 10 s, leave it for 40 s, and charge at 1 C for 10 s, then leave it.
3. The battery is discharged with a current value of 40 A for 6 minutes to decrease the battery SOC to the next SOC point and leave it for 30 minutes.
4. These steps are repeated nine times to obtain the complete data for the test.

According to the experimental procedure of the capacity test, the battery is filled with energy by using the constant-current–constant-voltage method. The standard current is then discharged to the discharging cut-off voltage of 2.75 V at a standard constant-current rate of 1 C. The voltage, capacity, current, and energy variation curves from the experiment are obtained. The steps involved in the capacity test are shown in Figure 5.17.

The HPPC test is conducted for parameterization and verification of the model accuracy. The HPPC test is carried out using simple steps to obtain the necessary

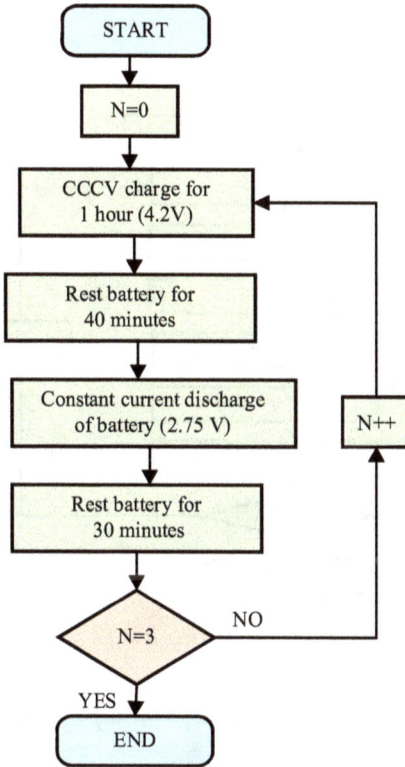

Figure 5.17 Capacity test process

values needed for parameterization and further computation as far as the research is concerned. The experimental current and voltage curves of the test can be obtained and presented subsequently. The experimental process of the HPPC test can be seen in Figure 5.18.

5.4 State of charge estimation

5.4.1 Simulated estimation results

The experimental results and parameterization are used for SOC estimation. Figure 5.19 shows the following estimate, which describes the decreasing slope over time as SOC reduces from 1 to 0.1. The result shows that the discharge time is longer than the charging time, and there is an alternating charge–discharge process during the experiment, so the discharging time is gradually decreasing and fluctuating.

According to the estimation effect analysis, the battery starts to discharge after being fully charged. As shown in the figure, the state of charge estimation depicts a decreasing trend of fluctuation as time increase in the whole process. This is due to the alternating charge–discharge process of the experiment, and the fact that

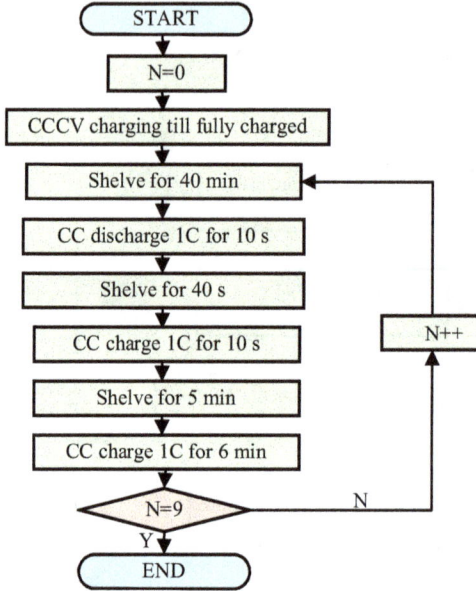

Figure 5.18   HPPC test process

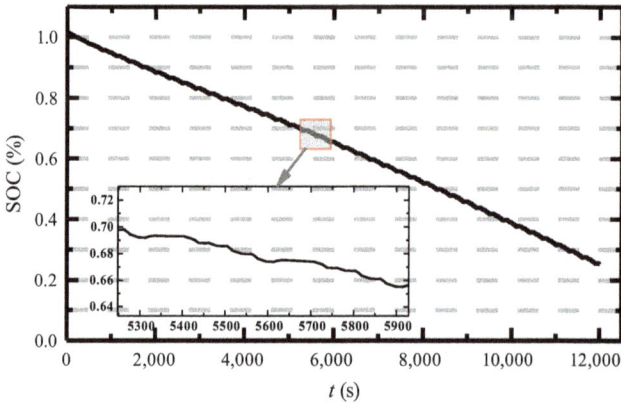

Figure 5.19   SOC variation curve

discharge time is longer than the charging time. Figure 5.20 shows simulation results using the algorithms in the model established.

Several simulations are conducted to achieve a perfect result. Some adjustments done are to make sure the error rate which is the difference between the actual SOC, in this case, SOC1, and the estimations using the algorithms are minimal, in which SOC2 is used for EKF and SOC3 representing AEKF. Figure 5.20(a) shows the initial results of experimental SOC against the estimation with the proposed AEKF algorithm. Figure 5.20(b) shows the initial results of the experimental SOC and the estimation from the use of the EKF and AEKF algorithms.

5.4.2 Voltage traction effect

The second-order Thevenin equivalent circuit model is established through parameterization with the HPPC experimental results. To verify the validity of the SOC estimation in the simulation, the results are compared with the results of the HPPC experiment, and then calculations are performed. The value of current I in the experimental data obtained by the test equipment is taken as the input condition, and the simulation terminal voltage is obtained through the simulation model, and the experimental terminal voltages are compared to obtain the results as shown in Figure 5.21.

Wherein U_1 is the change curve of experimental terminal voltage data obtained through the experiment, while U_2 is the output terminal voltage curve obtained through the simulation model. The figure shows the variation trend of both the experiment and simulation curves is like that of the actual test curve. This means

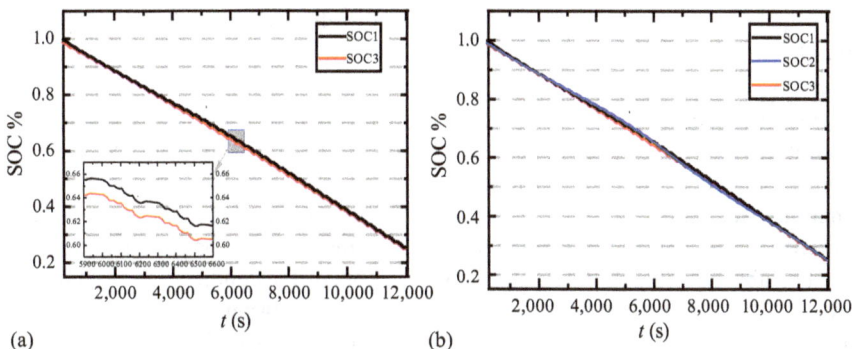

Figure 5.20   SOC variation curves: (a) variation for SOC1 and SOC3 and (b) comparison for SOC1, SOC2, and SOC3

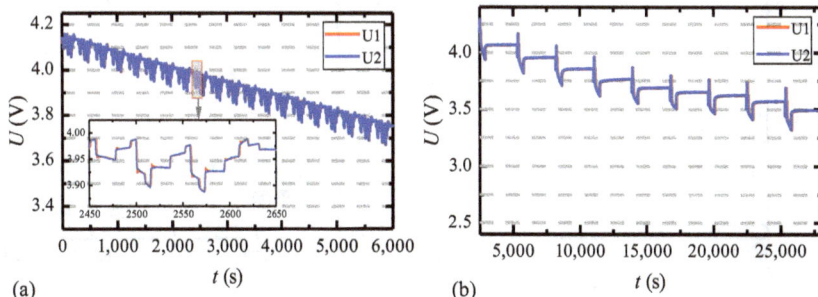

Figure 5.21   Comparison of voltage variation curves: (a) voltage simulation of BBDST test and (b) voltage simulation of HPPC test

that the results can be verified and authenticated as appropriate for use in any calculation toward the accurate estimation of SOC.

5.4.3   Pulse-current estimation verification

The experimental BBDST operating condition is intended to demonstrate the adaptability and efficiency of the proposed improved adaptive Kalman filtering algorithm for SOC estimation, and the efficacy of the algorithm for correct SOC estimation is compared to the EKF algorithm. To validate the simulation results, several simulations are performed to verify the effect. The result shows that the AEKF algorithm used is effective as the difference is minimal compared to the EKF algorithm. The SOC estimation results from the proposed model and the use of the two algorithms are then compared to assess and verify the validity and efficient performance of the model and algorithms implemented. The three methods are compared as shown in Figure 5.22.

It can be observed from Figure 5.22 that the three methods all follow the same trend and have a good convergence. The figure also shows the difference in error tolerance between the estimated curves using EKF and AEKF estimation curves and the use of real SOC curves. Figure 5.22(a) is the result of SOC estimation for three different methods. SOC1 is the true SOC value, and SOC2 is the SOC estimate using the EKF algorithm and SOC3 is the SOC estimation using the AEKF. Figure 5.22(b) is the error curve obtained through the difference in SOC value curves. The error of using the extended Kalman filtering algorithm to estimate the state of charge is around 5%, and the error of SOC estimation using the improved adaptive extended Kalman algorithm is less than 2%, which has a strong correction function. According to the simulation result, the main factor affecting the convergence speed of the algorithm is the initial value of the error covariance matrix.

The ability of the AEKF to adapt adjustments of the condition for initial value error toward the accurate SOC estimation is better and faster than that of the EKF. In the quest to improve the accuracy and reliability of the lithium-ion battery state of charge estimation, the second-order Thevenin model is established and the

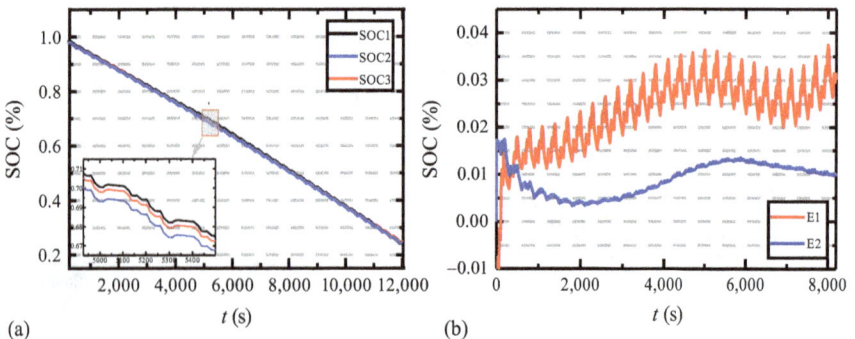

Figure 5.22   SOC variation curves for three different methods: (a) comparison of SOC estimation results and (b) SOC estimation error comparison

parameters are obtained. The selection of an adaptive law for the process noise covariance matrix shows an improvement in estimation performance. The result indicates that the maximum relative errors of both algorithms are less than 5%, which can generally satisfy the precision requirements for practical engineering calculation, such as algorithms based on EEC for advanced management.

The improved AEKF algorithm for SOC estimation used in this work can effectively and accurately estimate the battery SOC and has high precision compared with the EKF algorithm. The results are of great instructional significance to the application in practical control systems for the EEC of batteries. The improved AEKF algorithm has good convergence speed, higher estimation accuracy, and more stability, which is appropriate and convenient for SOC estimation. Therefore, the second-order RC model may be the first choice for stringent applications such as automobiles and high-power demand equipment, and the SOC estimation using an adaptive algorithm will obtain more accurate and timely information.

Measurement of new energy states such as the SOC of the lithium-ion battery is very important because it depends a lot on research to improve the safety and lifespan of the energy stored. It is therefore imperative to encourage the use of combined methods and especially the invention of new and improved algorithms towards the accurate estimation of battery states. Research in this area is not only improving state measures but also leading to discoveries about stored energy and how the BMS can be improved to give concise and timely notification.

The experiments conducted for this research did not consider the influence of temperature on the lithium-ion battery. It is worth noting that lithium-ion batteries are affected by temperature, which is one of the external effects of the battery. Further research in the area will focus on the effects of temperature on the battery by conducting the experiment at different temperatures and comparing the results for analysis and conclusive opinions.

5.4.4   BBDST estimation results

The EKF algorithm and PNC-AEKF algorithm are used to realize the estimation under the BBDST experiment. Figure 5.23(a) shows the comparison between simulation estimation results of the lithium-ion battery by two different algorithms and experimental results. Figure 5.23(b) and (c) shows the enlarged view of the estimation results. Figure 5.23 (d) shows the estimation error curves by two algorithms under BBDST.

In Figure 5.23(a), S_1 and S_3 are the estimated values by the EKF algorithm and improved algorithms, respectively. S_2 is the experimental test result. From the enlarged figures, one can conclude that the prior noise correction can enhance the tracking performance of the EKF algorithm. It is easy to find that the improved method can effectively reduce the fluctuation of the estimation. In Figure 5.23(d), E_1 is the AEKF estimation error and E_2 is the EKF estimation error. In BBDST working conditions, the maximum error is 0.59% by the AEKF algorithm, the maximum error is 1.19% of EKF. The simulation results show that the proposed improved method can improve the SOC estimation accuracy.

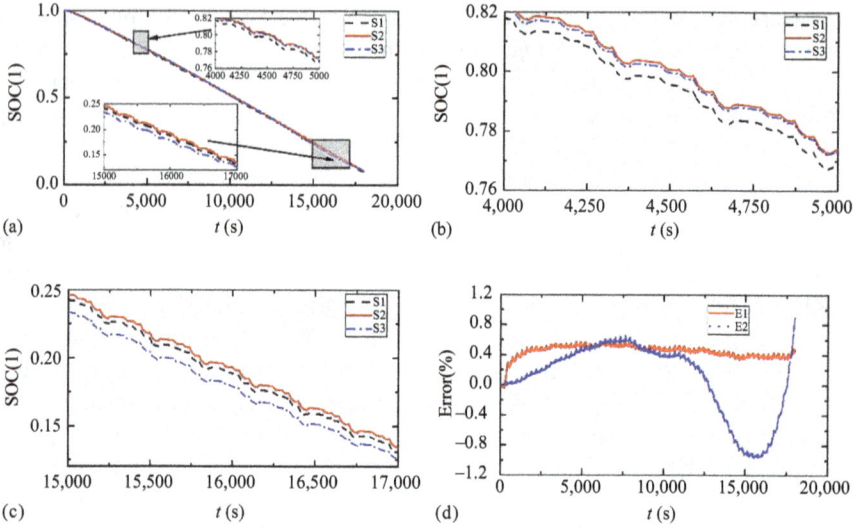

Figure 5.23   *State of charge estimation results: (a) state of charge estimation, (b) enlarged view of estimation effect, (c) enlarged view of estimation effect, and (d) state of charge estimaiton error*

During the discharging process of the lithium-ion battery, since the EFK algorithm uses the Taylor expansion to linearize the battery system, EKF and AEKF will cause a certain SOC estimation error due to linearization. Besides, one of the use conditions of EKF is that the system noise and the observation noise are Gaussian white noise. Since the noise conditions cannot be met, errors in the SOC estimation will occur. The error change of the EKF algorithm in Figure 5.23(d) can be verified well. In this figure, with the SOC decrease of the lithium-ion battery, the nonlinear degree of the lithium-ion battery will increase. The EKF algorithm will ignore the high-order term of the system, so it will further increase the estimation error.

By comparing the simulation results of the improved algorithm and the EKF algorithm, the advantages of the adaptive prior noise correction are illustrated. This method can effectively improve the estimation accuracy and make up for the lack of the EKF algorithm that does not consider noise characteristics.

5.5   State of power prediction

5.5.1   State of power characteristics

The state of power (SOP) is a critical parameter of the lithium-ion batteries, which reflects the maximum and minimum power during charging or discharging. In recent years, more and more SOP estimation methods is proposed by scholars around the world. The common estimation methods include the HPPC test estimation method, data-driven approaches, and model-based approaches; data-driven approaches include NN algorithms, fuzzy control algorithm, and so on. The model-based approaches mainly adopt equivalent circuits to realize the SOP estimation.

The HPPC test estimation algorithm mainly recognizes the interpolation table by the HPPC test so that peak power can be estimated by looking up the table. However, this method requires a large amount of experimental data, so that it is seldom used. The data-driven approaches require many training samples and calculations, which will greatly increase the computing load of BMS. Besides, their accuracy is also influenced by the accuracy of the training samples. The model-based approaches use the lithium-ion battery models to calculate peak current. The peak current during charging and discharging can be calculated accordingly based on the operating range of the open-circuit voltage and other parameters in the equivalent circuit model of the lithium-ion battery, and then the SOP can be calculated by the peak current and limited voltage.

5.5.2   SOC-based SOP estimation

State of charge is a state parameter related to SOP. The SOC can be estimated by some filtering algorithms or other methods. According to the definition of SOC, the relationship between SOC and peak current can be obtained. Consequently, the SOP can be computed by the current, which is shown in the following equation:

$$\begin{cases} I_{\max}^{dis} = Q(k) \cdot (\text{SOC}(k) - \text{SOC}_{\min})/\eta T \cdot N \\ I_{\max}^{chg} = Q(k) \cdot (\text{SOC}_{\max} - \text{SOC}(k))/\eta T \cdot N \end{cases} \tag{5.59}$$

where $Q(k)$ is the capacity of the battery and N is the number of sampling points.

5.5.3   EEC-based SOP estimation

Taking the Thevenin EEC as the research objection, the state of the model after N time points can be estimated, which is shown as

$$\begin{cases} X(k+N) = A^N(k)X(k) + \left(\sum_{j=0}^{N-1}A^{N-1-j}(k)B(k)\right)I(k) \\ U_{L,k+N} = U_{OC}(SOC_{k+N},Q_N) + \begin{bmatrix} 0 \\ 1 \end{bmatrix}^T (A^N(k) \cdot X(k)) \\ \quad + \left(\sum_{j=0}^{N-1}A^{N-1-j}(k)B(k)\right)I(k)\right) + R_{0,k}I(k) \end{cases} \tag{5.60}$$

where A is the state transition matrix, B is the control matrix, $U_{L,k+N}$ is the observation voltage. The OCV can be linearized by the Taylor expansion. Therefore, the peak current based on EEC is shown in the following equation:

$$
\begin{cases}
I_{\max}^{dis} = \dfrac{U_{L,\min} - U_{OC}(\text{SOC}_{k+N}, Q_N) - [\,0 \quad 1\,] \cdot A^N(k) \cdot X(k)}{\dfrac{\eta T \cdot N}{Q_N} \dfrac{\partial U_{OCV}}{\partial \text{SOC}}\Big|_{\text{SOC}=\text{SOC}_N} + [\,0 \quad 1\,]\displaystyle\sum_{j=0}^{N-1} A^{N-1-j}(k) \cdot B(k) + R_0} \\[4mm]
I_{\max}^{chg} = \dfrac{U_{L,\max} - U_{OC}(\text{SOC}_{k+N}, Q_N) - [\,0 \quad 1\,] \cdot A^N(k) \cdot X(k)}{\dfrac{\eta T \cdot N}{Q_N} \dfrac{\partial U_{OCV}}{\partial \text{SOC}}\Big|_{\text{SOC}=\text{SOC}_N} + [\,0 \quad 1\,]\displaystyle\sum_{j=0}^{N-1} A^{N-1-j}(k) \cdot B(k) + R_0}
\end{cases}
$$

$$(5.61)$$

where $N = L/T$, T is the sampling time, N is the total sampling points in L s, A is the state transition matrix, and B is the control matrix. When the state after L s is obtained, the equation to calculate the peak current can also be computed.

5.5.4 Multi-constraint SOP estimation

According to the SOC, EEC, and rated current, the maximum power can be calculated, and the computation equation is shown in the following equation:

$$
\begin{cases}
I_{\max}^{dis} = \min\{I_{\max}^{dis,V}, I_{\max}^{dis,SOC}, I^{dis}\} \\
I_{\max}^{chg} = \min\{I_{\max}^{chg,V}, I_{\max}^{chg,SOC}, I^{chg}\} \\
P_{\max}^{dis} = I_{\max}^{dis} \cdot U_{\min} \\
P_{\max}^{chg} = I_{\max}^{chg} \cdot U_{\max}
\end{cases}
$$

$$(5.62)$$

wherein $I_{\max}^{dis,\,V}$ and $I_{\max}^{chg,\,V}$ are the current calculated by EEC, $I_{\max}^{dis,\,SOC}$ and $I_{\max}^{chg,\,SOC}$ are the value computed by SOC, and I^{dis} and I^{chg} are the rated current.

5.5.5 BBDST estimation results

To verify the efficiency of the SOP estimation method, BBDST working conditions are used to calculate SOP, as shown in Figure 5.24.

In Figure 5.24, S1 and S2 are the estimated discharge and charging power, respectively. S3 and S4 are the reference discharge power and charge power,

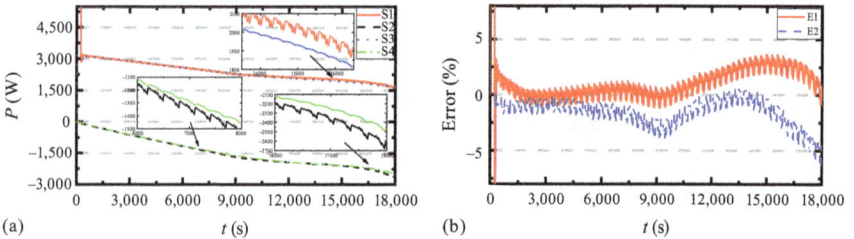

(a)   (b)

Figure 5.24   SOP prediction results for the BBDST condition: (a) SOP prediction and (b) SOP error

respectively. E1 is the estimation error of the discharge SOP, and its maximum is 3.41%. E2 is the charge SOP error, and its maximum value is 7.11%. The simulation results show that the proposed method can effectively estimate the SOP. The simulation model is designed and established. The implementation of the algorithms is introduced into the model.

5.6 Chapter summary

In this chapter, the square root algorithm of AEKF is improved, which is used to estimate the lithium-ion battery state and the estimation effect as well. Based on the previous research, it is used to establish the estimation model. The function of each component module is analyzed. The model is analyzed with measured experimental data onto various working conditions, such as constant-current discharge, cyclic discharge, and dynamic stress test. The feasibility of model estimation is verified by comparison. The results show that the proposed algorithm has a fast convergence speed, good tracking effect, stable deviation, and accurate overall performance of the battery state estimation. The work is accomplished through the experimental data acquired for parameterization and its use in the simulation. The second-order Thevenin equivalent circuit model is chosen because it is simple and has a minimal error, which can give the desired results. The OCV, HPPC, and volume testing have made a huge contribution to the data and results obtained.

Acknowledgment

The work is supported by the National Natural Science Foundation of China (No. 61801407), Sichuan Science and Technology Program (No. 2019YFG0427), China Scholarship Council (No. 201908515099), and Fund of Robot Technology Used for Special Environment Key Laboratory of Sichuan Province (No. 18kftk03).

Chapter 6
Dual extended Kalman filtering prediction for complex working conditions

Abstract

The Kalman filtering algorithm plays a vital role in battery state estimation. To solve the difficulties in real-time estimation and low precision under various working conditions for the battery state of charge and the state of health (SOH) estimation, the extended Kalman filtering algorithm is improved and a dual extended Kalman filtering method is proposed to realize the joint estimation of the model parameters and state of charge (SOC) in this chapter. Taking the experimental current and voltage as inputs, the model state-space equation is established, and the SOC estimation results are obtained. On this basis, the estimation method of battery internal resistance is established. Through experiment tests, the recursive least-square method is assisted by using the established simulation model to achieve the prediction of model parameters in various working conditions. Through the experimental results under complex operating conditions, the internal resistance of the battery under various operating conditions is estimated, according to which the SOC and SOH can be predicted accurately for lithium-ion batteries.

Keywords: State of charge; State of health; Real-time estimation; Extended Kalman filter; State-space equation; Dual extended kalman filter; System state; Nonlinear system; Recursive least square; Internal resistance; Lithium-ion batteries; Knee-Arrhenius algorithm; State of safety; End of life; Arrhenius model; Capacity fade; Rain-flow counting

6.1 Introduction

With the widespread application of lithium-ion batteries in the field of new energy sources, its health detection has been paid much attention [222]. Among them, it is of great significance to accurately estimate the state of charge (SOC) and state of health (SOH) for the lithium-ion batteries [63,64,223] in giving full play to the performance of batteries and realizing the real-time state detection and safety control of lithium-ion batteries [224].

Lithium-ion batteries are often used under complex working conditions [225], and their state detection is susceptible to environmental noise. Moreover, the internal electrochemical reactions of the lithium-ion batteries are complicated, often accompanied by polarization effects and ohmic effects [226], combined with variable discharge currents under complex conditions, internal battery temperature, self-discharge of the battery, and repeated use of materials to cause aging [65–68]. Due to the interference of other factors, accurate real-time SOC and SOH estimation of lithium-ion batteries is facing greater difficulties [227]. However, due to the complex internal structure of the lithium-ion batteries, it often exhibits strong nonlinear characteristics when used under complicated conditions [228], making it difficult for the traditional equivalent model to characterize the characteristics of the lithium-ion batteries fully and correctly [229]. There are still many problems and deficiencies in equivalent modeling and state estimation [230]. Therefore, how to establish an equivalent model for the working characteristics of lithium-ion batteries, estimate the battery state with the correct and appropriate algorithm, conduct real-time monitoring and safety control of lithium-ion batteries, and improve the efficiency of the battery is of great significance [37].

The current methods for estimating battery SOH mainly include empirical model-based methods, equivalent circuit model-based methods, and data-driven estimation methods [231]. The empirical model obtains the change of battery performance state through the analysis of large experimental data and summarizes the change law of the battery SOH [69,232,233], such as the incremental capacity analysis method [234]. It has the advantages of low modeling difficulty and wide application range, but the physical meaning of the empirical model is not clear [235], relying on experimental data, and lacks precision and accuracy in evaluating the results [70]. This method is based on the equivalent circuit model that simulates the relationship between the external parameters of the battery and the internal state quantities through the circuit. Based on the equivalent circuit model, the filter is used to realize the SOH estimation.

Filtering algorithms mainly include Kalman filter (KF), particle filter (PF), and least square (LS) [236]. Data-driven methods include autoregressive models, artificial neural networks (ANNs), support vector machines (SVM), and Gaussian process regression [72–76]. These methods have received increasing attention because they do not involve complex physical models. However, the effectiveness of these methods largely depends on the quality and quantity of the test data, and the derived models usually require a lot of computational intensity. To avoid dependence on data, an iterative algorithm based on the equivalent circuit model is selected to estimate the battery SOH. Since the state estimation relies heavily on the equivalent model, the model selection and construction are very important [80]. Therefore, the electrical equivalent circuit (EEC) modeling and state estimation of batteries under various complex working conditions need to be further developed.

Aiming at the goal of accurately describing the working state of the ternary lithium-ion battery, considering the accuracy of the characterization, the dual extended Kalman filter (DEKF) algorithm is used to collaboratively estimate the SOC and the SOH of the lithium-ion battery and to estimate both states simultaneously and improve the accuracy of the estimation [77,78,237]. In this modern era, batteries have become powerful sources of alternative energy, and the choice of lithium-ion chemistry battery technology is a highly efficient option for this kind of energy system [238]. However, the decent and secure application of lithium-ion batteries has always been an important issue. Studies have shown that as an important indicator of the battery management system, the safety status is firmly associated with the decent and secure application of batteries [239]. However, measuring a reliable state of safety (SOS) is difficult because its value cannot be directly calculated by the sensor, and there is no accurate solution yet [240].

Storing renewable energy at a competitive cost makes the battery energy source more reliable and comparable to the limited fossil fuel energy system [241]. Thus, the battery energy system is gaining more interest as alternative energy because of the rising environmental concern [242]. Additionally, as the most recent invention of the battery industry, lithium-ion technology has turned into a game changer. Over a decade, the lithium-ion batteries have become an essential source of power for portable and non-portable electronic devices including electric vehicles and smart grids. The long cyclic lifespan with high energy and power density characteristic of lithium-ion batteries makes this technology a reliable option for many industries [243]. However, the operation of the lithium-ion batteries required careful monitoring and regulation for special handling to avoid the deterioration of battery performance and prevent the situation that could result in severe damage or explosions. In this matter, the BMS plays an essential role in improving performance [244], prolonging life, and ensuring safety.

The BMS monitors the performance of the batteries by identifying some main state parameters of batteries such as SOC, SOH [245], and the state of power (SOP). Wherein the SOC represents the amount present charge, the SOH represents the status of aging, the SOP represents the amount of present power [246]. Additionally, some other minor parameters also have an effective relationship with the performance of batteries, such as the beginning of life (BOL), end of life (EOL), and remaining useful life (RUL) [247]. Although an accurate forecasting technique of lithium-ion batteries performance is very essential for efficient energy management, especially for safety to avoid the circumstance of fires or explosions [248,249], the identification of these state parameters is not easy. Because the internal characteristics of lithium-ion batteries are very dynamic, in a complete life cycle, many external and internal factors are also influencing this.

The main aim in this section is also introducing a reliable identification process of batteries state parameter, but in this research, the state parameter is going to study is not a common one [250]. The focus of this study is a stand-alone state parameter that is used for identifying the safety status of batteries and is known as the SOS. The SOS is a rarely studied state parameter of lithium-ion batteries and the research literature about this topic is also very few [251]. Most of the time, researchers documented the safety factors for the batteries based on the SOH value, but the SOH only describes the aging factor of the batteries, which is not enough for taking decisions about the safety of batteries. Additionally, there is no well-defined numerical quantification for the SOS like the SOC or SOP. A proper definition of the SOS is introduced by covering the numerical quantification [252].

The main scope of this section is the safety parameter study of lithium-ion batteries. The highlights of the proposed method are a newly developed numerical quantification of SOS and an adaptive Knee-Arrhenius algorithm to analyze the capacity degradation of lithium-ion batteries [253]. To cover the complete aging phenomenon of batteries, the structure of the proposed method is divided into two sub-models. The first sub-model is labeled as the battery aging model (BAM), and the second sub-model is labeled as the capacity fade model (CFM). Comprehensively, the two sub-models are designed that is based on two different internal characteristics of batteries [254]. Then, the output of each model is designed identically to extract the final SOS result without any extra mathematical analysis.

The output of both models is an EOL index derived from the EOL equation of both models [255]. Its concept is well documented for the lithium-ion batteries in the research domain, which is also frequently used by researchers [256]. EOL is defined under a general condition of capacity fade or internal resistance growth [257]. It represents a time threshold depending on the capacity fade or internal resistance growth, after that threshold the battery will no longer be able to use in any application [258]. According to the researchers, the threshold value of EOL for lithium-ion batteries is defined as the time either when it lost 20% of its nominal capacity or it gains 33% of its nominal internal resistance [259].

Three experimental processes are carried out in this research work to validate the proposed method [260], the cycle-life test, the hybrid pulse-power characterization (HPPC) test, and the capacity characterization test. The only aging phenomenon that is studied in this work is cyclic aging; the calendar aging is out of scope for this research work. The thermal safety of the lithium-ion batteries turned to be worse after the calendar aging [261]. The HPPC test is performed to analyze the internal resistance growth for the BAM, and the capacity characterization test is performed to analyze the capacity degradation mechanism of the CFM.

The scope of the BAM is the study of the internal resistance growth of lithium-ion batteries [262]. The internal resistance of lithium-ion batteries is a highly important parameter because of its frequently application, which is used for determining the battery power capability and aging status [263]. Accurate information about internal resistance growth during the aging of batteries can be very helpful for optimal energy management strategies. However, the internal resistance of the lithium-ion battery is not purely ohmic, and it might vary due to the polarization effects of the battery, such as impedance and capacity [264]. As a result, the battery modeling methods for characterizing the internal resistance growth always required a polarization parameter known as the RC networks, and the battery model included the RC network always provides better accuracy than the non-polarized battery models [265]. Hence, the second-order RC network Thevenin equivalent battery model is unitizing in this research work to characterize the accurate internal resistance characteristics of the experimented battery.

The scope of the CFM is the study of the capacity degradation mechanism of lithium-ion batteries [130]. Capacity degradation is a wide issue in the research field of lithium-ion batteries [266]. The researchers have mentioned it because of the internal chemical reaction of batteries which is begun from the very first charge-discharge cycle [267]. The cause of capacity degradation is also very different, Arora *et al.* [268] identified the side reactions as the fundamental cause of the lithium-ion battery capacity fade. They mention that inside of a lithium-ion battery, the lithium ions move from one electrode to another electrode during charging and discharging [269]. When they moved various side reactions took place inside the battery such as electrolyte decomposition [270], dissolution of active material [271], and passive film formation [272]. These side reactions can lead to the capacity fading of the battery and limit the battery lifespan. Similarly, Gantenbein *et al.* also identify two main reasons for capacity fade, either loss of active electrode material on both anode and cathode or loss of active lithium [273].

The scope of mathematical analysis for the CFM is a combined method of the knee-point algorithm and Arrhenius model quantification. The whole mathematical analysis is divided into three phases, including identification of knee-point and knee-values, Arrhenius model quantification, and extraction of EOL index from the combined knee-Arrhenius algorithm [274]. The idea of knee-point is very new in the battery domain and it is mainly utilizing in the dynamic characteristic of lithium-ion batteries capacity degradation mechanism. Outside of the battery domain, one of the first definitions is the knee-point. They developed an algorithm that used the mathematical concept of curvature and defined the maximum curvature point as the knee for continuous functions.

In the battery domain, the concept of knee points first came from the IEEE standard 485TM-2010, which defined the knee as the transformation to a phase of rapid degradation in the capacity for batteries. Also, many research references are documented for different methods of identifying the knee-point based on capacity degradation in recent times. The method identifies the knee point as the two tangent lines intersection point of the capacity fade curve [275]. An online knee detection algorithm based on quantile regression defines the knee-point as the intersection of two straight lines with different slopes. A different kind of knee feature on the voltage response curve is described [276] for the current pulse test and selected the knee points as the degradation features for the SOH estimation [277]. Most recently, the machine learning technique is introduced [278] to identify the knee-point and the knee-onset from the capacity fade curve [279].

The second phase of the mathematical analysis of the CFM is the Arrhenius model quantification of the identified knee-values. The application of the Arrhenius model on the battery domain is quite popular, and researchers have been numerously used this method to identify the effect of external forces on the degradation of batteries. The cylindrical 18,650 lithium-ion cells are tested [280] to analyze the changes of power capability, cell impedance, capacity rate by Arrhenius equation quantification. A lifetime prediction method [281] is developed for lithium-ion batteries based on the Arrhenius model. The model did not fit perfectly for all of their experimental data. One hundred seven aged lithium-ion batteries are tested [282] by using th USABC PHEV Battery Test Manual and realize that the capacity fade and internal resistance growth for the calendar aging at 30–50 °C follow the Arrhenius-like kinetics.

Zhou *et al.* [283] described that the number of charge–discharge cycles of lithium-ion batteries could change according to the working temperature condition due to the Arrhenius dependence of the reaction rate on temperature. NASA recommended a specific testing procedure for the prototype cells on the Aerospace Flight Battery Program to calculate the Arrhenius factors and activation energies for the capacity fade degradation. Recently, Li *et al.* [284] developed a SOH balancing control method for recycled batteries and mentioned that the Arrhenius model can express the capacity fade and internal resistance growth for a reference ambient temperature.

The third phase of the mathematical analysis of the CFM is the extraction of the EOL index from the combined knee-Arrhenius algorithm. The research literature has also documented the relation of knee point with EOL and frequently mentions that after the knee point batteries lost capacity in a rapid trend until the EOL. One of the earlier literature that introduces the relation between knee point and the EOL is Neubauer *et al.* [285], where they defined the EOL based on the knee point, but the definition of the EOL has not been widely accepted.

Later on, Yang *et al.* [286] proposed an RUL prediction method with a model-based Bayesian approach and observed that lithium-ion batteries show two-phase degradation behavior with evidence of an inflection point named knee point. Ecker *et al.* [287] studied Li(NiMnCo)O$_2$ based 18,650 lithium-ion batteries for both calendar and cycle aging and observed a rapid drop of capacity after a certain number of cycles. Saxena *et al.* expressed that the characterization of the knee point of batteries capacity degradation is necessary for accurate EOL prediction. Schuster *et al.* [288] mentioned that the capacity fade of cells exhibits a linear dependency on the charge throughput, but this linear characteristic turns into a nonlinear form after a turning point named knee point. Han *et al.* experimented three lithium titanium oxide anode battery under accelerated cyclic life and the results indicate that at 55 °C the batteries shows higher capacity fade with two-stage of degradation characteristics.

These are the overall scopes in this section and the techniques that are applied to make the structure of the proposed method. The next section will describe the problem formation and the realization of the proposed method.

6.2   Aging modeling methods

The evaluation processes of battery aging are not a few, and several of these methods are created to quantify the SOH and EOL by aging. As there is no standard definition for the term aging, researchers introduce different aspects of aging and utilizing them to introduce the different models of aging. So far, most of the researchers are agreed on two basic parameters for characterizing aging, which include the capacity fade and internal resistance growth. One of the well-documented research literature that defined different methods of modeling battery aging is Ecker *et al.* [281]. According to this literature, four different models could apply to estimate battery aging. There are electrochemical-based, analytical-based, equivalent-circuit-based, and statistical-based models.

6.2.1 Aging mechanisms

The aging characteristics of lithium-ion batteries are very dynamic. Generally, aging is defined as the degradation of the battery electrode which takes place in the chemical composition, and because of this, it is hard to categorize. Also, the degradation mechanism of the positive and negative electrodes is different. Researchers differentiate the origin of the aging mechanism as either chemical or mechanical and are strongly dependent on electrode composition. At the present time, most of the battery negative electrodes are composed of graphite and the main aging factor of the graphite electrode is the development of the interface of a solid layer which is also called the solid electrolyte interphase (SEI).

The SEI is a natural layer that is generated at the very beginning of the battery life cycle also provides a guarantee of security. However, the development of SEI over time also causes the loss of active lithium ions and electrolyte decomposition; as a result, the cell impedance increases and causes the aging phenomenon. On the other hand, the aging characteristics of the positive electrode are very different from the negative electrode. The positive electrode shows a lower decomposition rate over time and is similar to the negative electrode also depending on the composed materials. However, the SEI layer also exists on the positive electrode but more difficult to detect for the high voltage issues.

6.2.2 Electrochemical aging models

The core technology of electrochemical-based aging modeling is determining the physical model of the battery. The purpose of this method is to provide a sharp understanding of the specific physical and chemical phenomena happening during the battery working stage. Moreover, the electrochemical-based models are further divided into two parts by researchers such as the phenomenological approach and the atomistic and molecular approach.

6.2.3 Analytical aging models

Analytical-based aging models are the data dependable methods. These models are required as large as possible data from experiments to evaluate model parameters. The most popular analytical-based model is the "coulomb counting" method, which could estimate the SOH through a simple Ampere-hour (Ah) experiment method [289]. The accuracy is always depending on the accuracy of the measurements of the experiment process. The other approaches of analytical-based models are mainly based on the recursion algorithm such as fuzzy logic (FL), ANNs, EKF, and particle filter (PF). The accuracy of analytical-based models is relatively high, but the major drawback of this method is the requirements of doing large-scale recursion under the same condition and as a result, it cannot be done in real-time.

6.2.4 Equivalent circuit aging models

Equivalent circuit-based aging modeling is the simplest BAM that tries to clone the internal characteristics of batteries. The main technique of this approach is utilizing the equivalent battery model for quantifying the aging process. An EEC-based model may have three major parts. First, a static part that represents the thermodynamics properties of the battery, such as the nominal capacity and open-circuit voltage (OCV) as a function of the SOC. Second, a dynamic part that represents the kinetic aspects of the battery internal impedance behaviors such as polarization resistor and polarization capacity. Third, a source or load to complete the circuit for charge–discharge. The main goal of this approach is to reduce the need to understand detailed mechanisms and only require a few parameters, which are easily obtainable from the experiments.

6.2.5   Statistical aging models

The statistical-based models are very similar to the analytical-based model, but these methods do not require any prior knowledge of the aging mechanisms. Autoregressive moving average (ARMA) is one of the simplest modeling methods that consider the aging levels as a chronological series. Another implementation of this model is Eom *et al.* [290] where they consider the battery EOL criteria as a failure and model this EOL by Weibull's law.

6.2.6   Battery aging model

The proposed BAM in this section is based on the equivalent circuit-based aging modeling. This method is also called a model-based method and mainly focuses on the simple battery internal parameters to clone the overall characteristics of the battery. One of the benefits of using this aging model is that it allows us to estimate the battery aging from direct experimental data parameter identification or a complex algorithmic approach [291–293]. The algorithmic approach is mentioned as complex because such methods required a large and diverse data set from the time-consuming test, such as particle filtering, machine learning, SVM, and relevance vector machine. The proposed BAM is shown in Figure 6.1.

Figure 6.1 shows the complete BAM used in this research. The proposed model is designed based on the simplex technique of the equivalent circuit-based aging model (second-order Thevenin model), where the internal resistance growth is considered as the main model parameter. The experimental process used to identify the model parameter is a standard HPPC test. The mathematical analysis is conducted for the proposed BAM along with the required experimental process and the experimental setup to validate the proposed model.

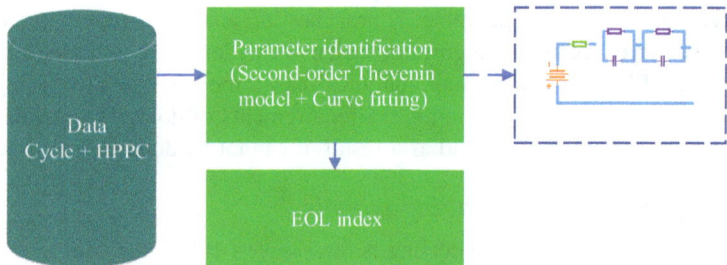

Figure 6.1   Proposed battery aging model

6.2.7 Internal resistance growth

As mentioned in the previous section, the BAM of the proposed method is mainly designed based on the internal resistance growth characteristic of batteries. This section briefly discusses the mechanism for the internal resistance growth of lithium-ion batteries. The internal resistance is an essential property of lithium-ion batteries for determining the available power, heat generation, and SOH. As a result, precise monitoring of this property is crucial for any battery management system. Generally, the internal resistance of lithium-ion batteries depends on the operating temperature, current rate, and SOC [294].

A lithium-ion battery experiences various levels of internal resistance development throughout its life cycle, depending on its causes. The structure of the SEI layer and the absence of active lithium ions are thought to be the key causes of this effect. The SEI layer worked as a safety barrier between the negative electrode and electrolyte, but it is permeable to the lithium ions and other charged elements to pass the layer. However, under a high-temperature condition, the SEI layer may dissolve and make the lithium salt less permeable to the lithium ions and it causes the increase of negative electrode impedance as well as the internal resistance [295]. On the other hand, low temperatures cause a decrease of the lithium-ion within the SEI layer and graphite [296] which can overlay the electrode with a lithium plating and reduce the active lithium-ion inside the battery.

6.2.8 Mathematical aging models

This section will describe the necessary mathematical analysis of the proposed BAM. The model parameters already listed in Table 6.4 can be identified by the experimental data of the HPPC test by using an exponential function fitting method on MATLAB. To initiate the calculative process, the fundamental equations of the second-order Thevenin equivalent circuit battery model need to be discretized first because the input data of the next calculation process will be discrete sets of points. After discretization, the derived equations can be listed as shown below:

$$U_t(k) = U_{oc} - U_1(k) - U_2(k) - I_k R_0 \tag{6.1}$$

$$U_1(k) = I_k R_1 \left(1 - e^{-t/\tau_1} \right) \tag{6.2}$$

$$U_1(k) = I_k R_2 \left(1 - e^{-t/\tau_2} \right) \tag{6.3}$$

The calculation of time constants will be done by performing the curve-fitting on HPPC test zero-input response data. In the HPPC test, the battery rests for a short time after each charge–discharge pulse. The time constants τ_1 and τ_2 can be represented by

$$\begin{cases} \tau_1 = R_1 C_1 \\ \tau_2 = R_2 C_2 \end{cases} \tag{6.4}$$

As a result, the current becomes zero and this results in the voltage response of branch U_1 to branch U_4 as a zero-input response in the HPPC test, as shown in Figure 6.2.

Herein, U_1 to U_2 is the drop of voltage for the discharge pulse current. U_2 to U_3 is the zero-input phase. U_3 to U_4 is the transition of the terminal voltage, and ΔU is the voltage variation from U_3 to U_4. According to Figure 6.2, the changing of terminal voltage in the rest period can be presented by curve fitting as shown below:

$$\begin{cases} U_t = a - b\left(1 - e^{-t/\lambda_1}\right) - c\left(1 - e^{-t/\lambda_2}\right) \\ y = a - b*(1 - \exp(-x/f)) - c*(1 - \exp(-x/g)) \end{cases} \quad (6.5)$$

where U_t is the terminal voltage, a, b, and c are the constant variables of the model, and λ_1 and λ_2 are the time constant. On the other hand, based on the discretized battery model equations, the terminal voltage U_t of the battery model for the rest period can be represented by an equation that shows the calculation process of R_1, R_2, C_1, and C_2. The mathematical relationship is described as shown in the following equation:

$$\begin{cases} U_t = U_{oc} - R_0 I - R_1 I e^{-t/\tau_1} - R_2 I e^{-t/\tau_2} \\ U_t = U_{oc} - R_0 I - R_1 I e^{-t/R_1 C_1} - R_2 I e^{-t/R_2 C_2} \\ R_1 = \dfrac{b}{I}, R_2 = \dfrac{c}{I}, C_1 = \dfrac{\tau_1}{R_1}, C_1 = \dfrac{\tau_2}{R_2} \end{cases} \quad (6.6)$$

Herein, the value of variables b, c, τ_1, and τ_2 are collected from the curve-fitting result. Finally, the ohmic resistance R_0 will be derived from the immediate voltage decline after the current pulse according to the following equation:

$$R_0 = \frac{\Delta U}{I} = \frac{U_2 - U_1}{I} \quad (6.7)$$

where ΔU denotes the terminal voltage step-variation as shown in Figure 6.2 when the discharging pulse process stops. In this research work, U_2 is selected from the voltage increase in the first second of the rest period [297] and U_1 is selected from the terminal voltage when the long rest period ends.

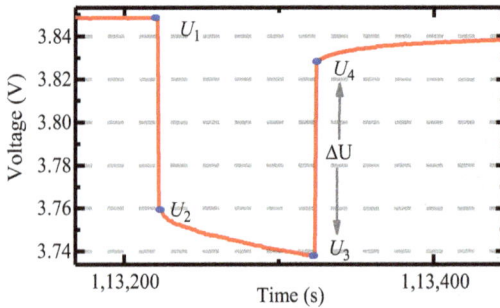

Figure 6.2   Zero-input response on the HPPC test

6.3 Iterative calculation algorithm

6.3.1 Cyclic aging expression

The aging of lithium-ion batteries is closely related to the number of cycles. Therefore, the statistics of the lithium-ion battery cycle times are also a way of life prediction. The rain-flow counting algorithm is proposed as the statistical method of cycle times, and the lithium-ion battery with the capacity of 70 Ah is taken as the research object.

6.3.2 Rain-flow counting

The rain-flow counting algorithm, also known as the tower top algorithm, was proposed by two British engineers M. Matsuishi and T. Endo [298]. The main function of this counting method is to simplify the actual measured load history into several load cycles, and each cycle is a damage accumulation. This method is widely used in fatigue life calculation, and its main manifestation is the load–time curve. At the same time, to reduce the number of half-cycles, it is necessary to reconstruct the time history of the data before counting and move the absolute value of the peak or valley to the starting point of the process. In the estimation of lithium-ion battery life, the load is SOC, both set of SOC–time curves are obtained, and then the entire coordinate system is rotated 90° clockwise, and the time coordinate axis is vertically downward. The roof goes down, as shown in Figure 6.3.

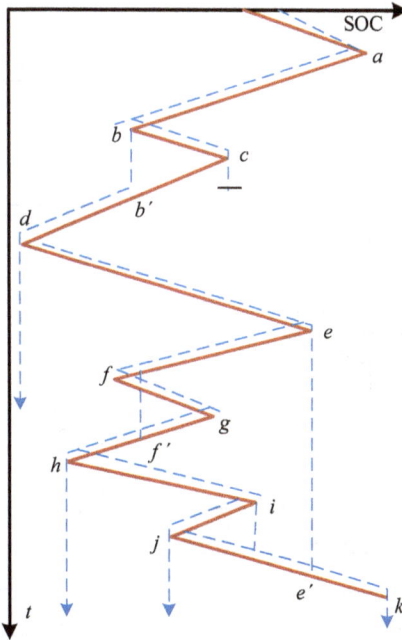

Figure 6.3   The time history of SOC

The counting rules are described as follows:

1. The rain-flow counting algorithm starts from the internal measurement of the peak position of the load time history and flows down the slope.
2. The rain current starts to flow from a certain peak point and stops flowing when it encounters a larger peak than its actual peak.
3. When the rain current meets the rain flowing down, it must stop flowing.
4. Take out all the full cycles and note the amplitude of each cycle.
5. The divergent convergent load time history remaining after counting in the first stage is equivalent to a convergent–divergent load time history, and the second stage rain-flow counting is performed. The total number of counting cycles is equal to the sum of the counting cycles of the two counting stages.

According to the above counting process and rules, the SOC time record includes three complete cycles b–c–b', f–g–f', i–j–e', and three half-cycles a–b–d, d–e–e', e–f–h. The flowchart of the improved rain-flow counting algorithm is shown in Figure 6.4.

First, it should be judged whether the starting value is the peak value. If not, the data before the peak value is cut to the end of the data segment, and the peak value is taken as the starting value. The maximum and minimum values are determined by the three-point method, and the data are updated at any time. The middle zone is used to store extreme points. According to the counting principle of the rain-flow counting

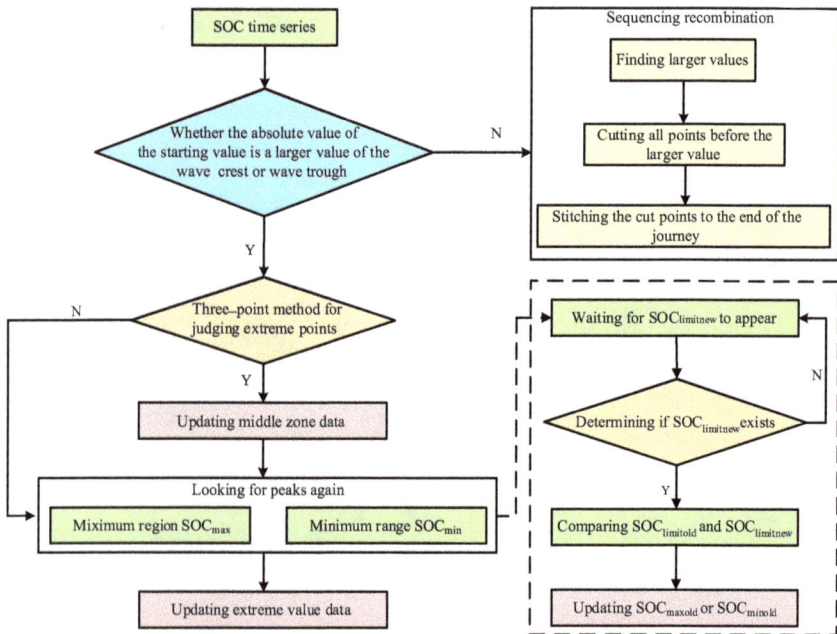

Figure 6.4   Improvement of rain-flow process

algorithm mentioned above, lithium-ion batteries may experience some small cycles during a deep cycle. These small cycles may be a small charge–discharge range of the oscillation caused by the electrochemical properties of lithium-ion batteries. Due to the advantages of lithium-ion batteries, the impact of these small cycles on battery life can be ignored. Besides, the traditional rain-flow counting algorithm is sensitive to the size of the cycle. Therefore, the improved rain-flow counting algorithm increases the middle region, which can filter out the small periods in these processes.

6.3.3 Cyclic charge–discharge variation

The research platform of this experiment is consistent. To test the cyclic life of the lithium-ion batteries, a lithium-ion battery with 4.2 V/70 Ah is used as the research object, and deep charge–discharge is conducted 80 times. The battery SOC change can be described as shown in Figure 6.5.

Rainfall can identify events similar to constant amplitude load data in complex load sequences. Figure 6.6 shows that the lithium-ion battery completes 80 cycles

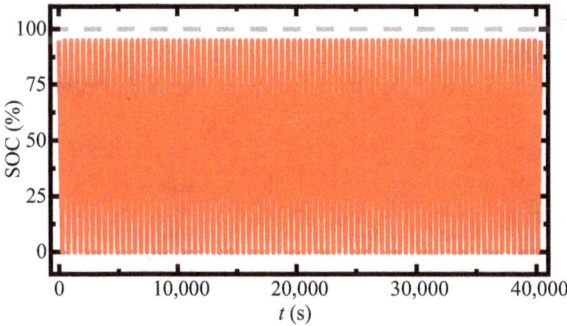

Figure 6.5   Cycle condition of SOC history

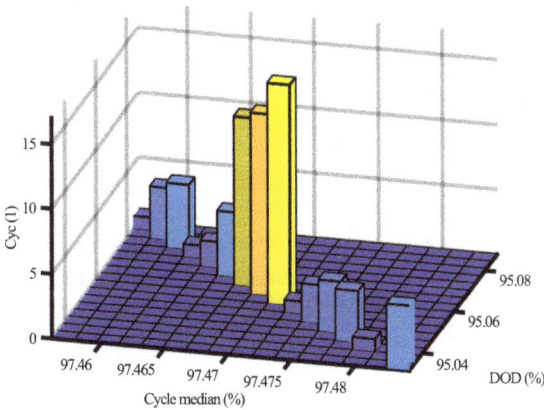

Figure 6.6   SOC cycle count of 95% DOD

at a DOD of 95%. This result is consistent with the actual number of cycles, which proves that the algorithm has good robustness.

In Figure 6.6, the median circulation depth is the average of the two circulation depths. The value can show the current load level of the battery, and the range of the value is 0%–100%. The larger the value, the heavier the current load of the battery, that is, the faster the aging speed will be. It can be seen from the figure that the rain-flow counting method can well track each cycle, and the statistical results of cycle times are quite accurate, which is consistent with the experimental data.

6.3.4   State of safety analysis

The main problem that is focused on this research work is the unavailability of a reliable SOS prediction method for the lithium-ion battery. Already mentioned in the previous section that the SOS is a relatively new state parameter for the lithium-ion battery and the numerical quantification of SOS is also not well defined. So far most of the research literature has already connected the SOS with the health and aging parameters of the batteries. That is why the problem formation for a reliable SOS prediction method required a deep study for the complex aging characteristics of lithium-ion batteries. Hence, in this research work, the problem structure is designed with the relevant aging mechanism of lithium-ion batteries for a reliable solution. For better understanding, the whole process is listed according to the following manner. This research provides a brief definition of the existing and newly designed key points that are going to utilize to solve existing problems or limitations and provides a framework of the problem-solution models of the proposed method which will help to realize the solution background.

6.3.5   Definitions of key points

6.3.5.1   State of safety

The safety of batteries for a battery-powered application means reliable use and no situation of hazards, like fire or explosion. The SOS is one of the state parameters that indicate the health and remaining usefulness status of batteries. A general derivation of the SOS can be obtained from the inverse function of battery abuse. That is described as shown below:

$$\begin{cases} \text{Safety} \propto \dfrac{1}{\text{Abuse}} \\[2mm] f_{Safety}(x) \propto \dfrac{1}{f_{Abuse}(x)} \end{cases} \tag{6.8}$$

The amount of battery abuse can be calculated from different degradation factors of batteries, such as internal resistance growth or capacity fade. In this research work, the abuse of batteries is calculated from both internal resistance growth and capacity fade in the form of the EOL Index.

6.3.5.2   State of safety index (SOS index)

The SOS index is a list of SOS status numerical values with the decision and required action according to the battery hazard level. The proposed SOS index in

Table 6.1   Proposed SOS index

SOS index	Description	Decisions
0	No hazard level	No replacement required
1	Modest hazard level	
2		
3	Above-average hazard level	Replacement required for high EOL index batteries
4		
5		
6	Significant hazard level	Replacement is required for all batteries
7		
8	Complete hazard level	Battery destroyed

this section has five status labels with a range of values from 0 to 8, which is described as shown in Table 6.1.

The value of the SOS index from 0 to 2 indicates that the battery cell is in no hazard or modest hazard level that comes with a decision that the battery cell should not require to replace with a new cell. The value of the SOS index from 3 to 5 indicates that the battery cell is at an above-average hazard level that comes with a decision that a high EOL index battery should replace with a new cell for avoiding any unwanted situation. Finally, the SOS index value from 6 to 8 indicates that the battery cell is in a significant hazard level or destroyed, that come with a decision that most of the battery cell need to replace for avoiding any kind of accidental situation.

6.3.5.3   End of life

EOL is a time threshold that represents the end of the useful life of a battery. After that, the battery will no longer be able to use in any application. EOL can be defined from both the capacity fade and internal resistance growth characteristics of batteries. According to the references, a lithium-ion battery reached its EOL when it lost 20% of its nominal capacity or get a 33% increase in its nominal internal resistance [174] as shown in the following equation:

$$\begin{cases} Q_{EOL} = 0.8 * Q_n \\ R_{EOL} = 1.33 * R_n \end{cases} \tag{6.9}$$

Equation (6.9) shows the EOL condition for both capacity fade and internal resistance growth. Here, Q_{EOL} is the capacity of the battery at the EOL, Q_n is the nominal capacity of the battery, R_{EOL} is the internal resistance of the battery at the EOL and R_n is the nominal internal resistance of the battery.

6.3.5.4   End of life index

EOL index is a list of EOL status numerical values with remarks. In this research work, two EOL indexes are defined. Table 6.2 shows the EOL index for the BAM.

Table 6.2   EOL index for the battery aging model

EOL index	Conditions	Remarks
0	If $R_0 < 1.08 * R_n$	No damage
1	If $1.16 * R_n > R_0 \geq 1.08 * R_n$	Little damage
2	If $1.25 * R_n > R_0 \geq 1.16 * R_n$	Above-average damage
3	If $1.33 * R_n > R_0 \geq 1.25 * R_n$	Severe damage
4	If $R_0 \geq 1.33 * R_n$	Complete damage

Table 6.3   EOL index for the capacity fade model

EOL Index	Conditions	Remarks
0	If $0.95 * Q_n < Q_r \leq Q_n$	No damage (BOL)
1	If $0.90 * Q_n < Q_r \leq 0.95 * Q_n$	Little damage
2	If $0.85 * Q_n < Q_r \leq 0.90 * Q_n$	Above-average damage
3	If $0.80 * Q_n < Q_r \leq 0.85 * Q_n$	Severe damage
4	If $Q_r \leq 0.80 * Q_n$	Complete damage

Table 6.3 shows the EOL index for the CFM. Both EOL indexes are designed with five decision statuses and the values ranged from 0 to 4.

For the BAM, the five conditions are designed for the 0.8%–33% increase of batteries internal resistance, and for the CFM, the five conditions are designed for the 5%–20% fade of batteries nominal capacity. Both EOL index is independent and developed based on two different internal characteristics of lithium-ion batteries.

Lithium-ion batteries show a rapid transition in their capacity degradation process after a certain point and generate a knee shape on its capacity fade curve. This certain point is named as knee point. Knee values are a set of normalized capacity values between the maximum and minimum slope-changing ratio point of the capacity fade curve. The concept of knee values is completely new and this research work is the very first literature that introduces the idea of knee values for calculating the remaining capacity of lithium-ion batteries.

6.3.6   Electrical equivalent circuit modeling

The equivalent circuit model includes the internal resistance model, Thevenin model, PNGV model, etc. The internal resistance model takes the operating characteristics of the battery into account and the structure is simple. Based on the internal resistance model, the Thevenin model introduces the parallel circuit of resistance and capacitance to describe the polarization effect in the battery. It also stimulates the dynamic characteristics of the battery. When compared with the PNGV model and the GNL model, the Thevenin model has a simple structure. Besides, it belongs to a nonlinear low-order model. It involves fewer parameters and its accuracy can meet the requirements of engineering applications. On this basis, the effects of polarization on the voltage and SOC are fully considered, and the Thevenin equivalent circuit model is established to characterize the battery characteristics, as shown in Figure 6.7.

In Figure 6.7, U_{oc} represents the open-circuit voltage. U_o represents the terminal voltage. R is the ohmic resistance, and U_R is the ohmic voltage, which is the battery voltage drop effect at the end of the discharge. The RC parallel loop is composed of a polarization resistor R_1 and a polarization capacitor C_1 for characterizing the polarization effect of the lithium-ion battery. U_1 is the polarization voltage.

6.3.7 Model parameter identification

Parameters of the BAM can be obtained. The BAM is designed based on the internal resistance growth characteristic of lithium-ion batteries. As a result, the main parameter of the BAM is internal resistance. To calculate the internal resistance of batteries, the corresponding battery model applied in this research work is a second-order Thevenin equivalent circuit battery model and the validation experiment is a HPPC test, as shown in Table 6.4.

Table 6.4 shows the definition of all parameters of the BAM derived from the corresponding battery model. The theoretical and mathematical analysis of the corresponding battery model that is used to validate the BAM is described.

Parameters of the CFM can be obtained. The CFM is designed based on the capacity fade characteristic for the cyclic aging of batteries. The key experimental analysis used for model validation is a standard capacity characterization test. Table 6.5 shows all of the parameters of the CFM.

The mathematical analysis of the CFM is divided into three phases. Phase one is the knee point and knee values identification, phase two is the Arrhenius model

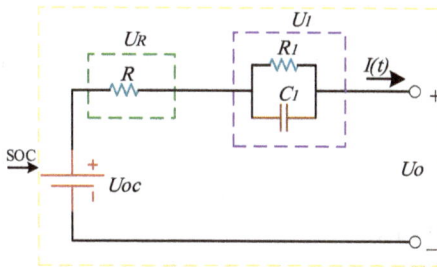

Figure 6.7   Equivalent circuit model

Table 6.4   Parameters of the battery aging model

U_{oc} (V)	R_0 (Ω)	R_1 (Ω)	R_2 (Ω)	C_1 (F)	C_2 (F)
Open-circuit voltage	Internal resistance	Polarization resistance	Polarization resistance	Polarization capacity	Polarization capacity

Table 6.5   Parameters of the capacity fade model

h	P_{QMAX}	$P_{Q,}$ knee	Q_{kneev} $[P_{QMAX},$ $P_{QMIN}]$	E_a (J/mol)	Q_1 (Ah)
Minimum slope-changing ratio	Maximum slope-changing ratio	Knee point	Knee values	Activation energy	Capacity loss

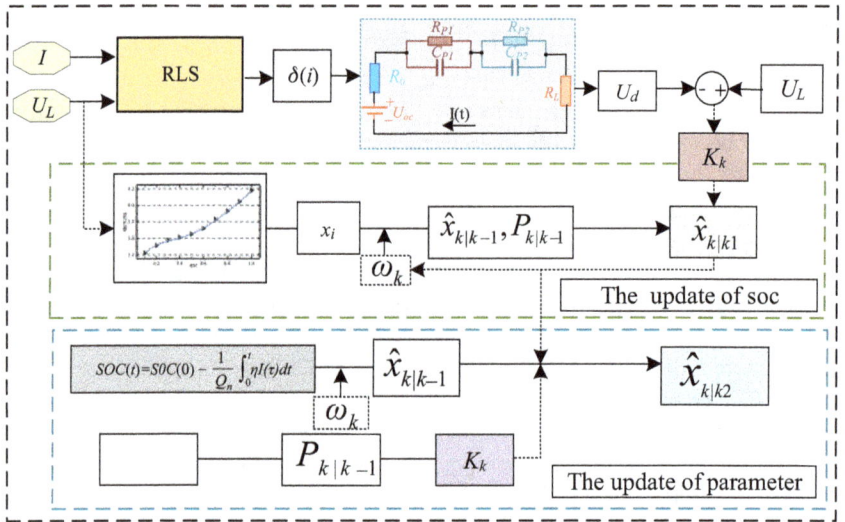

Figure 6.8   Structure diagram of double extended Kalman algorithm

quantification of knee values, and phase three is the EOL index extraction. The details of the CFM are discussed.

6.3.8   Dual extended Kalman filtering

The DEKF algorithm adds the EKF to update the battery equivalent model parameters in real-time. The DEKF structure can jointly estimate the model parameters and SOC that can improve the model accuracy and the SOC accuracy simultaneously. The structure diagram of the joint estimation of model parameters and SOC by the DEKF algorithm is shown in Figure 6.8.

From the voltage–current relationship of the Thevenin model, the LS form of the difference equation can be obtained, as shown in the following equation:

$$
\begin{cases}
U_m(k) = U_{OC}(k) - U_L(k) \\
U_m(k) = \dfrac{R_P C_P}{R_P C_P + T} U_m(k-1) - \left(\dfrac{T R_P}{R_P C_P + T} + R_0 \right) I(k) + \dfrac{R_0 R_P C_P}{R_P C_P + T} I(k-1)
\end{cases}
$$

$$(6.10)$$

In (6.10), U_{oc} represents the open-circuit voltage of the equivalent model, and U_L is the terminal voltage of the model. R_0, R_p, and C_p are the ohmic resistance, polarization internal resistance, and polarization capacitance of the model, respectively. T is the sampling interval of the working condition simulation experiment, and I is the experimental current. The differential equation of (6.10) is written in the standard least-square form as shown below:

$$y(k) = X^T(k)\theta \tag{6.11}$$

In Equation (6.11), $y(k)$ is the output matrix, $X^T(k)$ is the input matrix, and θ is the parameter matrix to be identified. The RLS algorithm uses the recursive equation to adaptively update the parameters at each sampling point, as shown in the following equation:

$$
\begin{cases}
\widehat{\theta}_{N+1} = \widehat{\theta}_N + y(n+1)P_N X(N+1)\left[y(n+1) - X^T(N+1)\widehat{\theta}_N\right] \\[2mm]
P_{N+1} = \dfrac{1}{\lambda}\left[P_N - y(n+1)P_N X(N+1)X^T(N+1)P_N\right] \\[2mm]
y(n+1) = 1/[X^T(N+1)P_N X(N+1)] \\[2mm]
P_N = \left(X_N^T X_N\right)^{-1}
\end{cases} \tag{6.12}
$$

In (6.12), λ is the forgetting factor, which is used to eliminate the influence caused by data saturation, and the value is 0.95. After obtaining the real-time model parameters of each sampling point through the above recursive steps, a state estimation model is built to perform two-stage filtering and correction on the SOC of the lithium-ion batteries. In the second stage of SOC filtering, the Ah integration method is used to update the SOC time and correct the SOC with the result of the first filtering. The recursive process of Kalman filtering is shown in the following equation:

$$
\begin{cases}
\widehat{X}_0 = E[x(0)], P_0 = Var[x(0)] \\[2mm]
\widehat{X}_{k+1|k} = f\left(\widehat{X}_k, u_k\right) = \begin{bmatrix} 1 & 0 \\ 0 & \exp(-\Delta t/\tau) \end{bmatrix} * \begin{bmatrix} SOC_k \\ U_{p,k} \end{bmatrix} + \begin{bmatrix} -\eta\Delta t/c \\ R_p(1 - \exp(-\Delta t/\tau)) \end{bmatrix} \\[4mm]
P_{k+1,k} = A_k \widehat{P}_k A^T_k + Q_k, A_k = \left.\dfrac{\partial f}{\partial x}\right|_{x=k} = \begin{bmatrix} 1 & 0 \\ 0 & \exp(-\Delta t/\tau) \end{bmatrix} \\[4mm]
K_{k+1} = P_k C^T_k (C_k P_k C^T_k + R_k)^{-1} \\[2mm]
\widehat{X}_{k+1|k+1} = X_{k+1} + K_{k+1}(y_{k+1} - g(X_{k+1}, u_{k+1}))^{-1} \\[2mm]
g(X_{k+1}, u_{k+1}) = f_k(soc) - I_k R_0 - U_{p,k} \\[2mm]
\widehat{P}_{k+1|k+1} = (1 - K_{k+1}C_{k+1})P_{k+1}
\end{cases} \tag{6.13}
$$

In (6.13), X represents the selected state value. In the adaptive extended Kalman filtering algorithm, $X = R_0$ in the first filter, and $X = [SOC\ U_p]$ in the second filter, in which P represents the corresponding error covariance. Q_k and R_k represent the expected values of process noise ω_k and observation noise v_k, respectively, y_k is the experimental value of the system observation variable at k time, and K_k represents the Kalman gain.

6.4   Parameter test and identification

6.4.1   *Experimental platform setup*

In this research work, all experimental process is performed at the New Energy Measurement Laboratory (DTlab) by using BTS200-100-104 made by Shenzhen Yakeyuan Technology Co., Ltd. To maintain the constant temperature, the cell is placed in a temperature chamber named DGBELL-BTKS (-70–150 °C) manufactured by Guangdong Bell Experiment Equipment Co., Ltd. The full experimental equipment setup used for the battery experiments is shown in Figure 6.9.

The temporal temperature deviation of the temperature chamber is $T < \pm 2$ °C with a spatial temperature accuracy of ± 0.5 °C. Additionally, a high configuration host computer is connected with the experimental equipment to store the experiment data and analyze the data for calculating the results.

6.4.2   *Whole-life-cycle HPPC test*

The HPPC test is a well-known experimental process to identify the unknown model parameters of the model-based battery modeling approach. In this research work, four ideal HPPC tests are performed to identify the unknown parameters of the proposed BAM. The complete profile of the HPPC test can be designed as shown in Figure 6.10.

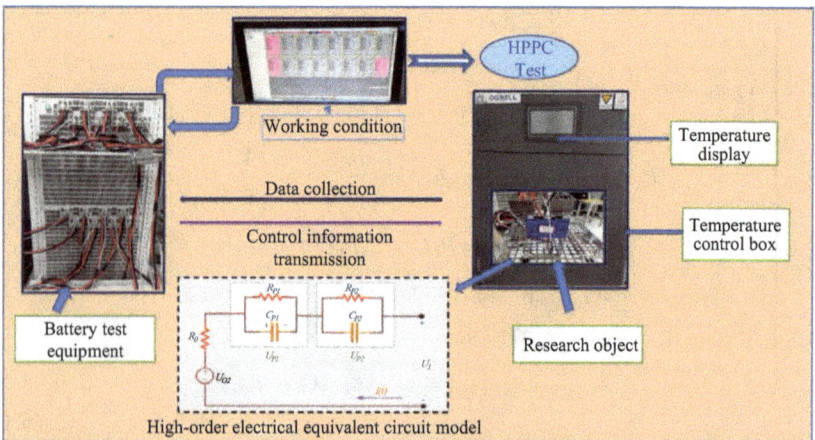

Figure 6.9   Experimental equipment setup

(a) Full voltage profile

(b) Full current profile

(c) Single cycle step

(d) Single pulse

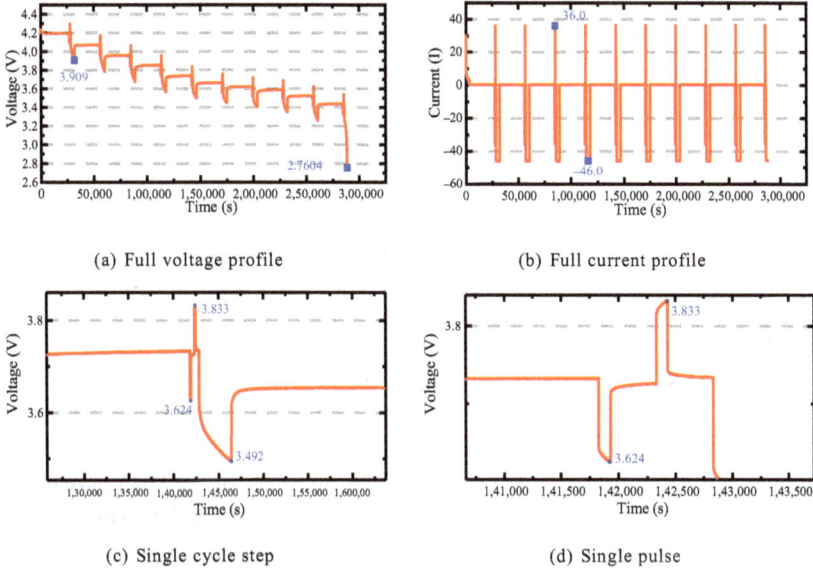

Figure 6.10   Complete hybrid pulse-power characterization test profile

The HPPC test curve contains a 720-s discharging cycle with a fixed current rate of −46 A (nearly 1 C) to achieve a 10% decrease in SOC and a rest period of 3,600 s before the next test cycle. Between every two steps, a discharge for 10 s at −36 A, rest for 40 s, and charge with 10 s at 36 A is performed to identify the transient response of terminal voltage. Ideally, the HPPC experiment pulse is performed at a charge–discharge rate of 3 C, but for more accurate data analysis, the charge–discharge pulse is done at about 0.90 C in this work.

6.4.3   Capacity characterization test

The capacity characteristic test is a simple experimental process to identify the present capacity status of a battery. An ideal capacity characterization test contains three steps such as standard charging, discharging, and rest. The designed capacity characterization test in this section contains a constant current and constant voltage charge up to 100% SOC at C/2 current rate and constant current discharge down to 0% SOC at C/10 current rate. The rest period between the charge–discharge step is 2 h, as shown in Figure 6.11.

Figure 6.11 shows the flowchart of a complete capacity characterization test. Figure 6.12(a) shows the voltage–time–current plot, (b) shows the voltage–time plot, (c) shows the current–time plot, and (d) shows the capacity–time plot of a single capacity characterization test.

The low current rate on the discharge capacity characterization test is to allow the system to measure the actual capacity fades. Because the higher current rate could be distorted, the capacity fades by causing the lower voltage limit to be

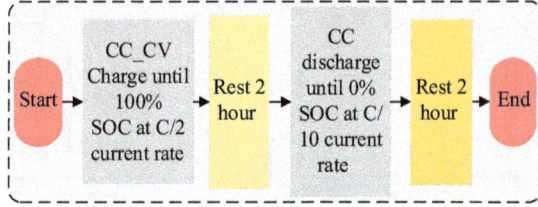

Figure 6.11   Flowchart of capacity characterization test

reached earlier, and the measurement will not reflect the actual capacity. A single capacity characterization test takes about 12 h.

6.4.4   Open-circuit voltage test

The pulse discharge test is conducted on a lithium-ion battery. The curves of current and terminal voltage are shown in Figure 6.13. During the experiment, the direction of the charging current is positive.

In Figure 6.13, the red line represents the current and the blue line represents the terminal voltage. It can be seen from the voltage curve of 40 min at the end of each constant current discharge that the battery voltage will gradually stabilize after a long period after the end of discharge, which means that the internal chemical reaction and thermal effect have reached equilibrium. The battery voltage is its OCV, so the relationship between OCV and SOC can be obtained as

$$U_{OC} = 2.91154SOC^5 - 11.15548SOC^4 + 15.57393SOC^3 - 9.12235SOC^2$$

$$+ 2.73594SOC + 3.23474$$

$$(6.14)$$

6.4.5 Recursive least-square method

To better observe the prediction of internal resistance, the online parameter identification method is used to identify the battery model parameters. The method used is the recursive least-square (RLS) method. First, according to the equivalent circuit model in Figure 6.7, the output equation of the circuit is

$$U_L(s) = U_{oc}(s) + I(s)\left(R_o + \frac{R_1}{1 + R_1 C_1 s}\right) \qquad (6.15)$$

The direction of the current in (6.15) is opposite to that in Figure 6.7. Substituting $s = 2(1 - z^{-1})/T(1 + z^{-1})$ into (6.15), the mathematical expression can be obtained as

$$\frac{E(s)}{I(s)} = \frac{\frac{R_o T + R_1 T + 2R_1 C_1 R_o}{T + 2R_1 C_1} + \frac{R_o T + R_1 T - 2R_1 C_1 R_o}{T + 2R_1 C_1} z^{-1}}{1 + \frac{T - 2R_1 C_1}{T + 2R_1 C_1} z^{-1}} \qquad (6.16)$$

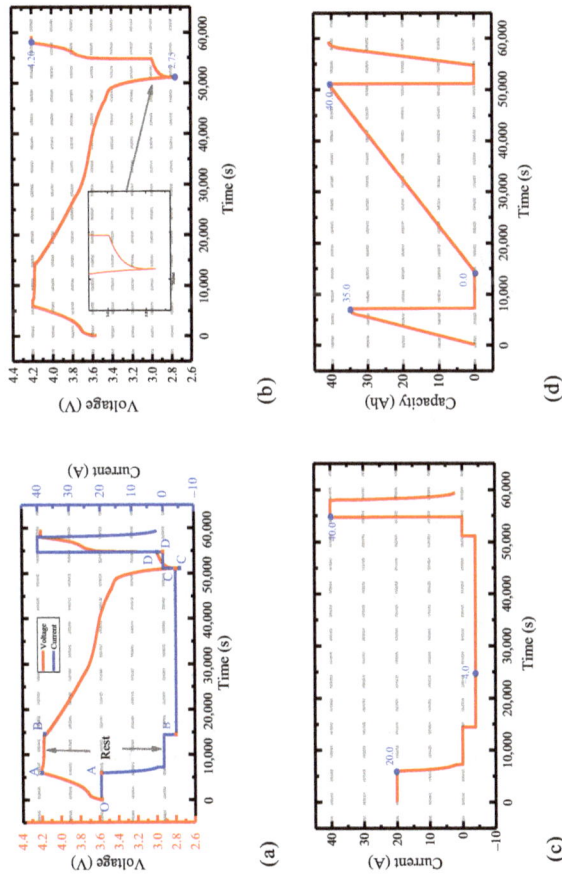

Figure 6.12   Complete capacity characterization test profile: (a) voltage–time–current curve, (b) full voltage profile, (c) full current profile, and (d) full capacity profile

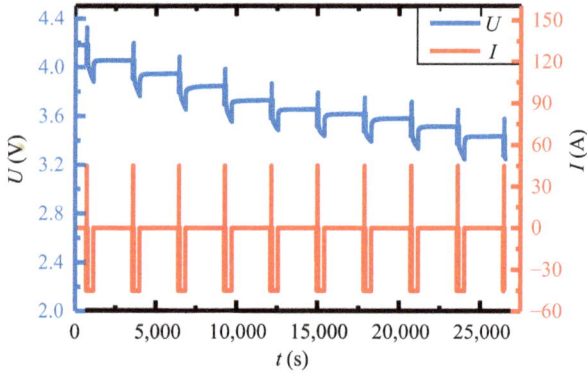

Figure 6.13   Current and terminal voltage curves of the HPPC test

In (6.16), $E(s) = U_L(s) - U_{oc}(s)$. According to the principle of RLS, the parameter calculation process can be obtained as

$$
\begin{cases}
\theta_1 = \dfrac{(T - 2R_1C_1)}{(T + 2R_1C_1)} \\[3mm]
\theta_2 = \dfrac{(R_oT + R_1T + 2R_1C_1R_o)}{(T + 2R_1C_1)} \\[3mm]
\theta_3 = \dfrac{(R_oT + R_1T - 2R_1C_1R_o)}{(T + 2R_1C_1)}
\end{cases}
\tag{6.17}
$$

Combining (6.16) and (6.17), the equation can be obtained that needs to be identified as

$$
E(k) = -\theta_1 E(k-1) + \theta_2 I(k) + \theta_3 I(k-1)
\tag{6.18}
$$

The RLS method can be used to estimate the values of θ_1, θ_2, and θ_3, and then according to (6.16), the parameters in the circuit can be calculated as

$$
\begin{cases}
R_o = \dfrac{(\theta_2 - \theta_3)}{(1 - \theta_1)} \\[3mm]
\tau = R_1C_1 = \dfrac{(1 - \theta_1)}{(2\theta_1 + 2)} \\[3mm]
R_1 = (1 + 2\tau)\theta_2 - 2R_o\tau - R_o \\[3mm]
C_1 = \dfrac{\tau}{R_1}
\end{cases}
\tag{6.19}
$$

Through this method, the internal resistance of the lithium-ion battery can be identified online, and then the SOH can be estimated in real time by the resistance definition method in the SOH evaluation process.

6.5 Complex condition experiment

To verify the estimation effect of the algorithm, the BBDST condition test is combined for analysis.

6.5.1 Test platform construction

An NMC battery with a rated capacity is selected for the test. The main specifications of the battery are presented in Table 6.6.

The new ternary lithium-ion battery with a nominal capacity of 45 Ah is selected for the test. The instruments used in the test include the power cell large-rate charge-discharge tester, a three-layer independent temperature control high- and low-temperature test chamber (BTT-331C), and other supporting experiments equipment.

As the parameters in the model are affected by temperature, the test is carried out at 25 °C. The battery will age due to recycling and other reasons, and the actual capacity of the battery will greatly deviate from the calibration capacity. The true discharge capacity of the battery is important for the estimation of the SOC of the lithium-ion battery. Therefore, the capacity calibration of the lithium-ion battery must be performed. In this study, to better observe the change of resistance, the online parameter identification method is adopted.

6.5.2 Cyclic aging procedure design

The main work of research is to study the cyclic aging of lithium-ion batteries. Both proposed models are also designed that is based on the cyclic aging of the battery. As a result, a large part of the complete experimental process contains the cyclic aging test. In this study, a lithium-ion battery cell with a nominal capacity of 40 Ah is investigated to validate the proposed method of the SOS prediction, as shown in Figure 6.14.

The battery cell is composed of graphite as an anode and LiCoMn and carbon black as a cathode. The upper voltage limit of the cell is 4.2 V, and the lower voltage limit of the cells is 2.6 V. Figure 6.14(a) shows the flowchart of the complete cycle-life test. The cyclic aging test profile is shown in Table 6.7.

The cyclic life test contained a full three hundred cycles of standard charge–discharge steps. After every one hundred cycles, the battery cell is rested for one

Table 6.6   Basic technical parameters of the battery

Factors	Specification
Rated voltage (V)	3.7
Maximum load current (A)	5 C
Rated capacity (Ah)	45
Charge cut-off voltage (V)	4.2
Discharge cut-off voltage (V)	2.75

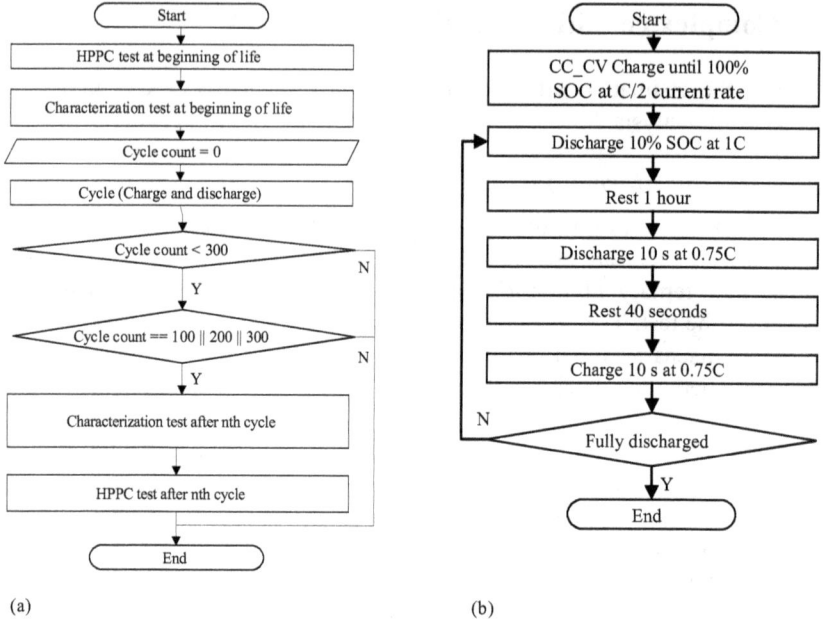

(a)   (b)

Figure 6.14   Experimental processes of battery aging model: (a) flowchart of cycle-life test and (b) flowchart of HPPC Test

Table 6.7   Cyclic aging test profile

Properties	Values
Temperature (°C)	25
Charge current rate	C/2
Discharge current rate	C/4
SOC range (%)	10-80

day to perform the HPPC test and a capacity characterization test. A constant 25 °C temperature condition is maintained for the complete cyclic life experiment. Table 6.7 shows the conditions of charging current rate, discharging current rate, and the SOC range of the cycle-life test.

The aging and capacity fade characteristics are analyzed for lithium-ion batteries. Several commonly used models for battery aging evaluation are summarized, including electrochemical-based models, analytical-based models, equivalent circuit-based models, and statistical-based models, and their advantages and disadvantages are summarized. Besides, methods for identifying the capacity fade of lithium-ion batteries are introduced. The model-based method effectively solves the dynamic problem of the lithium-ion battery safety prediction. Through research and use of BAM and CFM, a reliable prediction method for the SOS of lithium-ion batteries is proposed. The SOS is defined as a function of battery aging and capacity degradation. Based on the BAM with increased internal resistance, the HPPC test is performed, and the parameter identification is performed.

6.5.3 Battery aging modeling results

As mentioned in the previous chapters, to develop and experimentally validate the proposed BAM, four HPPC tests are carried out where one of them is performed at the beginning of the cycle-life test and called the BOL test. Another three HPPC tests are performed after every 100 cycles of the cycle-life test. Finally, the values of the internal resistance calculated from the HPPC tests data analysis are used to identify the output (the EOL index) for the BAM. The results of these HPPC tests, the defined EOL indexes, and the final output of the BAM are presented in Table 6.8.

Table 6.8 shows all identified parameter values of the BAM from the HPPC test data based on the corresponding mathematical analysis. Herein, the value of R_0 at the 50% SOC is taken as the reference value of internal resistance for the HPPC test. This value of R_0 is used to define the nominal internal resistance R_n to evaluate the conditions of the EOL index for the BAM, as shown in Table 6.9.

Table 6.9 shows the defined EOL indexes for the BAM. According to the BAM EOL index, if the value of R_0 is less than $3{,}237.87\Omega$, then the battery EOL index will be 0. If R_0 is between $3{,}477.68\Omega$ and $3{,}237.87\Omega$, then the battery EOL index will be 1. If R_0 is between $3{,}747.50\ \Omega$ and $3{,}477.68\ \Omega$, then the EOL index will be 2. If R_0 is between $3{,}987.34\Omega$ and $3{,}747.50\Omega$, then batteries EOL index will be 3. Finally, if R_0 is more than $3{,}987.34\Omega$, then batteries EOL index will be 4. The internal resistance and EOL index values for all HPPC test is shown in Table 6.10.

Finally, Table 6.10 shows the values of internal resistance of all HPPC tests with the EOL indexes. As can be seen from Table 6.10, after 100 cycles, the battery cell is still on the EOL index 0, and after 200 cycles, the battery cell reached EOL index 1 and after the 300 cycles, the battery cell finally reached the EOL index 2. These EOL index values are used to get the value of the SOS index which will be discussed in the following section.

6.5.4 Parameter identification results

To verify the characterization of the battery voltage in the actual operating conditions of the constructed second-order Thevenin equivalent circuit model, the experimental voltage and current data under the condition of cyclic discharge are imported into the equivalent battery model constructed, and the model is verified by combining with the online parameter identification results. The estimated value is compared with the actual terminal voltage value, and the comparison result and the corresponding error are shown in Figure 6.15.

Figure 6.15 shows the comparison between the estimated value of the battery terminal voltage and the true value with the discharge of the battery under the condition of the cyclic discharge hold test. The red solid line is the estimated value based on the constructed model, and the black solid line is the actual battery

Table 6.8 *Battery aging model parameters values for BOL HPPC test*

Parameter	U_{oc} (V)	R_0 (μΩ)	R_1 (Ω)	R_2 (Ω)	C_1 (F)	C_2 (F)
Values (SOC = 100%)	4.1899	2,845.50	7.6111×10^{-4}	3.0250×10^{-5}	2.2848×10^{4}	4.4298×10^{4}
Values (SOC = 90%)	4.0656	2,888.19	5.6500×10^{-4}	9.2024×10^{-4}	3.6938×10^{4}	2.2222×10^{4}
Values (SOC = 80%)	3.9533	2,876.61	3.5556×10^{-5}	8.8556×10^{-4}	3.7153×10^{4}	1.9256×10^{4}
Values (SOC = 70%)	3.8488	2,936.58	1.0275×10^{-3}	4.2611×10^{-5}	1.8637×10^{4}	4.7617×10^{4}
Values (SOC = 60%)	3.7320	2,965.27	3.1667×10^{-5}	8.7917×10^{-4}	2.0807×10^{4}	1.9598×10^{4}
Values (SOC = 50%)	3.6536	2,998.36	7.7889×10^{-4}	8.9694×10^{-5}	3.7001×10^{4}	3.6546×10^{4}
Values (SOC = 40%)	3.6145	2,948.83	1.9211×10^{-4}	3.2639×10^{-5}	1.2857×10^{5}	7.1969×10^{4}
Values (SOC = 30%)	3.5819	2,895.83	6.3750×10^{-4}	4.5111×10^{-5}	2.9412×10^{4}	3.1988×10^{4}
Values (SOC = 20%)	3.5159	2,856.83	7.6167×10^{-4}	9.9944×10^{-6}	2.6219×10^{4}	1.1797×10^{5}
Values (SOC = 10%)	3.4306	2,863.61	1.0897×10^{-3}	3.6833×10^{-5}	1.8280×10^{4}	4.4742×10^{4}

Table 6.9   *Defined EOL index from the nominal internal resistance value*

EOL index	Condition	Remarks
0	If $R_0 < 3{,}237.87$	No damage (BOL)
1	If $3{,}477.68 > R_0 \geq 3{,}237.87$	Little damage
2	If $3{,}747.50 > R_0 \geq 3{,}477.68$	Above-average damage
3	If $3{,}987.34 > R_0 \geq 3{,}747.50$	Severe damage
4	If $R_0 \geq 3{,}987.34$	Complete damage

Table 6.10   *Internal resistance and EOL index values for all HPPC test*

Cycle number	0	100	200	300
HPPC tests	BOL HPPC test	HPPC test after 100 cycles	HPPC test after 200 cycles	HPPC test after 300 cycles
R_0 (μΩ)	2,998.36	3,156.18	3,368.55	3,587.37
EOL index	0	0	1	2
Remarks	No damage (BOL)	No damage (BOL)	Little damage	Above-average damage

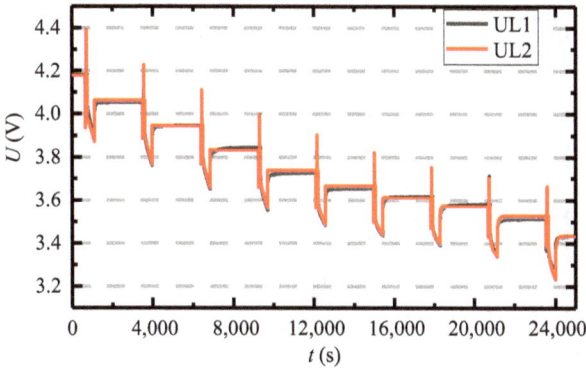

Figure 6.15   *Comparison of model voltage and the actual voltage*

terminal voltage value. It can be seen from Figure 6.15 that the estimated value has a good tracking effect on the true value, and the average estimated deviation is about 0.03 V, which can characterize the value of the terminal voltage of the battery during operation.

6.5.5   BBDST verification

The BBDST condition is obtained by processing the data collected from the starting, acceleration, sliding, braking, rapid acceleration, and stopping of the Beijing bus. According to the actual situation, the power of each step is reduced to simulate this condition. The experimental BBDST data can be obtained as shown in Figure 6.16.

6.5.6   Estimation result verification

The BBDST condition data are substituted into the algorithm to obtain the estimation results. The SOC estimation results are shown in Figure 6.17.

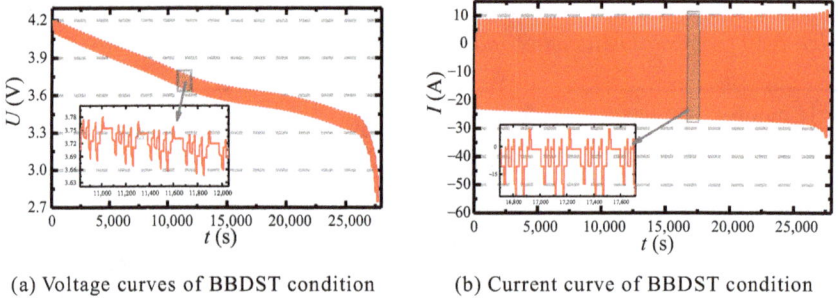

(a) Voltage curves of BBDST condition   (b) Current curve of BBDST condition

Figure 6.16   Experimental results for the BBDST conditions

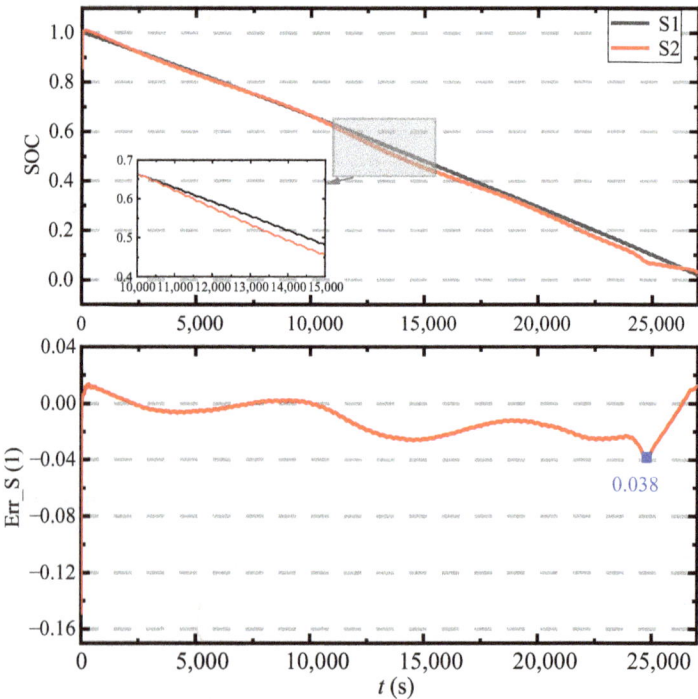

Figure 6.17   SOC estimation under the BBDST condition

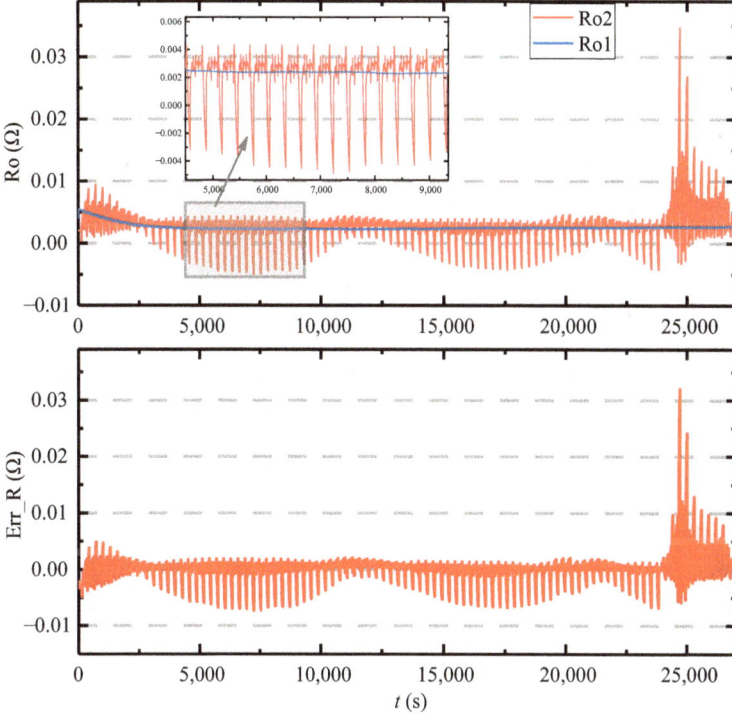

Figure 6.18   Internal resistance estimation under the BBDST condition

In Figure 6.10, S_1 is the true SOC, and S_2 is the SOC estimated by the DEKF algorithm. Under the BBDST condition, DEKF tracks the SOC well, and the estimation error is within 0.038, so it can track the SOC value of the battery well under this condition. The estimated resistance results are shown in Figure 6.18.

In Figure 6.18, R_{o1} is the measuring internal resistance, and R_{o2} is the estimated internal resistance by the DEKF algorithm. As can be known from the experimental results, it has good convergence in estimating internal resistance, and the internal resistance can be estimated. After convergence, the error is kept within 0.005Ω except for the end of discharge, but there are certain fluctuations, and the algorithm needs further improvement. However, SOH can be estimated by calculating the average value of ohmic resistance. The error of estimating SOC is within 3.8%.

6.6   Chapter summary

The accurate estimation of SOC and SOH provides an important guarantee for efficient and safe management. The aging characteristics of lithium-ion batteries are analyzed. Based on the second-order Thevenin equivalent circuit model and online parameter identification method, a DEKF is proposed to estimate SOC and SOH simultaneously. The results show that the algorithm can well estimate both SOC and SOH. The algorithm presents good convergence and can give a good reference to the battery management system. This chapter also proposes a rain-flow counting algorithm to calculate the number of lithium-ion battery cycles. Without considering the dynamic characteristics of the lithium-ion batteries, this algorithm can accurately count the number of charge–discharge cycles under the cycle condition, which is helpful to the SOH prediction.

Acknowledgment

The work is supported by the National Natural Science Foundation of China (No. 61801407), Sichuan Science and Technology Program (No. 2019YFG0427), China Scholarship Council (No. 201908515099), and Fund of Robot Technology Used for Special Environment Key Laboratory of Sichuan Province (No. 18kftk03).

Chapter 7

Unscented particle filtering of safety estimation considering capacity fading effect

Abstract

To get a more accurate state of charge estimation for lithium-ion batteries, this chapter introduces an improved particle filtering algorithm named unscented particle filtering. This algorithm adopts the framework of the unscented Kalman filter. With an appropriate probability density function, the unscented transformation is used in the traditional particle filter algorithm. Through the unscented transformation, the improved algorithm can realize the mean and variance calculation more accurately and solve the shortcoming of particle exhaustion in traditional particle filter algorithms. To verify the effectiveness of the algorithm, the ternary lithium-ion battery is selected as the research object, and the Thevenin equivalent circuit model is constructed. Finally, the experimental analysis is carried out under the Beijing bus dynamic stress test condition. The verification results show that there exists a great variation when the improved algorithm is used to estimate the state of charge value for lithium-ion batteries. The effect is very good, and the algorithm shows extremely strong tracking and robustness. The prediction error is stabilized within 1.5%, which brings good performance to the lithium-ion batteries.

Keywords: Ternary lithium-ion batteries; Thevenin equivalent circuit model; State of charge; Probability density function; Unscented particle filter; Unscented transformation

7.1 Introduction

In recent years, with the increasing demand for environmentally friendly and sustainable energy development, the development of new energy has increasingly become the mainstream of energy development strategies [299]. In the development wave of the new energy automobile industry, the power battery industry has also maintained a momentum of sustained and rapid growth, and the enterprise capacity construction scale expands rapidly [300]. At present, the development of new energy vehicles is still in its infancy [301]. Although its vehicle and electric drive control technology has matured, the power batteries are still a key factor hindering the development of the automobile industry [302]. There are problems such as short battery life and short battery cyclic life [121]. Among various electronic products, lithium-ion batteries have the advantage of high energy density, which makes their applications very extensive and gradually move towards the power field of new energy vehicles [303]. In the context of increasingly severe global environmental problems, vehicles have switched to storage batteries as their main power source, and lithium-ion batteries have been included as an ideal choice [304].

To meet the power requirements of electric vehicles for high-voltage and large-capacitance batteries, lithium-ion batteries are often combined in series and parallel [305]. However, due to the phenomena of over-charging, over-discharging, and over-heating, often occurring during the charging and discharging of the battery [306], there will be inconsistencies between the batteries, which is inevitable and will reduce its efficiency and shorten the battery life [307,308]. The inconsistency problem will aggravate with the increase in the number of uses, and relevant measures can only be taken to restrain its aggravation, but it cannot be eliminated fundamentally [309].

The battery management system (BMS) can detect battery physical parameters and estimate the state of charge (SOC), state of health (SOH), state of power (SOP), state of balanced (SOB), which are all born in the development field of the electric vehicle. Therefore, BMS can monitor the working status of the lithium-ion batteries [310], balance and control the inconsistency, and prevent the polarization phenomenon such as ohmic polarization [311], concentration polarization, and electrochemical polarization in the charging process of the battery pack [312–315]. Among them, the lithium-ion battery SOC can characterize the remaining power of the battery, and accurate estimation can provide a reference for the BMS to determine the timing of equilibrium and provide a reference for the driver to predict the remaining mileage more accurately. The ternary lithium-ion batteries are selected as the research object for their state estimation process [316].

7.2   Capacity fade modeling methods

7.2.1   Capacity fade mechanism

The battery capacity represents the amount of energy that can be stored in a full charge condition [317]. It can be defined by integrating the current drawn from the battery over time from fully charge condition to discharge condition as shown in the following equation:

$$Q = \int_{t_{ch}}^{t_{disch}} i dt \tag{7.1}$$

Here, Q (Ah) is the capacity; t_{ch} (s) and t_{disch} (s) can be used to express the time for the full charge–discharge condition, respectively; and i (A) is the battery current. However, the capacity of a battery does not remain the same, and it gradually fades whether a battery is used or not. As a result, there is a threshold limit for the capacity fade of lithium-ion batteries, after which the battery will no longer be used in applications. The state parameter that defines this threshold is called the end of life (EOL) for capacity fading.

The capacity fades are defined as the loss of the cell energy storage capacity due to degradation. Normally, the amount of capacity attenuation is calculated by the percentage. The percentage of capacity attenuation can be defined by

$$CF = 100\% \times \frac{Q_n - Q_r}{Q_r} \tag{7.2}$$

where Q_n (Ah) is the cell nominal capacity and Q_r (Ah) is the actual capacity after several cycles. In the past few decades, researchers have discovered different causes of capacity fading, most of which have mentioned that the loss of reversible lithium ion and the loss of active materials are the two main reasons for the capacity fading of batteries. As mentioned above, during the cycling process different side reactions occur in the battery system and an amount of recyclable active material became irreversible due to the formation of the solid electrolyte interphase (SEI) layer. Another explanation for this degradation process is that if the cathode material becomes unstable at a high potential, it can no longer store lithium, causing the active lithium ions to get stuck on the SEI layer for irreversible electroplating.

7.2.2   Capacity fade modeling methods

Identification of the battery capacity degradation mechanism is very important for designing a reliable battery-powered device for its connection with aging. A full understanding of this mechanism enables action for proper maintenance and reduces safety hazards. There are different methods to identify capacity degradation, but the model-based approach is the most suitable for the easy implementation processes. Researchers have introduced the model-based capacity degradation identification process of a lithium-ion battery by two methods [318,319]. One method is mainly based on the chemical change inside the cell, such as the loss of lithium ions, the growth of the SEI layer, and the variation law of the cell impedance.

Another method called the empirical life model is mainly based on accelerated life experiments, which directly reflect how external stresses influence the capacity to fade [320]. Moreover, the capacity degradation process of lithium-ion batteries and their consequence is also very interesting. In a complete life cycle, a battery could face two types of capacity fading such as capacity fading for calendar aging and capacity for cyclic aging. The capacity fade for calendar aging is defined as a function of time. On the contrary, the capacity fading for cyclic aging is defined as a function of usages, and not only that, the impact of both capacity fade on batteries is also different. Generally, the capacity fades for the calendar happened to an application where the operation period is shorter than the rest period such as an electric vehicle. However, the capacity fades for cyclic aging happened to an application that charged and discharged so frequently such as mobile phones. However, if the cycle depth and the current rate are relatively low, the calendar aging could also simultaneously occur in cyclic aging.

7.2.3   Capacity fade modeling

The capacity fade model proposed in this research work is mainly designed based on the loss theory of the active lithium ions. Therefore, the whole model characterized only the capacity fades mechanism for cyclic aging. Because the core idea of the proposed method is a state of safety prediction, which is connected to thermal safety, the capacity fade for calendar aging has a positive effect on thermal safety. The complete diagram of the proposed capacity fade model is presented in Figure 7.1.

The input of the proposed model is the experiment result data set of the cyclic aging test and the capacity characterization test. The mathematical analysis of the model is designed based on the "knee point" analysis and "Arrhenius model." The model parameters are already listed in Table 6.5. The algorithm used to calculate the model parameters is named "knee-Arrhenius algorithm" which will be described in the next section. The output of the proposed model is an EOL index, as presented in Table 6.3.

7.2.4   Mathematical knee point expression

The mathematical analysis for capacity fades mode is divided into three phases. The first phase is the knee-point analysis that begins with the capacity fade curve characterization. According to basic algebra, the capacity fading curve is characterized as a parabolic mathematical equation as shown below:

$$\begin{cases} y = f(x) = 1 - Ax^B - Cx^D \\ NC = 1 - a \times N^b - c \times N^d \end{cases} \tag{7.3}$$

Then, the equation of normalized capacity based on the model [321] will be derived, where NC is the normalized capacity and N is the number of the cycle. a, b, c, and d are the model coefficients. The calculation process can be described as shown in Figure 7.2.

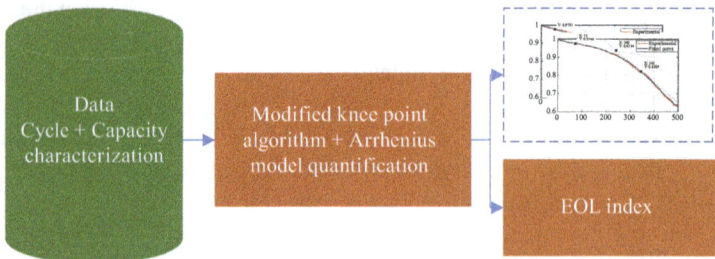

Figure 7.1   Proposed capacity fade model

Experiment data

Capacity fade curve characterization

$$NC = 1 - a \times K^b - c \times K^d$$

Calculating slope changing ratio

Find P_{QMIN}, P_{QMAX} and draw tangent lines.

Find $P_{Q,knee}$ and Q_{kneev} $[P_{QMAX}, P_{QMIN}]$

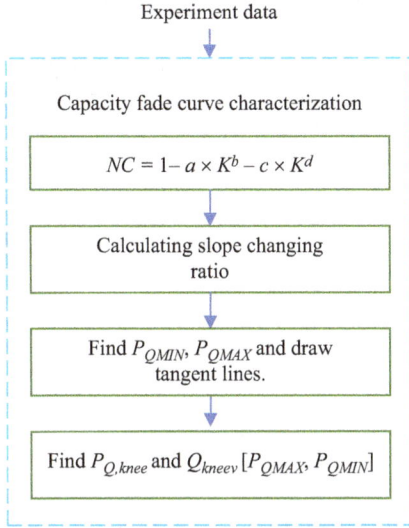

Figure 7.2   First phase of capacity fade model mathematical analysis

The next step of the first phase is calculating the slope-changing ratio of the fitted capacity fades curve and find the maximum point of the slope-changing ratio. The third step of the first phase is finding the minimum absolute slope-changing ratio (P_{QMIN}) and the maximum absolute slope-changing ratio (P_{QMAX}) of the fitted capacity fade curve, drawing the corresponding tangent lines. The final step of the first phase is finding the knee point ($P_{Q,knee}$) value, which is the value of cycle number at the intersection point of the tangent lines, and finding the knee values ($Q_{kneev}[P_{QMAX}, P_{QMIN}]$), which are the values of the normalized capacity from point P_{QMIN} to point P_{QMAX}. Figure 7.2 shows the step-by-step process of the first phase of the mathematical analysis.

7.2.5   Arrhenius model quantification

The second phase of the capacity fade model mathematical analysis is fitting the calculated knee values by using the Arrhenius equation. The capacity fades of a lithium-ion battery can be quantified by using the Arrhenius equation at a constant charge or discharge rate. The capacity degradation of lithium-ion batteries follows the Arrhenius law under the external stress of ambient temperature (T). Based on these studies and according to the Arrhenius model of Bloom *et al.* [322] and Ramadass *et al.* [323]. The derived Arrhenius equation is shown in the following equation:

$$Q_l = A \exp\left(\frac{-E_a}{RT}\right) t^z \tag{7.4}$$

Equation (7.4) is used to quantify the capacity fades of the experimented battery of this work. Here, Q_l is the capacity loss (Ah), A is the pre-exponential factor,

E_a is the activation energy (J), R is the gas constant (8.3145 J/mol.K), T is the external ambient temperature in Kelvin, t is time (week), and z is the adjustable factor. For simple diffusion control or layer growth, $z = 1/2$ [322].

The values of unknown parameters A, E_a, and z in (7.5) can be calculated by the covariance matrix adaptation evolution strategy technique [324], but this method required a large dataset of capacity characterization tests and additional computing complexity. As a result, the value of A and E_a is calculated by the following mathematical derivation with help of the linear least-square curve fitting method.

$$\begin{cases} ln\ Q_l = ln\ A\ e^{\dfrac{-E_a}{RT}} \\ ln\ Q_l = ln\ A - \dfrac{E_a}{RT} ln\ e \\ ln\ Q_l = -\dfrac{E_a}{R}\cdot\dfrac{1}{T} + ln\ A \end{cases} \tag{7.5}$$

Then, (7.5) is transformed into a general equation of a straight line ($y = mx + b_1$). Therefore, E_a can be calculated from the negative multiplication of gas constant and the slope of the fitted line by (7.5), and the natural logarithmic of A is calculated from the y-intercept. The final step of the second phase is finding the capacity loss Q_l by plugin the values of A, Ea, R, T, t, and z on (7.6).

$$\begin{cases} E_a = -R.m \\ ln\ A = b_1 \end{cases} \tag{7.6}$$

The second phase of the capacity fade model mathematical analysis is conducted as shown in Figure 7.3.

The third phase of the mathematical analysis contains the remaining capacity and its fading percentage calculation. The first step of the third phase is the

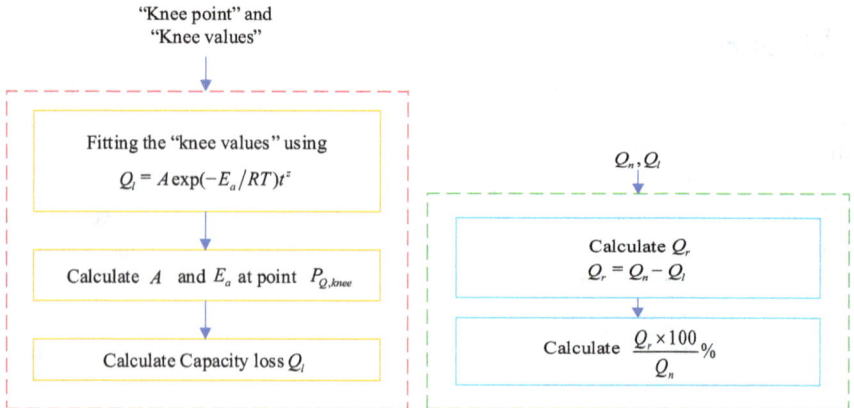

Figure 7.3   Second phase of capacity fade model mathematical analysis

calculation of the remaining capacity Q_r (Ah) according to (7.7). Where Q_n (Ah) is the nominal capacity and Q_l (Ah) is the amount of capacity loss.

$$Q_r = Q_n - Q_l \tag{7.7}$$

The final step of the third phase is calculating the value of what the remaining capacity Q_r is of the nominal capacity Q_n, to get a prediction of the EOL for the battery. Figure 7.3 shows the step-by-step calculation process of the second and third phases of the mathematical analysis for the capacity fading model.

7.2.6 Knee-Arrhenius expression

The novelty in this section is introducing the knee-Arrhenius algorithm. This section will describe the whole proposed algorithm with the step-by-step calculation process. The algorithm starts with the capacity fade curve characterization and normalized capacity calculating according to (7.8).

$$f(x) = 1 - Ax^B - Cx^D \tag{7.8}$$

The next step of the algorithm is the initialization of equation parameters $A = a, B = b, C = c,$ and $D = d$. After that, the algorithm will calculate the first and second derivatives of the model equation according to the following equation:

$$\begin{cases} f'(x) = -a * b * x^{b-1} - c * d * x^{d-1} \\ f''(x) = -a * b * (b-1) * x^{b-2} - c * d * (d-1) * x^{d-2} \end{cases} \tag{7.9}$$

Here, $f'(x)$ is the first derivative, and $f''(x)$ is the second derivative of the model equation. In the next step, the algorithm will find the maximum and minimum slope-changing ratio of the capacity fades cure according to

$$\begin{cases} P_{QMIN} = MIN[ABS(f''(x))] \\ P_{QMAX} = MAX[f''(x)./f'(x)] \end{cases} \tag{7.10}$$

Here, P_{QMIN} is the point of minimum slope-changing ratio and P_{QMAX} is the point of maximum slope-changing ratio. The next step of the algorithm is drawing the tangent lines on the derived P_{QMIN} and P_{QMAX} point according to (7.11) for calculating the corresponding gradients.

$$\begin{cases} P_{QMIN}f(x) = 1 - a * P_{QMIN}^b - c * P_{QMIN}^d \\ P_{QMAX}f(x) = 1 - a * P_{QMAX}^b - c * P_{QMAX}^d \end{cases} \tag{7.11}$$

Here, $P_{QMIN}f(x)$ is the tangent line on P_{QMIN}, and $P_{QMAX}f(x)$ is the tangent line on P_{QMAX}. The next step of the algorithm is to calculate the gradient of the tangent line for the data points of P_{QMIN} and P_{QMAX} by (7.12).

$$\begin{cases} P_{QMIN}f'(x) = 1 - a * b * P_{QMIN}^{b-1} - c * d * P_{QMIN}^{d-1} \\ P_{QMIN}f'(x) = 1 - a * b * P_{QMIN}^{b-1} - c * d * P_{QMIN}^{d-1} \end{cases} \tag{7.12}$$

The next step of the algorithm is the calculation of knee-point and knee-values according to (7.13), where Q_{kneev} is the set of normalized capacity as knee-values from P_{QMIN} to P_{QMAX}.

$$\begin{cases} S = Solve\left[P_{QMIN}f' * (x - inf) + P_{QMIN}f - \left(P_{QMAX}f' * (x - Mratio) + P_{QMAX}f\right) = 0, x\right] \\ P_{Q,knee} = Round[Double(S)] \\ Q_{kneev}\left[P_{QMAX}, P_{QMIN}\right] = \{y_{PQMIN}, \ldots, y_{PQMIN}\} \end{cases}$$

(7.13)

Finally, the secondary initialization part will come for the Arrhenius Model quantification. Where the x-axis will contain the natural log of the knee-values (x-axis $= \ln Q_{kneev}[i]$) and the y-axis will contain the inverse temperature values in kelvin (y-axis $= 1/T$). The algorithm ends with the calculation of the gradient and y-intercept from the Arrhenius plot for calculating the capacity loss and remaining capacity according to the following equation:

$$\begin{cases} E_a = -R.m \\ \ln A = b_1 \\ Q_r = Q_n - Q_l \end{cases}$$

(7.14)

7.3   Estimation modeling methods

7.3.1   *Equivalent circuit modeling*

Lithium-ion batteries have strong nonlinear dynamic characteristics due to the multi-parameter coupling processes. To describe the relationship between the SOC influencing factors (voltage, ambient temperature, charge–discharge rate, etc.) and the internal nonlinear operating characteristics due to parameter coupling, a battery model is constructed to simulate the internal complex electrochemical reactions, charge–discharge charge transfer, and energy conversion when replacing the lithium-ion batteries. The complexity of the working environment and application situation aggravates the difficulty in the mathematical modeling process of lithium-ion batteries.

Considering its aging and the environment complexity and variability, the existing studies have constructed equivalent circuit models to simulate the voltage response characteristics under different load conditions. In this paper, the Thevenin equivalent circuit model is selected, which is simple in structure, involves fewer parameters, and is easy to identify. This model also takes into account the battery polarization effect, which satisfies the model accuracy requirements well, as shown in Figure 7.4.

Wherein, U_{oc} represents the open-circuit voltage and U_L represents the terminal voltage. R_o, R_p, and C_p represent ohmic resistance, polarization resistance, and polarization capacitance respectively. Among them, the first-order RC loop can accurately describe the dynamic characteristics and express the polarization effect

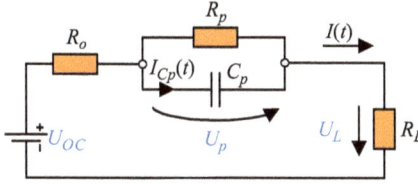

Figure 7.4   Thevenin equivalent circuit model

during the operation of lithium-ion batteries. According to the Thevenin equivalent model, its mathematical expression is described by Kirchhoff's law, as shown in (7.15). The discharge direction is taken as the reference direction

$$\begin{cases} U_L = U_{OC} - U_p - R_o I \\ I = \dfrac{U_p}{R_p} + C_p \dfrac{dU_p}{dt} \end{cases} \tag{7.15}$$

The SOC is defined uniformly by the *Electric Vehicle Battery Experiment Manual* from the power consumption perspective, which is issued by the United States Advanced Battery Federation. It is expressed as multiplying the lithium-ion batteries discharge loss rate by the lithium-ion batteries remaining capacity ratio to its rated capacity at a certain discharge rate, as shown below:

$$SOC(t) = SOC(t_0) - \frac{\eta}{Q_C} \int_{t_0}^{t} i(\tau)d\tau \tag{7.16}$$

In (7.16), η represents the Coulomb coefficient and Q_c represents the lithium-ion batteries nominal capacities. The state-space model of the lithium-ion batteries can be obtained as follows:

$$\begin{bmatrix} SOC_{k+1} \\ U_{P|k+1} \end{bmatrix} = \begin{bmatrix} 1 & 0 \\ 0 & e^{-\frac{T}{\tau}} \end{bmatrix} \times \begin{bmatrix} SOC_k \\ U_{P|k} \end{bmatrix} + \begin{bmatrix} -\dfrac{\eta T}{Q_C} \\ R_P\left(1 - e^{-\frac{T}{\tau}}\right) \end{bmatrix} \times I_{t|k} + \begin{bmatrix} w_{1|k} \\ w_{2|k} \end{bmatrix} \tag{7.17}$$

$$U_{t|k} = U_{OC}(SOC_k) - U_{p|k} - R_o I_{t|k} + v_k \tag{7.18}$$

In (7.17) and (7.18), $\tau_p = R_p C_p$. The state variable is $x_k = [SOC_k, U_{p/k}]$. The control variable is $u_k = I_{t/k}$. The observation variable is $y_k = U_{t/k}$. The system noise and the observation noise can be described by w_k and v_k, respectively.

7.3.2   Improved PNGV circuit modeling

In this research work, the utilized battery model for validating the proposed battery aging model is an improved PNGV circuit battery model. It is a stand-alone

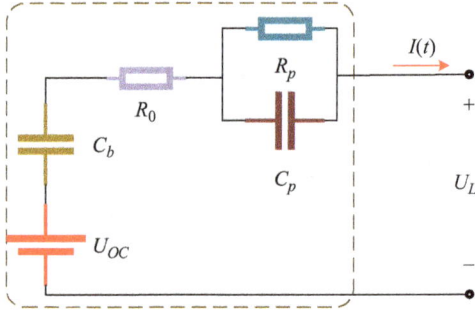

Figure 7.5   PNGV circuit model

model-based battery modeling technique and updated from a previous version. The primary version of this improved model is known as the PNGV circuit model. It has relatively higher accuracy than other electrical-based models such as the first-order Thevenin model. The circuit elements in the model are not very complex and have distinct physical explanations. The fundamental PNGV circuit model is shown in Figure 7.5.

Herein, U_{oc} (V) denotes a nonlinear voltage source as a function of the SOC and the first-order RC network consists of a capacitor C_p (F) and a resistor R_p (Ω). This capacitor and the resistor here are the polarization capacitance and polarization resistance. R_0 (μΩ) denotes the internal resistance. I (A) denotes the changing current. U_L (V) denotes the battery terminal voltage and C_b (F) denotes the bulk capacity. The PNGV model is a new model based on the improvement of the Thevenin model. In the process of discharge, the charge state of the battery will change, and at the same time, the open-circuit voltage of the lithium-ion battery will also change. Therefore, the capacitor C_b (F) is connected in series next to the open-circuit voltage U_{OC} (V). C_b (F) is used to describe the effect of time accumulation of load current I on open-circuit voltage U_{oc} (V). When considering the voltage of the RC network as U_p, according to Kirchhoff's voltage law, the following equations are obtained for the first-order Thevenin equivalent circuit, where τ_p is the time constant. The calculation procedures are shown in the following equations:

$$\begin{cases} U_{oc} = U_L + IR_o + U_p + U_b \\ I = \dfrac{U_p}{R_p} + C_p\dfrac{dU_p}{dt} \\ I = C_b\dfrac{dU_b}{dt} \end{cases} \qquad (7.19)$$

$$U_L = U_{oc} - IR_pe^{-\frac{t}{\tau_p}}, U_L = U_{oc} - IR_0 - IR_p - \left(U_p - IR_p\right)e^{-t/\tau_p} \qquad (7.20)$$

Although the PNGV circuit model seems relatively accurate as to the experimental battery model, it cannot reflect the dynamic characteristics of the battery

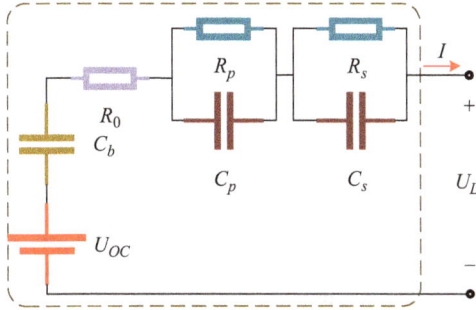

Figure 7.6   Improved PNGV circuit model

well and the fitting error is a little large. Because of the shortcomings of this model, the polarization circuit is changed based on this equivalent circuit treatment. The specific method is to connect an RC circuit in series to form two RC circuits. This model is called an improved PNGV circuit model. As a result, it can increase the efficiency of the battery model against the influential battery response and electrochemical polarization. The improved PNGV circuit model is presented in Figure 7.6.

Here, the new RC network also consists of a polarization capacitor C_s (F) and a polarization resistor R_s (Ω). When considering the voltage in the second RC network of U_s, the equation obtained from the circuit model will be derived by

$$\begin{cases} U_{oc} = U_L + IR_o + U_p + U_s + U_b \\ I = \dfrac{U_p}{R_p} + C_p \dfrac{dU_p}{dt} = \dfrac{U_s}{R_s} + C_s \dfrac{dU_s}{dt} \\ I = C_b \dfrac{dU_b}{dt} \end{cases} \tag{7.21}$$

$$U_L = Uoc - IR_p e^{-t/\tau_p} - IR_L e^{-t/\tau_L} \tag{7.22}$$

wherein τ_p and τ_s represent the time constant for U_1 and U_2, respectively. The current of the circuit will be derived. The battery model presented in Figure 4.7 is the main battery model that is used in this research work to validate the proposed battery aging model. They are the fundamental equations for this battery model.

7.3.3   High-order modeling realization

An ordinary differential equation is usable in the time domain of the high-order equivalent circuit. The equation for the resulting voltage response has been fixed [200,325,326]. In Simulink, the process of discretization is required before the model to obtain a discrete state-space equation for the high-order model. The largest block in the diagram is the corresponding circuit model. The inputs contain current I, internal ohmic resistance R_0, electrochemical polarization resistance R_{p1}, concentration polarization resistance R_{p2}, electrochemical polarization capacitor

C_{p1}, concentration polarization capacitor C_{p2}, open-circuit voltage U_{OC}, terminal voltage U_L, and load current I_L.

Besides, the current is taken as the input parameter, and the other parameters are the internal factors of the model. They are SOC functions as the independent variable. If the status of the charge varies, the beneficial relationship is taken from the identification of the previous segment. The SOC is the result of the latest changes at every moment. The simulation model is given in Figure 7.7.

The real-time SOC update can be accessed from a built-in ampere-time integration module. It is related to the end of the parameter described above. Then, the parameters of the model are obtained corresponding to the constant shift of the current input. To get the SOC with discharge current, an Ampere-hour (Ah) integral model is approached in Figure 7.8.

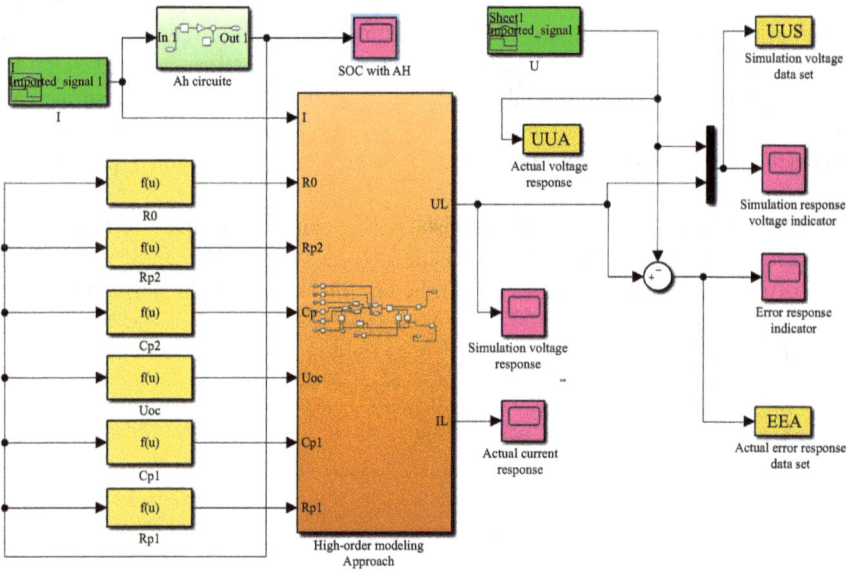

Figure 7.7   Top view structure of terminal voltage verification model

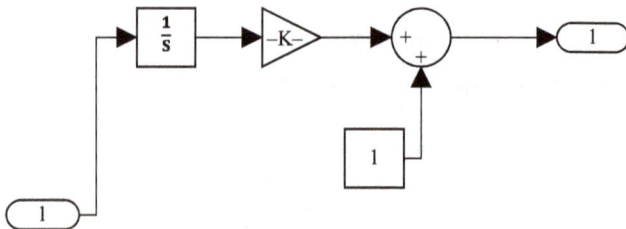

Figure 7.8   Inner structure of Ah integral model

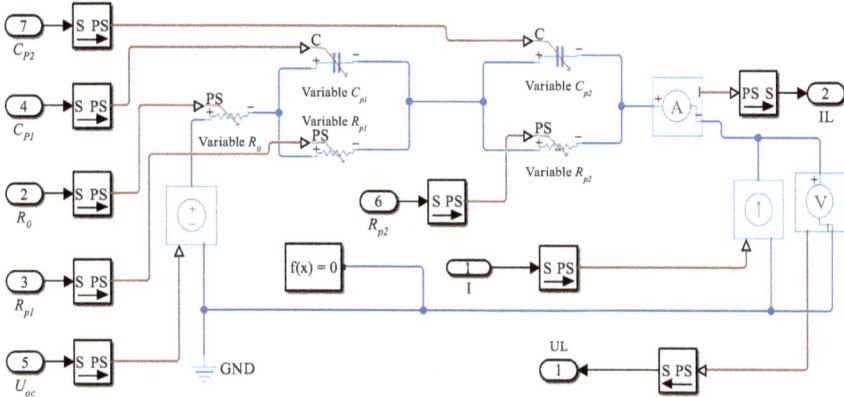

Figure 7.9   The central part of inside the high-order model in Simulink

In Figure 7.8, the central component of the whole module is the high-order internal circuit. The device configuration is used directly to construct the module, containing the internal ohmic resistance, two parallel resistance systems, a controllable voltage source, a controllable source of energy, voltage, current sensors, and the input–output interfaces. The internal structure of the high-order equivalent circuit model is shown in Figure 7.9.

Wherein, each circuit component is a controllable parameter that changes over time. The high-order model has seven inputs and one output, which indicates the inputs are I_L, R_0, R_{P1}, R_{P2}, C_{P1}, C_{P2}, and U_{OC}. The one output is the terminal voltage U_L. The controllable voltage source and current source are signal interfaces that can transform the signal into a material port. Convert external input voltage inputs to the voltage and current supply that can be attached to the circuit. Voltage sensors and current sensors are both signal transducers, which transform physical interfaces to signal interfaces.

7.3.4   Internal resistance estimation

According to the SOH estimation based on internal resistance, its variation of lithium-ion batteries under various operating conditions can be obtained by requiring the variation of internal resistance in the operating process. Based on the improvement of the EKF algorithm, the overall flow chart is shown in Figure 7.10.

The detailed steps to estimate the internal resistance of the lithium-ion batteries using the improved KF algorithm are described as follows. First, the state and measurement equations of battery internal resistance update are shown as follows:

$$\begin{cases} R_k = R_{k-1} + r_{k-1} \\ U_k = U_{OC}(SOC) - U_{1,k} - U_{2,k} - i_k R_k - w_{k,1} \end{cases} \tag{7.23}$$

wherein $U_{OC}(SOC)$ is the linear function of open-circuit voltage U_{OC} with *SOC* on the battery charge condition. The zero-state and zero-input response generated

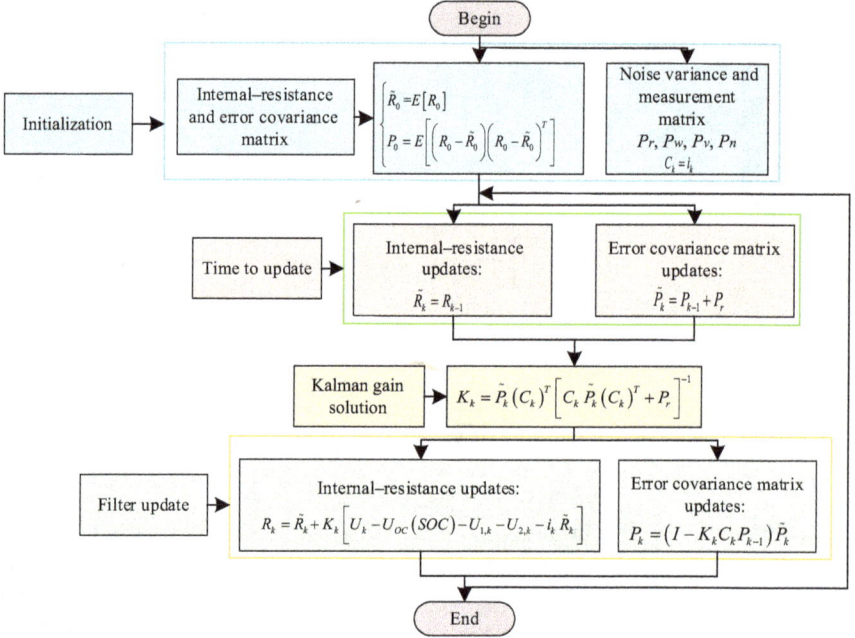

Figure 7.10   The flow chart of the improved extended Kalman filtering algorithm

by system excitation at both terminal voltages of $U_1(k)$ and $U_2(k)$ of two RC circuits are shown as follows:

$$
\begin{cases}
U_1(k) = U_1 e^{-t/R_1 C_1} + R_1 i_k \left(1 - e^{-t/R_1 C_1}\right) \\
U_2(k) = U_2 e^{-t/R_2 C_2} + R_2 i_k \left(1 - e^{-t/R_2 C_2}\right)
\end{cases}
\tag{7.24}
$$

(1) Initialization. The initial value of internal resistance needs to be obtained through the preliminary estimation by related experiments, while the initial value of error covariance is set according to actual needs. For $A_k = 1$ and $C_k = -i_k$ (incoming current), B_k and D_k are constant. When $k = 0$, the initial values of R_0 and P_k are shown in the following equation:

$$
\begin{cases}
\widetilde{R}_0 = E[R_0] \\
P_0 = E[(R_0 - \widetilde{R}_0)(R_0 - \widetilde{R}_0)^T]
\end{cases}
\tag{7.25}
$$

Here, $\widetilde{R}_0$ is the filtering initial value and R_0 is the original value of the internal resistance. The R_0 is taken as the initial value of internal resistance at the beginning of filtering. The initial value of the error covariance matrix needs to be adjusted according to the actual needs so that the final estimated internal resistance is truthfulness.

(2) Time to update. Attributing to the internal resistance changes very little in a short time, it is completely reasonable to take the accurately estimated internal

resistance value obtained at the time point of $k-1$ as the updated value of the time point k. The time update of state variable R_k and error covariance P_k is shown in the following equation:

$$\begin{cases} \widetilde{R}_k = R_{k-1} \\ \widetilde{P}_k = P_{k-1} + P_r \end{cases} \tag{7.26}$$

Wherein, these two core parameters are respectively the time update values of R_k and P_k, P_r is the variance of r_k, which denotes the fluctuation range of internal resistance under the influence of noise, and can be initialized and set as small as possible in the specific setting.

(3) Kalman gain solution. Kalman gain can be used to update the internal resistance and error covariance of the battery, and its calculation accuracy directly determines whether the algorithm can effectively obtain a more accurate internal resistance estimation value. The results are shown in the following equation:

$$K_k = \widetilde{P}_k(C_k)^T \left[C_k \widetilde{P}_k(C_k)^T + P_r \right]^{-1} \tag{7.27}$$

(4) Filter updates. The filtering update values obtained here include the internal resistance and error covariance matrix, which can be used as the initial value for the next time update. Kalman gain is used to filter and update state variable and error covariance, and the results are shown in the following equation:

$$\begin{cases} R_k = \widetilde{R}_k + K_k \left[U_k - U_{OC}(SOC) - U_{1,k} - U_{2,k} - i_k \widetilde{R}_k \right] \\ P_k = (I - K_k C_k) \widetilde{P}_k \end{cases} \tag{7.28}$$

Wherein, R_k and P_k are internal resistance and error covariance estimates after filter updates. This is the whole process of estimating the internal resistance of the battery by using the EKF algorithm under improvement. Whether the internal resistance can be accurately estimated depends on the setting of the initial value and the reasonable adjustment of the noise variance according to the actual situation. By introducing the current and voltage under various operating conditions, the estimation curve of internal resistance can be obtained.

7.3.5 Parameter identification

System model parameter identification is mainly used to describe system behavior characteristics. It is determined by the functional relationship of system input and output parameters over time [13]. The purpose of parameter identification is to make the error between the output data and the actual data within the specified range when the input data is transmitted into the system, which can judge the accuracy of the equivalent circuit model and provide a reliable basis for the subsequent SOC estimation. The offline identification method uses HPPC to test the performance of lithium-ion batteries at an ambient temperature of 25 °C [327]. Through the research on the lithium-ion batteries working characteristics, the

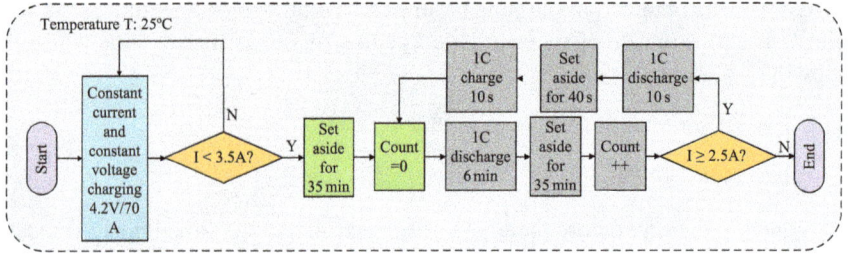

Figure 7.11   Experiment process

identified parameters of the Thevenin equivalent circuit model are obtained [328]. The detailed steps of the HPPC experiment designed are shown in Figure 7.11.

According to the HPPC experimental steps set in 0, during the cycle test, select the SOC to be 1–0.1 equidistant points for the experimental operation, until the battery capacity drops to 0, or when the current drops to 0.05 C, stop the experiment [329–331]. In this process, each time the lithium-ion battery releases 10% of its capacity [152], and the HPPC cycle test is performed.

7.3.6   Particle filtering algorithm

The particle filtering (PF) algorithm is also known as sequential filtering, which is used to realize the recursive Bayesian filtering by the non-parametric Monte Carlo simulation. It applies to any nonlinear system described by the state-space model, and its accuracy approaches the optimal estimation. It can adapt to nonlinear and non-Gaussian systems well which has no restrictions on measurement and process noises, and it is a Bayesian filtering technology based on Monte Carlo treatment. Its principle is to find a group of random particles propagating in the state space to describe the state of the system.

It is used to deal with the integral operation in Bayesian estimation, and the minimum mean square error estimation of the system state can be then obtained. When the number of particles tends to infinity, it can approach the system state which obeys any probability distribution. Compared with other filtering methods, it does not need to make any system assumptions. Its structure can be described as shown in Figure 7.12.

After sampling each particle for estimation, the previous distribution and weight of the next moment will be obtained. The re-sampling process from green to yellow particles removes the smaller weights, which retains the weight of resistance and then redistributes to normalize the particles.

In theory, it can be applied to any stochastic system which can be described by the state-space model. However, two problems limit the further development of the PF algorithm. The first problem is that the PF has a large amount of computation. The other problem is that particle degradation leads to the waste of computing resources and the deviation from computing results. Subsequently, the cost performance from computing resources is getting higher along with the development

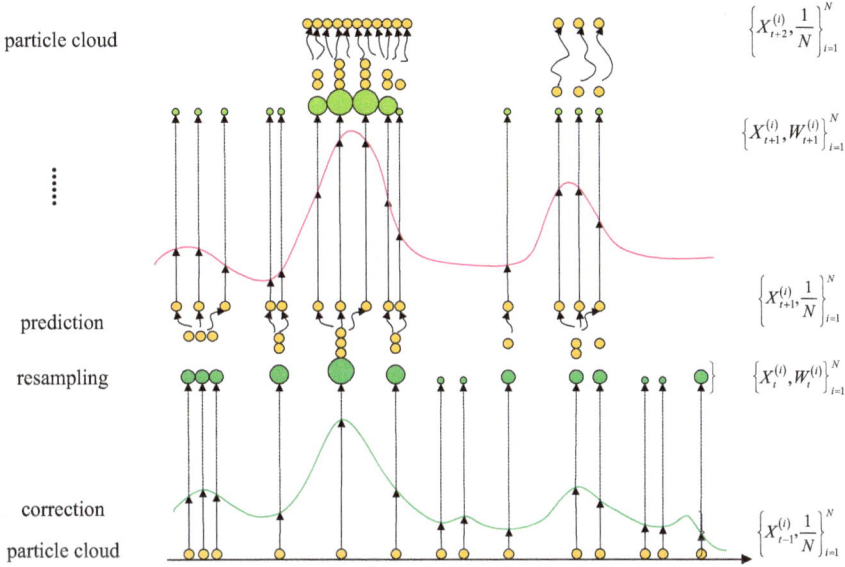

Figure 7.12   Schematic diagrams of algorithm implementation

of semiconductor technology. These problems have been effectively solved, and it has gradually become one of the key technologies. The PF algorithm is a probability density function based on Monte Carlo simulation, and its implementation process is mainly based on the Bayesian filter operation criterion to weight discrete random samples to complete the integral operation (sample mean). The algorithm can not only accurately predict and track dynamic parameters but also estimate and predict time-varying systems.

The algorithm can extract some discrete random particles, which use the probability density function to estimate the mean value in samples. This algorithm has two advantages. (1) It does not need to be integrated. (2) It can predict and estimate the dynamic parameters well, and the effect is good. It can be applied in more complex environments and achieves remarkable prediction and tracking effects. It can be considered as the best estimation algorithm and effective filtering algorithm. The algorithm theory is deduced in detail as follows. The iterative calculation process is obtained and described by the following steps.

Initialization: using the prior distribution to generate SOC initial particles and particle weight.

The algorithm cycle process is as follows.

1.   Updating particle weight

$$w_k^i = w_{k-1}^i p\left(U_{L(k)} \big| SOC_k^i\right) = w_{k-1}^i p\left(U_{L(k)} - h\left(SOC_k^i\right)\right), I = 1, 2, ..., N \tag{7.29}$$

2. Calculating the normalized weight value

$$w_k^i = \frac{w_k^i}{\sum\limits_{i=1}^{N} w_k^i} \tag{7.30}$$

3. Re-sampling: Obtaining the number of effective particles and several particle sets.
4. Calculating the least mean square estimation value

$$\widehat{SOC}_k \approx \sum_{i=1}^{N} w_k^i SOC_k^i \tag{7.31}$$

5. Prediction: The state equation is used to predict the unknown parameter.
6. The end condition of the program is judged. If it is not over at time $k = k + 1$, it jumps to step (1).

In the estimation simulation process, the state-space model of the lithium-ion batteries is obtained from the management process and observation model. When the model boundary strip is set, the observation variable is equal to the lithium-ion batteries load voltage, and the lithium-ion batteries state variable is the SOC, as shown in the following equation:

$$\begin{cases} x_{k+1} = f(x_k, i_k, w_k) = x_k - \dfrac{n i_k \Delta t}{\eta_i \eta_T \eta_n Q_n} + w_k \\[4mm] y_{k+1} = f(y_k, i_k, w_k) = k_0 - R i_k - \dfrac{k_1}{x_k} - k_2 x_k + k_3 ln(x_k) + k_4 ln(1 - x_k) + v_k \end{cases} \tag{7.32}$$

Wherein, w_k is the system process noise and v_k is the systematic observation noise, in which $w_k \sim N(0, Q)$, $v_k \sim N(0, R)$, and Δt is the system sampling period. When PF is applied, the accuracy of the model can improve the accuracy of SOC estimation. Under the condition of ensuring accuracy, the amount of calculation can be reduced.

The PF algorithm has an important relationship with particles, and the prediction is based on particles. This method is superior to other dynamic parameter prediction and tracking methods that can track dynamic parameters more accurately. The probability of random events in particle filtering is very low, which is equivalent to a significant advantage of the algorithm. The dynamic parameters and estimation effects of time-varying systems are the best.

This algorithm has been applied in many fields. The PF algorithm is to extract some discrete random particles and then use them to realize the probability density function, instead of extracting the mean value of the sample, so no integration operation is required. In this way, the dynamic parameters of the system can be estimated and a better estimation result can be obtained. Particle filtering can track many complex situations with high accuracy and effective estimation.

7.3.7 Unscented Kalman filtering

The unscented Kalman filtering (UKF) process is to complete the core content realization steps from unscented transformation (UT), prior estimation to posterior estimation correction. The algorithm derivation process is as follows.

7.3.7.1 Unscented transformation

UT is the core part of the UKF algorithm. The current state variables are constructed with statistical sampling points, and a limited number of sampling points and their corresponding weights are constructed. These sampling points mean and variance is equal to the state variable mean and variance at the current moment or there is only an acceptable limited tolerance, then the set of these sampling points is called the Sigma point set. For the nonlinear system given by the following equation, the UT process is performed on it.

$$\begin{cases} x_{k+1} = f(x_k, u_k) + w_k \\ z_k = h(x_k, u_k) + v_k \end{cases} \tag{7.33}$$

The $2n + 1$ sampling points of the state variable at the current moment are calculated by symmetrically distributed sampling method, where n is the state-space x dimension. The mean and variance of the Sigma point are set correspondingly equal to the state variable at the special time point. Supposing the mean value of x is its average value, the sampling point is determined by

$$\begin{cases} x_0 = \bar{x} \\ x_i = \bar{x} + (\sqrt{(n+P)})_i, i = 1, \dots, n \\ x_i = \bar{x} - (\sqrt{(n+P)})_i, i = n+1, \dots, 2n \end{cases} \tag{7.34}$$

In (7.34), x_i represents the i-th sampling point and i represents the i-th column of the matrix. Corresponding to each x_i, its weight is given by (7.35) when calculating the mean value.

$$\begin{cases} \omega_0{}^m = \dfrac{\lambda}{n+\lambda} \\ \omega_0{}^c = \dfrac{\lambda}{n+\lambda} + 1 - \alpha^2 + \beta, i = 1, \dots, 2n \\ \omega_i{}^m = \omega_i{}^c = \dfrac{1}{2(n+\lambda)} \\ \lambda = \alpha^2(n+\gamma) - n \end{cases} \tag{7.35}$$

In Equation (7.35), m represents the weight of the i-th sampling point in calculating the mean value and c represents the weight of the i-th sampling point in calculating the covariance. The n is the dimension degree of state variable x. The factor α is the distribution state of the sampling points around the mean, which controls the distance between the sampling points and the mean and generally takes a constant of 10^{-6} to 1. γ is required to satisfy the functional relationship of $\gamma + n \neq 0$. Since $n \neq 0$, $\gamma = 0$ can generally be selected. β is a priori distribution factor. Since it has been assumed that the system is given by obeys Gaussian distribution, $\beta = 2$ is optimal. The λ parameter represents the zoom ratio.

7.3.7.2   Sigma point set update

Sigma point set update is the process of using the state-space sampling points obtained by the UT at the previous moment to predict the sampling points at the current moment. It is also the Sigma further prediction, given by

$$x_{i,k+1|k} = f\left(x_{i,k|k}, u_k\right) \tag{7.36}$$

7.3.7.3   State variable *x* and covariance *P* prediction

Multiply the Sigma points predicted values by the corresponding weight to obtain the x predicted value at the next moment. Then, the covariance is calculated from the x predicted value and the covariance weight, and the calculation equation is given by

$$\begin{cases} \widehat{x}_{k+1|k} = \sum_{i=0}^{2n} \omega_i^m x_{i,k+1|k} \\ P_{k+1|k} = Q_k + \sum_{i=0}^{2n} \omega_i^c \left[\widehat{x}_{k+1|k} - x_{i,k+1|k}\right] \times \left[\widehat{x}_{k+1|k} - x_{i,k+1|k}\right]^T \end{cases} \tag{7.37}$$

7.3.7.4   Observations further prediction

The KF algorithm not only needs to predict the state and covariance at $k + 1$ from the state at time k but also needs to predict the state of the observation space at time $k + 1$. In the UKF algorithm, to predict the observation space at time $k + 1$, the observation predicted value at time $k + 1$ needs to be calculated from the state at time k, as shown below:

$$z_{i,k+1|k} = h\left(x_{i,k+1|k}, u_k\right) \tag{7.38}$$

7.3.7.5   System observation matrix and covariance matrix prediction

Observations obtained by nonlinear transfer based on sampling points are averaged to obtain the predicted amount of the system observation matrix. Then, the variance of the observed measurement at $k + 1$ given by the predicted value and the covariance between it and the state variable are calculated. The equation gives the calculation method of the prediction system observations, and the variance of the observations is given. Finally, the covariance between the observed quantity and the state quantity is shown below:

$$
\begin{cases}
\widehat{z}_{k+1|k} = \sum_{i=0}^{2n} \omega_i^m z_{i,k+1|k} \\[2ex]
P_{z_k z_k} = R_k + \sum_{i=0}^{2n} \omega_i^c \left[z_{i,k+1|k} - \widehat{z}_{k+1|k} \right] \left[z_{i,k+1|k} - \widehat{z}_{k+1|k} \right]^T \\[2ex]
P_{x_k z_k} = \sum_{i=0}^{2n} \omega_i^c \left[x_{i,k+1|k} - \widehat{z}_{k+1|k} \right] \left[z_{i,k+1|k} - \widehat{z}_{k+1|k} \right]^T
\end{cases}
\tag{7.39}
$$

7.3.7.6 KF gain calculation

The calculation method can be realized for estimating the state and observation at $k + 1$ from the information at time k. Next, the data at time $k + 1$ are required to perform a posterior estimation, that is, to modify the prior estimation to get a more accurate estimation value. Before the correction process starts, calculating the Kalman gain is a necessary step, as shown below:

$$
K_{k+1} = P_{x_k z_k} P_{z_k z_k}^{-1}
\tag{7.40}
$$

7.3.7.7 Status update and covariance update

The previous six steps are prepared for the status update in step (7). The covariance update is to prepare for the estimation at time $k + 2$. The status update and covariance update are shown in the following equation:

$$
\begin{cases}
\widehat{x}_{k+1|k+1} = \widehat{x}_{k+1|k} + K_{k+1} \left[z_{k+1} - \widehat{z}_{k+1|k} \right] \\[2ex]
P_{k+1|k+1} = P_{k+1|k} - K_{k+1} P_{z_k z_k} K_{k+1}^T
\end{cases}
\tag{7.41}
$$

Wherein, z_{k+1} is the observation measured by the instrument at $k + 1$. The observation is obtained from the state estimation at time k. The posterior state estimate at $k + 1$ and its optimal state estimation at the current time can be obtained accordingly. The estimating realization process can be realized by the nonlinear system state given by the UKF algorithm given above. The optimal estimation based on the minimum variance criterion can be obtained, and the estimation accuracy is better than the EKF algorithm processed by linearization.

7.3.8 Improved unscented particle filtering

The UKF is utilized to improve the PF algorithm, using unscented transformation. This method can theoretically calculate the posterior variance accuracy to the third order. The algorithm has higher accuracy and is also an effective means to calculate the mean and covariance. Taking a particle $u \sim U [0, 1]$ in the $[0, 1]$ uniform distribution, which satisfies x_k and forms a new particle set with the obtained particles as shown below:

$$\sum_{j=1}^{j-1} w_k(j) \le u \le \sum_{j=1}^{j} w_k(j) \tag{7.42}$$

The particle polynomial is resampled, and the weights of the particles are rechecked and screened to form a new particle library. In the particle set obtained, the weights are evenly distributed. The process at this stage is as follows.

(a) Initialization. Particles are extracted from the prior distribution $P(X_0)$ as the initial state of a new particle set. These particles are obtained from the important density function of the system.

$$\begin{cases} X_0^i = E(X_0^i) \\[2mm] P_0^i = E\left[\left(X_0^i - \bar{X}_0^i\right)\left(X_0^i - \bar{X}_0^i\right)^T\right] \\[2mm] X_0^{i,a} = E\left(\bar{X}_0^{i,a}\right) = \left[\left(\bar{X}_0^i\right)^T \quad 0 \quad 0\right]^T \\[2mm] P_0^{i,a} = E\left[\left(X_0^{i,a} - \bar{X}_0^{i,a}\right)\left(X_0^{i,a} - \bar{X}_0^{i,a}\right)^T\right] = \begin{bmatrix} P_0^i & 0 & 0 \\ 0 & Q & 0 \\ 0 & 0 & R \end{bmatrix} \end{cases} \tag{7.43}$$

(b) Importance sampling stage. Generate Sigma point set.

$$X_{k-1}^{i,a} = \left[X_{k-1}^{i,a} \quad X_{k-1}^{i,a} \pm \sqrt{(n_a + \lambda)P_{k-1}^{i,a}}\right] \tag{7.44}$$

A further prediction of the Sigma point set can be then obtained.

$$\begin{cases} \bar{X}_{k|k-1}^{i,a} = f\left(X_{k-1}^{i,x}, X_{k-1}^{i,v}\right) \\[2mm] \bar{X}_{k|k-1}^{i} = \sum_{j=0}^{2n_a} W_j^m X_{j,k|k-1}^{i,x} \\[2mm] P_{k|k-1}^{i} = \sum_{j=0}^{2n_a} W_j^c \left[X_{j,k|k-1}^{i,x} - \bar{X}_{k|k-1}^{i}\right]\left[X_{j,k|k-1}^{i,x} - \bar{X}_{k|k-1}^{i}\right]^T \\[2mm] Z_{k|k-1}^{i} = h\left(X_{k|k-1}^{i,x}, X_{k-1}^{i,n}\right) \\[2mm] \bar{Z}_{k|k-1}^{i} = \sum_{j=0}^{2n_a} W_j^c Z_{k|k-1}^{i} \end{cases} \tag{7.45}$$

Through the obtained observations, the system status is updated.

$$
\begin{cases}
P_{\overline{Z}_k} = \sum_{j=0}^{2n_a} W_j^c \left[Z_{j,k|k-1}^i - Z_{k|k-1}^i \right] \left[Z_{j,k|k-1}^i - Z_{k|k-1}^i \right]^T \\[2em]
P_{X_k,Z_k} = \sum_{j=0}^{2n_a} W_j^c \left[X_{j,k|k-1}^i - X_{k|k-1}^i \right] \left[X_{j,k|k-1}^i - X_{k|k-1}^i \right]^T \\[2em]
K = P_{\overline{Z}_k} P_{X_k,Z_k} \\[1em]
\overline{X}_k^i = \overline{X}_{k|k-1}^i + K \left(Z_k - \overline{Z}_{k|k-1}^i \right) \\[1em]
\widehat{P}_k^i = P_{k|k-1}^i - K P_{\overline{Z}_k} K^T
\end{cases}
\tag{7.46}
$$

(c) Normalized weight.

$$
w_k^i = \frac{w_k^i}{\sum_{i=0}^{N} w_k^i}
\tag{7.47}
$$

(d) State estimation.

$$
\bar{x}_k = \sum_{i=1}^{N} w_k^i \chi_x^i
\tag{7.48}
$$

7.4 Experimental result analysis

7.4.1 Test platform construction

To study the working characteristics of lithium-ion batteries, it is necessary to carry out lithium-ion battery test experiments under different working conditions. The ternary lithium-ion battery with the type of CFP70AH is used as the experimental object, and its rated capacity is 70 Ah, the charge cut-off voltage is 4.2 V, and the discharge cut-off voltage is 2.75 V. The test equipment is the sub-source BTS 750-200-100-4, which has a maximum charge–discharge power of 750 W, a maximum current of 100 A, and a maximum voltage of 200 V.

The charge–discharge experiments at different rates for capacity testing. The thermostat model is BTT-331C. The model is connected to the battery test equipment BTS into the incubator, connecting the BTS to the industrial computer. To carry out experiments, it is necessary to build an experimental test platform, and the platform structure is constructed. The battery test platform includes the following:

1. A lithium-ion battery with a rated capacity of 70 Ah.
2. A battery test system (NEWARE BTS-4000) for charging and discharging lithium-ion batteries. This device can detect the voltage, current, and temperature of the battery, and the sampling interval is 1 s.
3. A temperature box (TT-5166-7) that provides a constant temperature environment (25 °C) for the battery.

Based on the above test platform, all the lithium-ion battery test experiments required in it can be completed, and relevant experimental data can be obtained.

7.4.2   Open-circuit voltage characterization

OCV is the terminal voltage of the lithium-ion battery that has been left for a long time. In the perception of this, the lithium-ion batteries OCV and SOC have an excellent mapping relationship. The SOC–OCV step by step is to study the OCV characteristic of the lithium-ion battery voltage. The SOC–OCV calibration experiment for lithium-ions is performed using a battery test-time battery test device [332].

The OCV has a strong nonlinear relationship with the SOC. The relationship between them is determined by performing battery discharge and charging. For the charging test, charge the battery from 0% SOC at the recommended C/2 rate for 12 min, followed by a rest period of 12 min to allow the cell to return to a load balance condition before the next cycle is applied. Then, the OCV is measured [333]. The steps for designing the open-circuit voltage flow chart are shown in Figure 7.13.

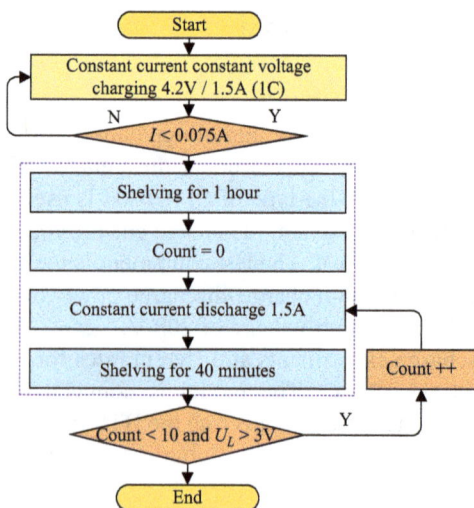

Figure 7.13   Open-circuit voltage and state of charge calibration

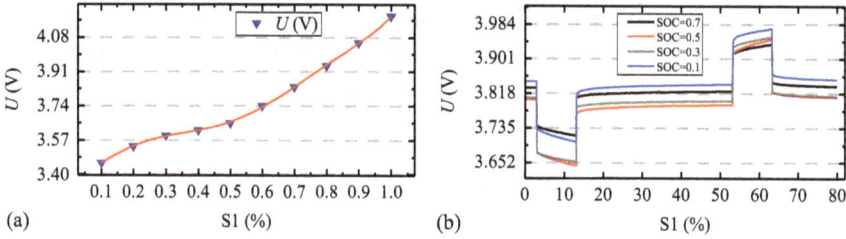

Figure 7.14 Specific open-circuit voltage at different SOC levels

In Figure 7.13, this process is repeated until the battery reached 0% from 100% of the SOC. As shown in Figure 7.14, these discrete points are extracted from the laboratory data to obtain an OCV-SOC relationship scatter plot. The recursive least-square (RLS) method is a mathematical optimization technique that determines the best function matching of the data by reducing the error square. It is suitable for curve fitting, which uses this method to fit the OCV-SOC relationship curve and relational polynomial.

Through curve fitting, it has been found that the batteries OCV of the batteries as a function of the SOC function can be expressed as a sixth-order polynomial equation. The experiment shows that the voltage stability of the battery after 40 min is equal to the open-circuit voltage of the battery. The curve fitting procedures are shown as follows:

$$U_{oc,k} = f(S_k) = P1 * S_k^6 + P2 * S_k^5 + P3 * S_k^4 + P4 * S_k^3 + P5 * S_k^2 + P6 * S_k + P7$$

$$(7.49)$$

Given the above consideration, after repeated experiments to compare the fitting effect, it is observed that the six-order polynomial has a strong fitting effect and is moderately reasonable to the processor. Thus, the functional relationship between the OCV and the SOC is accomplished by fitting to the fifth-order polynomial. In (7.49), P_1 to P_7 is the co-efficient obtained with 95% conviction limits using the least-square method, as shown below:

$$U_{oc,k} = f(S_k)$$

$$= 20.08 * S_k^6 - 61.22 * S_k^5 + 68.12 * S_k^4 - 32.07 * S_k^3 + 5.234 S_k^2$$

$$+ 0.6794 * S_k + 3.36$$

$$(7.50)$$

In (7.50), S_k is the battery SOC value at the time point of k, and $U_{OC,k}$ is the corresponding open-circuit voltage value. To obtain the battery OCV characteristics in the charge–discharge experiment. The OCV is 3.36 V when the SOC of the lithium-ion battery used in this experiment is zero. The SOC–OCV relationships provide a basis for the measure of the OCV in the revised mathematical help of the following parameters. The relation between the SOC and the OCV of the battery can be used to estimate the actual battery power. The approximate static battery capacity can be calculated by the measured capacity if it is left for an extended period.

7.4.3   Capacity fade modeling effect

Like the battery aging model, a four-capacity characterization test is performed to validate the capacity fade model. One of them is performed at the beginning of the cycle-life test which is called the BOL capacity characterization test and another three are performed after every 100 cycles of the cycle-life test. All capacity characterization test results are shown in Table 7.1.

The calculated values of Q_p from the capacity characterization test of BOL and after every 100 cycles are shown in Table 7.1. Like the previous model, the value of Q_p of the BOL capacity characterization test is defined as Q_n and used to evaluate the conditions of the EOL index for the capacity fade model, as presented in Table 7.2.

According to the capacity fade model EOL index, if the value of Q_r remains between 37.5715 and 39.5492 Ah, the EOL index will be 0. If Q_r is between 35.5941 and 37.5715 Ah, the EOL index will be 1. If Q_r is between 33.6166 and 35.5941 Ah, the EOL index will be 2. If Q_r is between 31.6392 and 33.6166 Ah, the EOL index will be 3. Finally, if Q_r is less than 31.6392 Ah, the EOL index will be 4, as shown in Figure 7.15.

After that, the values of Q_p and values of cycles are plotted to draw the capacity fade curve. Figure 7.15(a) shows the capacity fade curve with nominal

Table 7.1   All capacity characterization test results

Cycle number	0 (BOL)	100	200	300
Capacity characterization test	BOL capacity characterization test	Capacity characterization test after 100 cycles	Capacity characterization test after 200 cycles	Capacity characterization test after 300 cycles
Capacity Qp (Ah)	39.54925	37.91068	35.8463	34.82622

Table 7.2   Defined EOL index from the nominal capacity value

EOL index	Condition	Remarks
0	If $37.5715 < Q_r \leq 39.5492$	No damage (BOL)
1	If $35.5941 < Q_r \leq 37.5715$	Little damage
2	If $33.6166 < Q_r \leq 35.5941$	Above-average damage
3	If $31.6392 < Q_r \leq 33.6166$	Severe damage
4	If $Q_r \leq 31.6392$	Complete damage

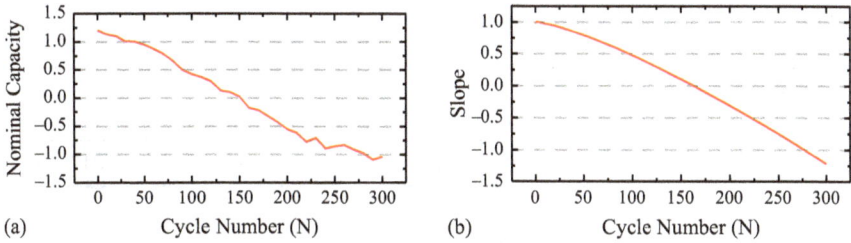

(a)   Cycle Number (N)

(b)   Cycle Number (N)

Figure 7.15   *Capacity fade results of lithium-ion batteries: (a) capacity fade curve
and (b) slope-changing ratio*

Table 7.3   *Curve fitting results and error for capacity fade curve characterization*

Parameter	a	b	c	d	R-squared	SSE
Value	−0.1983	−24.44	0.006881	1.004	0.9858	0.04247

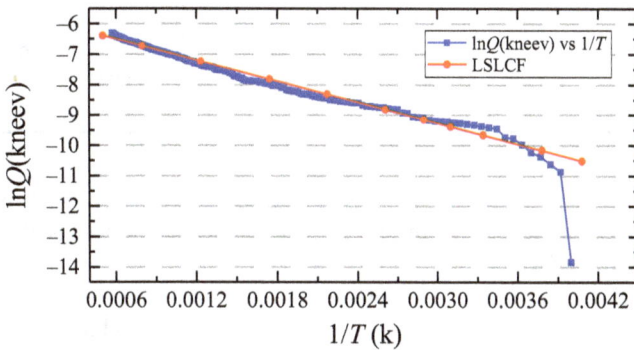

Figure 7.16   *Arrhenius plot linear least-square curve fitting*

capacity and cycle and Figure 7.15(b) shows the slope-changing ratio of the
capacity fade curve derived from the curve fitting. Table 7.3 shows the curve fitting
results and error analysis of the capacity fade curve characterization. The R-squared
error is 0.9858 and sum of squares due to error (SSE) is 0.04247 which evidence
that the fitted curve agrees well with the experimental data, as shown in Table 7.3.

The next step in the result calculation is to implement the proposed knee-
Arrhenius algorithm on capacity fading characterization result. Figure 7.16 shows
the Arrhenius plot of the knee values and temperature with linear least-square
curve-fitting treatment.

Table 7.4 shows the output of the knee-Arrhenius algorithm. According to it,
the experimented battery shows a capacity loss of 4.7701 Ah after 300 cycles and

Table 7.4   Capacity fade model parameters value

Parameter	P_{QMIN}	P_{QMAX}	$P_{Q,knee}$ (Cycle)	Gradient (m)	Ea (J/mol)	lnA	Ql (Ah)
Value	9.5106×10^{-09}	0.001804	415	1157	9619.876	−5.817	4.7701

Table 7.5   Arrhenius plot curve fitting result

Parameter	*R*-squared	Adjusted *R*-squared	RMSE	SSE
Value	0.9834	0.9833	0.1182	0.966

Table 7.6   Capacity fade model EOL value with error

Cycle number	300
Remaining capacity Q_r (Ah) by knee-Arrhenius algorithm	$Q_r = Q_n - Ql = 34.77915$
Remaining capacity from capacity characterization test	34.82622
Algorithm error	0.13%
EOL index	2
Condition	$33.6166 < 34.77915 \leq 35.5941$
Remarks	Above-average damage

the knee-Arrhenius algorithm predicted the knee point at 415 which is quite reasonable because the typical estimated life of a lithium-ion battery is about 300–500 cycles, as shown in Table 7.4.

Table 7.5 shows the linear least-square curve-fitting results of the Arrhenius plot. Wherein, both the root mean squared error (RMSE) and sum of square error (SSE) are obtained within the reasonable error margin, also the *R*-squared value and adjustable *R*-squared is in optimal point. That indicates that the fitted result validates the experiment data with the model quite accurately.

Finally, Table 7.6 shows the value of the remaining capacity according to the proposed knee-Arrhenius algorithm, remaining capacity estimated from capacity characterization tests, algorithm error, the EOL index for the battery, and remarks. After the 300 cycles according to the proposed knee-Arrhenius algorithm, the remaining capacity is 34.77915 Ah and the remaining capacity after the 300 cycles by the capacity characterization test is 34.82622 Ah. That means the algorithm error is about 0.13%. According to the EOL index, the battery cell is at 2, which indicates an above-average damage level.

7.4.4   State of safety evaluation

The result in this section is the prediction of the SOS status or its index for the experimented battery. The value of the predicted SOS status is generated by

EOL index from
aging model

~

EOL index from
capacity fade model

State of safety index

Figure 7.17   Calculation process of SOS from EOL indexes

Table 7.7   State of safety index for experimented battery

EOL index of battery aging model	EOL index of capacity fade model	State of safety index	Description	Decision
2	2	4	Above-average hazard level	Replacement required for high EOL index batteries

merging the value of the EOL index of the battery aging model and the EOL index value of the capacity fade model, as shown in Figure 7.17.

Figure 7.17 shows the calculation process of the state of safety index from the EOL indexes of the battery aging model and capacity fade model. Table 7.7 shows the present status of the state of safety of the experimented battery.

It can be seen that after 300 cycles, an index of the battery cell is 4, indicating that the battery cell is already in the above average-hazard level. The decision that can be taken from this is the higher EOL index batteries should be replaced from the pack or system to avoid any kinds of hazard situations.

In the commercial use of batteries, the efficient use of energy and safety is widely dependent on the proper maintenance and it requires an effective indication method of necessary parameters, including SOS, EOL, RUL, and SOH. As a result, the main focus in this section is to introduce a reliable SOS prediction method to improve the present SOS standards for lithium-ion battery applications, such as electric vehicles and smart power grids. For better understanding and easy implementation, the proposed method is divided into two parts. The only aging phenomenon studied in this research work is the cyclic aging of batteries.

The calendar aging of batteries is not within the scope of this work, because the thermal safety of batteries always increases after the calendar aging. This research work covers two important degradation parameters of lithium-ion batteries, such as internal resistance growth and capacity fade. The battery aging model is mainly designed based on the internal resistance growth mechanism and the capacity fade model is designed as it sounds like based on the capacity fade mechanism. A proper derivation of the proposed models is listed in the related chapters including the mathematical analysis, validation experimental analysis, and data analysis of the experimental results. To validate the proposed models, three experimental processes are carried out the cycle-life test, the hybrid pulse power characterization test, and the capacity characterization test. All experiments are carried out at the New Energy Measurement and Control Laboratory by the researchers.

7.4.5   Parameter identification tests

Figure 7.18 shows the terminal voltage and current variation curve of the lithium-ion battery during the experiment, and the HPPC experiment is partially enlarged.

According to the HPPC experiment results, U_1–U_2 and U_3–U_4 are the beginning and ending discharging stages of the lithium-ion battery. The reason for the sharp drop/rise in the process is mainly due to the ohmic resistance. In the period corresponding to U_4–U_5, since the circuit contains an RC network, when the circuit structure or parameters change suddenly, the system energy is only excited by the initial energy storage of the capacitor in the RC network, so the terminal voltage has a gentle upward trend. The parameters that need to be identified in this identification model are ohmic resistance R_0, polarization resistance R_P, and polarization capacitance C_P. The calculation process is shown in the following equations:

Ohm resistance R_0:

$$R_0 = \frac{|\Delta U_{12}| + |\Delta U_{34}|}{2I} \tag{7.51}$$

Polarization resistance R_P:

$$R_P = \frac{U_1 - U_3 - IR_0}{I\left(1 - e^{-(t3-t2)/\tau}\right)} \tag{7.52}$$

Polarization capacitance C_P:

$$C_P = \frac{\tau}{R_P}, \tau = R_P C_P = -\frac{t_5 - t_4}{\ln\left(\frac{U_1 - U_5}{U_1 - U_4}\right)} \tag{7.53}$$

According to the lithium-ion battery HPPC experiment results, the offline parameter identification method is used to derive the calculation process of the open-circuit voltage U_{OCV}, ohmic resistance R_0, polarization resistance R_P, and polarization capacitance C_P through the above derivation. The parameter identification results of the Thevenin model are obtained from the equally spaced points SOC = 1 to SOC = 0.1, as shown in Table 7.8.

The experimental data are fitted on curve-fitting treatment, and it is found that the six-order curve fitting effect is good, and the curve passes through the scattered

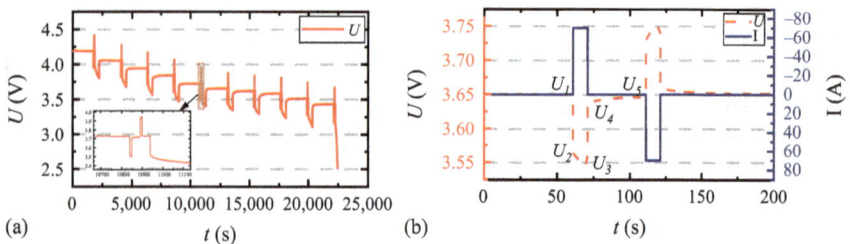

Figure 7.18   Terminal voltage and current change curve for the HPPC test

Table 7.8   Model parameters under different SOC states

SOC/%	U_{OC}/V	R_0/Ω	R_p/Ω	C_p/F
100	4.1877	0.00124	0.000385571	21503.14931
90	4.0535	0.00125	0.000414714	19570.0999
80	3.9363	0.00125	0.000458429	18286.38205
70	3.8268	0.00124	0.000463714	18328.09612
60	3.7406	0.00124	0.000454286	18303.45912
50	3.6504	0.00125	0.000321	25476.63551
40	3.6151	0.00127	0.000310857	24483.91544
30	3.5863	0.0013	0.000335571	23470.41294
20	3.5270	0.00134	0.000405286	20775.46704
10	3.4486	0.00142	0.000648714	11952.87382

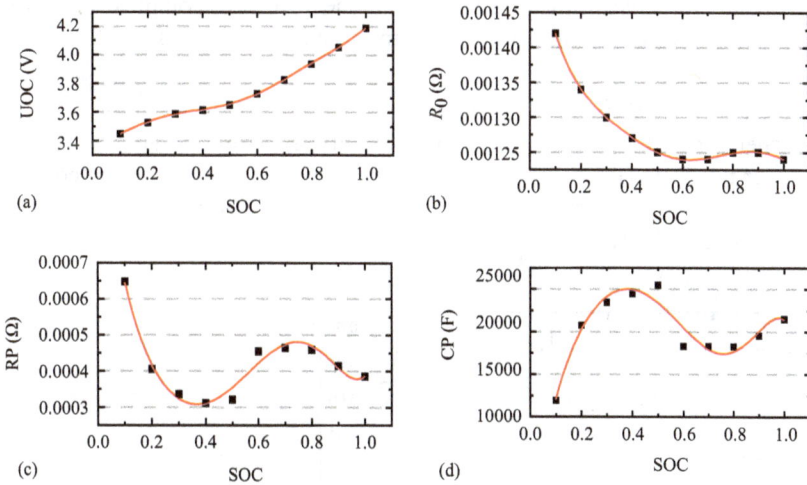

Figure 7.19   Parameter identification results

points smoothly. On the premise of ensuring the fitting accuracy, to avoid over-fitting, the subsequent code running speed, and the burden on the processor, the six-order polynomial is selected. The data in the table are visualized, and the scatter points of the relationship between each parameter and SOC are obtained, as shown in Figure 7.19.

Through the above research conclusions, the independent variable is SOC, and OCV, R_0, R_P, and C_P are taken as dependent variables, and the scatter diagram of the existing relationship among them is obtained. According to the identification results, when the SOC takes different values, the parameters to be identified by the model all fluctuate in a small range within a certain range.

After the test results of each identification parameter obtained above, check whether the Thevenin model is accurate and feasible for the parameter identification. The verification process is carried out/Simulink. The key to verifying whether it is accurate is that the simulation input must be consistent with the above-mentioned HPPC experimental process so that the error between the simulated output voltage and the actual voltage reflects the accuracy of the model. The Thevenin model simulation is shown in Figure 7.20.

It can be seen from the error curve as shown in Figure 7.20 (b) that the simulation model does not diverge under the whole working condition. The larger errors will occur in the pulse charging and discharging process because the sudden large current change in the lithium-ion batteries at these moments will cause a sharp drop/surge in the terminal voltage. It can be seen from the figure that the average error is within 0.04 V, while the rated voltage of the lithium-ion battery is 4.20 V, so the accuracy of this model can reach over 98%, which proves that the Thevenin model and the parameter identification method can accurately characterize the working process of the lithium-ion battery. On the other hand, the existence of errors means that the Thevenin model cannot be completely equivalent to the complex electrochemical reactions in lithium-ion batteries.

7.4.6 State estimation effect verification

To further verify the UPF algorithm practicability, experimental analysis is carried out under the BBDST. The SOC estimation test under BBDST working conditions is carried out for different algorithms. The SOC estimation and error tracking are shown in Figure 7.21.

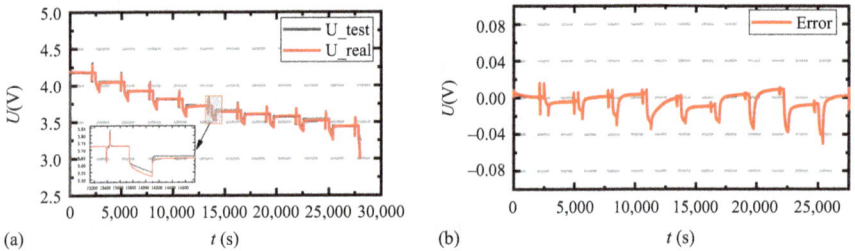

Figure 7.20 Thevenin model simulation results

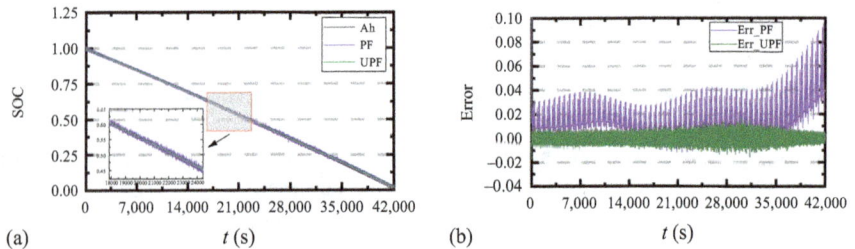

Figure 7.21 SOC estimation and error tracking curve under BBDST conditions

Through the BBDST working condition analysis, the charge tracking error has been in a jitter state, and the tracking error of the UPF algorithm is stable within 1.5%, which has high estimation accuracy, fast convergence speed, and reliable adaptability.

The knee-Arrhenius algorithm is proposed for estimating the knee-point and calculating the remaining capacity at the capacity fade model. The key technique of this algorithm is a combined calculation method of knee-point analysis and Arrhenius model quantification. Based on the mathematical analysis the knee-Arrhenius algorithm can divide into two phases. The first phase of the algorithm is estimating the knee-point and knee-values from the normalized capacity fade curve and the second phase of the algorithm is the quantification of knee-values according to the Arrhenius model and the parameters of the Arrhenius model are estimated by the least-square linear curve-fitting method. Moreover, the accuracy of the proposed method is quite reliable. The values of the R-square of the all-curve fittings are within 0.98–0.99, which means the experimental design and the mathematical analysis of the battery models agree with the dynamic experimental characteristics of lithium-ion batteries.

The simulation results show that the Thevenin model accuracy under HPPC conditions can reach more than 98%. The SOC accuracy of lithium-ion batteries based on the UPF algorithm is estimated to reach 98.5%. The knee-Arrhenius algorithm is also proved very efficient to predict an earlier knee-point within the range of the typical life span of lithium-ion batteries with an error rate of less than 0.15%. The possible future research work related to this study could be the implementation of the third-order Thevenin model for battery modeling to gain more accuracy on the internal resistance growth estimation process. Also, the knee-Arrhenius algorithm could be further modified through some machine learning techniques to improve identification accuracy.

7.5   Chapter summary

To verify the effectiveness of the UPF algorithm in estimating the SOC of the lithium-ion batteries, the parameters of the constructed lithium-ion battery Thevenin model are identified, the identified experimental result results are imported into the SOC estimation algorithm, and the battery SOC is simulated and estimated under the BBDST condition. Compared with the PF algorithm, it has higher accuracy, faster convergence, better stability, and higher robustness characteristics, which provides a new idea for SOC estimation.

Acknowledgment

This work is supported by the National Natural Science Foundation of China (No. 61801407), Sichuan Science and Technology Program (No. 2019YFG0427), China Scholarship Council (No. 201908515099), and Fund of Robot Technology Used for Special Environment Key Laboratory of Sichuan Province (No. 18kftk03).

References

[1] Yang, J.F., Xia, B., Huang, W.X., Fu, Y.H., and Mi, C., Online state of health estimation for lithium-ion batteries using constant-voltage charging current analysis. *Applied Energy*, 2018. **212**: p. 1589–1600.

[2] Zhang, J.N., Yang, X.G., Sun, F.C., Wang, Z.P., and Wang, C.Y., An online heat generation estimation method for lithium-ion batteries using dual-temperature measurements. *Applied Energy*, 2020. **272**: p. 1–12.

[3] Yuan, H., Dai, H.F., Wei, X.Z., and Ming, P.W., Model-based observers for internal states estimation and control of proton exchange membrane fuel cell system: A review. *Journal of Power Sources*, 2020. **468**: p. 1–17.

[4] Peng, J.K., Luo, J.Y., He, H.W., and Lu, B., An improved state of charge estimation method based on cubature Kalman filter for lithium-ion batteries. *Applied Energy*, 2019. **253**(113520): p. 1–10.

[5] Chen, K.H. and Ding Z.D., Lithium-ion battery lifespan estimation for hybrid electric vehicle. *2015 27th Chinese Control and Decision Conference (CCDC)*, 2015. p. 5602–5605.

[6] Chen, X.K. and Sun, D., Modeling and state of charge estimation of lithium-ion battery. *Advances in Manufacturing*, 2015. **3**(3): p. 202–211.

[7] Lai, X., Zheng, Y.J., and Sun, T., A comparative study of different equivalent circuit models for estimating state of charge of lithium-ion batteries. *Electrochimica Acta*, 2018. **259**: p. 566–577.

[8] Chen, W.C., Li, J.D., Shu, C.M., and Wang, Y.W., Effects of thermal hazard on 18650 lithium-ion battery under different states of charge. *Journal of Thermal Analysis and Calorimetry*, 2015. **121**(1): p. 525–531.

[9] Tang, S.X., Camacho-Solorio, L., Wang, Y.B., and Krstic, M., Digital battery management design for point of load applications with cell balancing. *IEEE Transactions on Industrial Electronics*, 2020. **67**(8): p. 6365–6375.

[10] Yan, Y.D., Hu, K.Y., and Tsai, C.H., State of charge estimation from a thermal–electrochemical model of lithium-ion batteries. *Automatica*, 2017. **83**: p. 206–219.

[11] Wang, S.L., Fernandez, C., Zou, C.Y., et al., Open circuit voltage and state of charge relationship functional optimization for the working state monitoring of the aerial lithium-ion battery pack. *Journal of Cleaner Production*, 2018. **198**(1): p. 1090–1104.

[12] Chemali, E., Kollmeyer, P.J., Preindl, M., and Emadi, A., State of charge estimation of Li-ion batteries using deep neural networks: A machine learning approach. *Journal of Power Sources*, 2018. **400**: p. 242–255.

[13] Xuan, D.J., Shi, Z.F., Chen, J.Z., Zhang, C.Y., and Wang, Y.X., Real-time estimation of state of charge in lithium-ion batteries using improved central difference transform method. *Journal of Cleaner Production*, 2020. **252**: p. 1–14.

[14] Wang, S.L., Shi, J.Y. Fernandez, C. Zou, C.Y., Bai, D.K., and Li, J.C., An improved packing equivalent circuit modeling method with the cell-to-cell consistency state evaluation of the internal connected lithium-ion batteries. *Energy Science & Engineering*, 2019. **7**(2): p. 546–556.

[15] He, Y.L., He, R., Guo, B., *et al.*, Modeling of dynamic hysteresis characters for the lithium-ion battery. *Journal of the Electrochemical Society*, 2020. **167**(9): p. 1–14.

[16] Mayur, M., Yagci, M.C., Carelli, S., Margulies, P., Velten, D., and Bessler, W.G., Identification of stoichiometric and microstructural parameters of a lithium-ion cell with blend electrode. *Physical Chemistry Chemical Physics*, 2019. **21**(42): p. 23672–23684.

[17] Meng, J.H., Cai, L., Stroe, D.I., Ma, J.P., Luo, G.Z., and Teodorescu, R., An optimized ensemble learning framework for lithium-ion battery state of health estimation in energy storage system. *Energy*, 2020. **206**: p. 1–14.

[18] Chaoui, H., Golbon, N., Hmouz, I., Souissi, R., and Tahar, S., Lyapunov-based adaptive state of charge and state of health estimation for lithium-ion batteries. *IEEE Transactions on Industrial Electronics*, 2015. **62**(3): p. 1610–1618.

[19] Zhang, Y.Z., Xiong, R., He, H.W., and Shen, W.X., Lithium-ion battery pack state of charge and state of energy estimation algorithms using a hardware-in-the-loop validation. *IEEE Transactions on Power Electronics*, 2017. **32**(6): p. 4421–4431.

[20] Chaoui, H. and Gualous, H., Adaptive state of charge estimation of lithium-ion batteries with parameter and thermal uncertainties. *IEEE Transactions on Control Systems Technology*, 2016. **25**: p. 1–8.

[21] Liu, K.L., Li, K., Peng, Q., and Zhang, C., A brief review on key technologies in the battery management system of electric vehicles. *Frontiers of Mechanical Engineering*, 2019. **14**(1): p. 47–64.

[22] Cacciato, M., Nobile, G., Scarcella, G., and Scelba, G., Real-time model-based estimation of SOC and SOH for energy storage systems. *IEEE Transactions on Power Electronics*, 2017. **32**(1): p. 794–803.

[23] Zhang, Z.L., Gui, H.D., Gu, D.J., Yang, Y., and Ren, X.Y., A hierarchical active balancing architecture for lithium-ion batteries. *IEEE Transactions on Power Electronics*, 2017. **32**(4): p. 2757–2768.

[24] Tang, X.P., Liu B.Y., and Gao F.R., State of charge estimation of LiFePO$_4$ battery based on a gain-classifier observer. *Energy Procedia*, 2017. **105**: p. 2071–2076.

[25] Wei, X., Mo, Y.M., and Feng, Z., Lithium-ion battery modeling and state of charge estimation. *Integrated Ferroelectrics*, 2019. **200**(1): p. 59–72.

[26] Wang, Q.Q., Wang, J., Zhao, P.J., Kang, J.Q., Yan, F., and Du, C.Q., Correlation between the model accuracy and model-based SOC estimation. *Electrochimica Acta*, 2017. **228**: p. 146–159.

[27] Yang, Q.X., Xu, J., Cao, B.G., and Li, X.Q., A simplified fractional order impedance model and parameter identification method for lithium-ion batteries. *PloS one*, 2017. **12**(2): p. e0172424–e0172424.

[28] Xia, B., Zhao, X., de Callafon, R., Garnier, H., Truong, N., and Mi, C., Accurate lithium-ion battery parameter estimation with continuous-time system identification methods. *Applied Energy*, 2016. **179**: p. 426–436.

[29] Corno, M., Bhatt, N., Savaresi, S.M., and Verhaegen, M., Electrochemical model-based state of charge estimation for Li-ion cells. *IEEE Transactions on Control Systems Technology*, 2015. **23**(1): p. 117–127.

[30] Ma, Y., Zhou X.W., Li, B.S., and Chen, H., Fractional modeling and SOC estimation of lithium-ion battery. *IEEE/CAA Journal of Automatica Sinica*, 2016. **3**(3): p. 281–287.

[31] Peng, S.M., Chen, C., Shi, H.B., and Yao, Z.L., State of charge estimation of battery energy storage systems based on adaptive unscented Kalman filter with a noise statistics estimator. *IEEE Access*, 2017. **5**: p. 13202–13212.

[32] Partovibakhsh, M. and Liu, G.J., An adaptive unscented kalman filtering approach for online estimation of model parameters and state of charge of lithium-ion batteries for autonomous mobile robots. *IEEE Transactions on Control Systems Technology*, 2015. **23**(1): p. 357–363.

[33] Tian, Y., Xia, B.Z., Sun, W., Xu, Z.H., and Zheng, W.W., A modified model based state of charge estimation of power lithium-ion batteries using unscented Kalman filter. *Journal of Power Sources*, 2014. **270**: p. 619–626.

[34] Barai, A., Uddin, K., Widanage, W.D. McGordon, A., and Jennings, P., A study of the influence of measurement timescale on internal resistance characterization methodologies for lithium-ion cells. *Scientific Reports*, 2018. **8**(1): p. 21.

[35] El Din, M.S., Abdel-Hafez, M.F., and Hussein, A.A., Enhancement in Li-ion battery cell state of charge estimation under uncertain model statistics. *IEEE Transactions on Vehicular Technology*, 2016. **65**(6): p. 4608–4618.

[36] Zhou, W.D. and Hou, J.X., A new adaptive high-order unscented kalman filter for improving the accuracy and robustness of target tracking. *IEEE Access*, 2019. **7**: p. 118484–118497.

[37] Zhang, Z.L., Cheng, X., Lu, Z.Y., and Gu, D.J., SOC estimation of lithium-ion batteries with AEKF and wavelet transform matrix. *IEEE Transactions on Power Electronics*, 2017. **32**(10): p. 7626–7634.

[38] Aung, H., Low, K.S., and Goh, S.T., State of charge estimation of lithium-ion battery using square root spherical unscented kalman filter (Sqrt-UKFST) in nanosatellite. *IEEE Transactions on Power Electronics*, 2015. **30**(9): p. 4774–4783.

[39] Chang, T.F., Watteyne, T., Pister, K., and Wang, Q., Adaptive synchronization in multi-hop TSCH networks. *Computer Networks*, 2015. **76**: p. 165–176.

[40]   Camci, F., Ozkurt, C., Toker, O., and Atamuradov, V., Sampling based state of health estimation methodology for Li-ion batteries. *Journal of Power Sources*, 2015. **278**: p. 668–674.

[41]   Su, J., Lin, M.S., Wang, S.L., Li, J., Coffie-Ken, J., and Xie, F., An equivalent circuit model analysis for the lithium-ion battery pack in pure electric vehicles. *Measurement and Control*, 2019. **52**(3–4): p. 193–201.

[42]   Tian, N., Wang, Y.B., Chen, J., and Fang, H.Z., On parameter identification of an equivalent circuit model for lithium-ion batteries. *2017 IEEE Conference on Control Technology and Applications (CCTA)*. 2017.

[43]   Qin, D.C., Li, J.J., Wang, T.T., and Zhang, D.M., Modeling and simulating a battery for an electric vehicle based on modelica. *Automotive Innovation*, 2019. **2**(3): p. 169–177.

[44]   Zhang, X.Q., Zhang, W.P., and Lei, G.Y., A review of Li-ion battery equivalent circuit models. *Transactions on Electrical and Electronic Materials*, 2016. **17**: p. 311–316.

[45]   Saxena, S., Raman, S.R., Saritha, B., and John, V., A novel approach for electrical circuit modeling of Li-ion battery for predicting the steady-state and dynamic I–V characteristics. *Sādhanā*, 2016. **41**: p. 479–487.

[46]   Camara, M.B. and Dakyo, B., Energy management for hybrid electric vehicles using load power fluctuation compensation—Ultracapacitors and lithium-battery. *2015 International Aegean Conference on Electrical Machines & Power Electronics (Acemp), 2015 International Conference on Optimization of Electrical & Electronic Equipment (Optim) & 2015 International Symposium on Advanced Electromechanical Motion Systems (Electromotion)*, 2015: p. 46–51.

[47]   Mendoza, S., Rothenberger, M., Hake, A., and Fathy, H., Optimization and experimental validation of a thermal cycle that maximizes entropy coefficient fisher identifiability for lithium iron phosphate cells. *Journal of Power Sources*, 2016. **308**: p. 18–28.

[48]   Rothenberger, M.J., Docimo, D.J., Ghanaatpishe, M., and Fathy, H.K., Genetic optimization and experimental validation of a test cycle that maximizes parameter identifiability for a Li-ion equivalent-circuit battery model. *Journal of Energy Storage*, 2015. **4**: p. 156–166.

[49]   Zhang, C., Allafi, W., Dinh, Q., Ascencio, P., and Marco, J., Online estimation of battery equivalent circuit model parameters and state of charge using decoupled least squares technique. *Energy*, 2018. **142**: p. 678–688.

[50]   Xu, Y.D., Hu, M.H., Zhou, A.J., *et al.*, State of charge estimation for lithium-ion batteries based on adaptive dual Kalman filter. *Applied Mathematical Modelling*, 2020. **77**: p. 1255–1272.

[51]   Rahimi-Eichi, H., Baronti, F., and Chow, M.Y., Online adaptive parameter identification and state of charge coestimation for lithium-polymer battery cells. *IEEE Transactions on Industrial Electronics*, 2014. **61**(4): p. 2053–2061.

[52]   Writer, B., Models, SOC, maximum, time, cell, data, parameters, in *Lithium-Ion Batteries: A Machine-Generated Summary of Current Research*, B. Writer, Editor. 2019, Springer: Cham. p. 195–247.

[53] Barai, A., Widanage, W.D., Marco, J., McGordon, A., and Jennings, P., A study of the open circuit voltage characterization technique and hysteresis assessment of lithium-ion cells. *Journal of Power Sources*, 2015. **295**: p. 99–107.

[54] Cai, J. and Zhong, X.L., An adaptive square root cubature kalman filter based SLAM algorithm for mobile robots. *2015 IEEE International Conference on Mechatronics and Automation*, 2015: p. 2215–2219.

[55] Bizeray, A.M., Zhao, S., Duncan, S.R., and Howey, D.A., Lithium-ion battery thermal-electrochemical model-based state estimation using orthogonal collocation and a modified extended Kalman filter. *Journal of Power Sources*, 2015. **296**: p. 400–412.

[56] Ai, S.J., Zong, C.X., Wu, W., *et al.*, Effect of derivatives of glycerol sulfite as electrolyte additive on electrochemical performance of lithium ion battery. *Chemical Journal of Chinese Universities-Chinese*, 2018. **39**(11): p. 2520–2528.

[57] Birkl, C.R., McTurk, E., Roberts, M.R., Bruce, P.G., and Howey, D.A., A parametric open circuit voltage model for lithium ion batteries. *Journal of the Electrochemical Society*, 2015. **162**(12): p. A2271–A2280.

[58] Appiah, W.A., Park, J., Byun, S., *et al.*, A coupled chemo-mechanical model to study the effects of adhesive strength on the electrochemical performance of silicon electrodes for advanced lithium ion batteries. *Journal of Power Sources*, 2018. **407**: p. 153–161.

[59] Wang, C.W., Ma, X.L., Cheng, J.G., Sun, J.T., and Zhou, Y.H., Synthesis and electrochemical properties of Ca-doped $LiNi_{0.8}Co_{0.2}O_2$ cathode material for lithium ion battery. *Journal of Solid State Electrochemistry*, 2007. **11**(3): p. 361–364.

[60] Li, X.Y., Yuan, C.G., and Wang, Z.P., State of health estimation for Li-ion battery via partial incremental capacity analysis based on support vector regression. *Energy*, 2020. **203**: p. 1–14.

[61] Bester, J.E., El Hajjaji, A., and Mabwe, A.M., Modelling of lithium-ion battery and SOC estimation using simple and extended discrete kalman filters for aircraft energy management. *IECON2015—41st Annual Conference of the IEEE Industrial Electronics Society*, 2015: p. 2433–2438.

[62] Tian, J.P., Xiong, R., and Shen, W.X., State of health estimation based on differential temperature for lithium ion batteries. *IEEE Transactions on Power Electronics*, 2020. **35**(10): p. 10363–10373.

[63] Zhao, M.H., Zhong, S.S., Fu, X.Y., Tang, B.P., Dong, S.J., and Pecht, M., Deep residual networks with adaptively parametric rectifier linear units for fault diagnosis. *IEEE Transactions on Industrial Electronics*, 2021. **68**(3): p. 2587–2597.

[64] Xu, T.T., Peng, Z., and Wu, L.F., A novel data-driven method for predicting the circulating capacity of lithium-ion battery under random variable current. *Energy*, 2021. **218**(119530): p. 1–13.

[65] Xiao, F., Li, C.R., Fan, Y.X., Yang, G.R., and Tang, X., State of charge estimation for lithium-ion battery based on Gaussian process regression

with deep recurrent kernel. *International Journal of Electrical Power & Energy Systems*, 2021. **124**: p. 1–15.

[66]   Wang, C., Yang, Y.Y., and Zhou, P.Z., Towards efficient scheduling of federated mobile devices under computational and statistical heterogeneity. *IEEE Transactions on Parallel and Distributed Systems*, 2021. **32**(2): p. 394–410.

[67]   Tian, J.Q., Wang, Y.J., and Chen, Z.H., An improved single particle model for lithium-ion batteries based on main stress factor compensation. *Journal of Cleaner Production*, 2021. **278**(123456): p. 1–12.

[68]   Sun, B.X., He, X.T., Zhang, W.G., Ruan, H.J., Su, X.J., and Jiang, J.C., Study of parameters identification method of Li-ion battery model for EV power profile based on transient characteristics data. *IEEE Transactions on Intelligent Transportation Systems*, 2021. **22**(1): p. 661–672.

[69]   Seruga, D., Gosar, A., Sweeney, C.A., Jaguemont, J., Van Mierlo, J., and Nagode, M., Continuous modelling of cyclic ageing for lithium-ion batteries. *Energy*, 2021. **215**(119079): p. 1–14.

[70]   Ma, J., Shang, P.C., Zou, X.Y., *et al.*, A hybrid transfer learning scheme for remaining useful life prediction and cycle life test optimization of different formulation Li-ion power batteries. *Applied Energy*, 2021. **282**(116167): p. 1–17.

[71]   Lui, Y.H., Li, M., Downey, A., *et al.*, Physics-based prognostics of implantable-grade lithium-ion battery for remaining useful life prediction. *Journal of Power Sources*, 2021. **485**(229327): p. 1–12.

[72]   Liu, K.L., Shang, Y.L. Ouyang, Q., and Widanage, W.D., A data-driven approach with uncertainty quantification for predicting future capacities and remaining useful life of lithium-ion battery. *IEEE Transactions on Industrial Electronics*, 2021. **68**(4): p. 3170–3180.

[73]   Li, Y.H., Li, K., Liu, X., Wang, Y.X., and Zhang, L., Lithium-ion battery capacity estimation—A pruned convolutional neural network approach assisted with transfer learning. *Applied Energy*, 2021. **285**(116410): p. 1–12.

[74]   Li, X.Y., Huang, Z.J., Tian, J.D., and Tian, Y., State of charge estimation tolerant of battery aging based on a physics-based model and an adaptive cubature Kalman filter. *Energy*, 2021. **220**(119767): p. 1–13.

[75]   Li, W.H., Sengupta, N., Dechent, P., Howey, D., Annaswamy, A., and Sauer, D.U., Online capacity estimation of lithium-ion batteries with deep long short-term memory networks. *Journal of Power Sources*, 2021. **482** (228863): p. 1–11.

[76]   Lee, C., Jo, S., Kwon, D., and Pecht, M.G., Capacity-fading behavior analysis for early detection of unhealthy Li-ion batteries. *IEEE Transactions on Industrial Electronics*, 2021. **68**(3): p. 2659–2666.

[77]   Kwak, E., Jeong, S., Kim, J.H., and Oh, K.Y., Prediction of compression force evolution over degradation for a lithium-ion battery. *Journal of Power Sources*, 2021. **483**(229079): p. 1–11.

[78]   Chen, L., Wang, H.M., Liu, B.H., Wang, Y.J., Ding, Y.H., and Pan, H.H., Battery state of health estimation based on a metabolic extreme learning

machine combining degradation state model and error compensation. *Energy*, 2021. **215**(119078): p. 1–10.

[79] Cao, X.B., Xu, W.Z., Liu, X.X., Peng, J., and Liu, T., A deep reinforcement learning-based on-demand charging algorithm for wireless rechargeable sensor networks. *Ad Hoc Networks*, 2021. **110**(485): p. 1–13.

[80] Zhu, Y., Fang, Y.D., Su, L., and Ye, F., Experimental investigation of start-up and transient thermal performance of pumped two-phase battery thermal management system. *International Journal of Energy Research*, 2020. **44** (14): p. 11372–11384.

[81] Zhu, S.X., Han, J.D., Wang, Y.N., *et al.*, In-situ heat generation measurement of the anode and cathode in a single-layer lithium ion battery cell. *International Journal of Energy Research*, 2020. **44**(11): p. 9141–9148.

[82] Zhu, R.J., Zhu, C.Y., Sheng, N., Rao, Z.H., Aoki, Y., and Habazaki, H., A widely applicable strategy to convert fabrics into lithiophilic textile current collector for dendrite-free and high-rate capable lithium metal anode. *Chemical Engineering Journal*, 2020. **388**: p. 1–13.

[83] Zhu, J.L., Nie, W., Wang, Q., Wen, W., Zhang, X.H., and Li, F.J., A competitive self-powered sensing platform based on a visible light assisted zinc-air battery system. *Chemical Communications (Cambridge)*, 2020. **56**(43): p. 5739–5742.

[84] Zhu, J.G., Darma, M.S.D., Knapp, M., *et al.*, Investigation of lithium-ion battery degradation mechanisms by combining differential voltage analysis and alternating current impedance. *Journal of Power Sources*, 2020. **448** (227575): p. 1–12.

[85] Zhu, H.J., Song, Z.Y., Hou, J., Hofmann, H.F., and Sun, J., Simultaneous identification and control using active signal injection for series hybrid electric vehicles based on dynamic programming. *IEEE Transactions on Transportation Electrification*, 2020. **6**(1): p. 298–307.

[86] Zhu, C., Shang, Y.L., Lu, F., Jian, Y., Cheng, C.W., and Mi, C., Core temperature estimation for self-heating automotive lithium-ion batteries in cold climates. *IEEE Transactions on Industrial Informatics*, 2020. **16**(5): p. 3366–3375.

[87] Zhou, Y., Wu, J.Z., Long, C., and Ming, W.L., State of the-art analysis and perspectives for peer-to-peer energy trading. *Engineering*, 2020. **6**(7): p. 739–753.

[88] Zhou, Y., Li, H., Ravey, A., and Pera, M.C., An integrated predictive energy management for light-duty range-extended plug-in fuel cell electric vehicle. *Journal of Power Sources*, 2020. **451**: p. 1–14.

[89] Zhou, Y.K. and Cao, S.L., Coordinated multi-criteria framework for cycling aging-based battery storage management strategies for positive building–vehicle system with renewable depreciation: Life-cycle based techno-economic feasibility study. *Energy Conversion and Management*, 2020. **226**: p. 1–14.

[90] Zhou, W., Huang, R.J., Liu, K., and Zhang, W.G., A novel interval-based approach for quantifying practical parameter identifiability of a lithium-ion

battery model. *International Journal of Energy Research*, 2020. **44**(5): p. 3558–3573.

[91]   Zhou, S.P., Hu, A.P., Liu, D.N., *et al.*, Building three-dimensional carbon nanotubes-interwoven Ni_3S_2 micro-nanostructures for improved sodium storage performance. *Electrochimica Acta*, 2020. **339**: p. 1–14.

[92]   Zhou, S.G., Liu, G., Ding, N.W., Shang, L.L., Dang, R., and Zhang, J.W., Improved performances of lithium-ion batteries by graphite-like carbon modified current collectors. *Surface & Coatings Technology*, 2020. **399**: p. 1–14.

[93]   Zhou, L., He, L., Zheng, Y.J., Lai, X., Ouyang, M.G., and Lu, L.G., Massive battery pack data compression and reconstruction using a frequency division model in battery management systems. *Journal of Energy Storage*, 2020. **28**(101252): p. 1–9.

[94]   Zhou, H.K., Dai, C.H., Liu, Y., Fu, X.T., and Du, Y., Experimental investigation of battery thermal management and safety with heat pipe and immersion phase change liquid. *Journal of Power Sources*, 2020. **473**: p. 1–14.

[95]   Zhou, G.H., Tian, Q.X., Leng, M.R., Fan, X.Y., and Bi, Q., Energy management and control strategy for DC microgrid based on DMPPT technique. *IET Power Electronics*, 2020. **13**(4): p. 658–668.

[96]   Zhou, D., Zheng, W.B., Fu, P., and Pan, X.L., Research on online estimation of available capacity of lithium batteries based on daily charging data. *Journal of Power Sources*, 2020. **451**(227713): p. 1–16.

[97]   Zheng, N.B., Fan, R.J., Sun, Z.Q., and Zhou, T., Thermal management performance of a fin-enhanced phase change material system for the lithium-ion battery. *International Journal of Energy Research*, 2020. **44**(9): p. 7617–7629.

[98]   Zheng, Q.F., Yamada, Y. Shang, R., *et al.*, A cyclic phosphate-based battery electrolyte for high voltage and safe operation. *Nature Energy*, 2020. **5** (4): p. 291–298.

[99]   Zheng, B.B., Yu, L.H., Li, N.R., and Xi, J.Y., Efficiently immobilizing and converting polysulfide by a phosphorus doped carbon microtube textile interlayer for high-performance lithium-sulfur batteries. *Electrochimica Acta*, 2020. **345**: p. 1–14.

[100]   Zhen, Y.H. and Li, Y.D., A high-performance all-iron non-aqueous redox flow battery. *Journal of Power Sources*, 2020. **445**: p. 1–14.

[101]   Zhao, Y.M., Liu, X.X., Zhang, S., *et al.*, Preparation and kinetic performances of single-phase $PuNi_3$-, Ce_2Ni_7-, Pr_5Co_{19}-type superlattice structure La–Gd–Mg–Ni-based hydrogen storage alloys. *Intermetallics*, 2020. **124**: p. 1–14.

[102]   Zhao, X.X., Gu, Z.Y., Li, W.H., Yang, X., Guo, J.Z., and Wu, X.L., Temperature-dependent electrochemical properties and electrode kinetics of Na3V2(PO4)(2)O2F cathode for sodium-ion batteries with high energy density. *Chemistry—A European Journal*, 2020. **26**(35): p. 7823–7830.

[103]   Wang, Y.J., Sun Z.D., and Chen Z.H., Development of energy management system based on a rule-based power distribution strategy for hybrid power sources. *Energy*, 2019. **175**: p. 1055–1066.

[104] Fan, Y.C., Wang, T.S., Legut, D., and Zhang, Q.F., Theoretical investigation of lithium ions' nucleation performance on metal-doped Cu surfaces. *Journal of Energy Chemistry*, 2019. **39**: p. 160–169.

[105] Pavkovic, D., Krznar, M., Komljenovic, A., Hrgetic, M., and Zorc, D., Dual EKF-based state and parameter estimator for a LifePO$_4$ battery cell. *Journal of Power Electronics*, 2017. **17**(2): p. 398–410.

[106] Qiu, Y., Li, X., Chen, W., Duan, Z.M., and Yu, L., State of charge estimation of vanadium redox battery based on improved extended Kalman filter. *ISA Transactions*, 2019. **94**: p. 326–337.

[107] Dai, K.W., Wang, J., and He, H.W., An improved SOC estimator using time-varying discrete sliding mode observer. *IEEE Access*, 2019. **7**: p. 115463–115472.

[108] Rosewater, D., Ferreira, S., Schoenwald, D., Hawkins, J., and Santoso, S., Battery energy storage state of charge forecasting: Models, optimization, and accuracy. *IEEE Transactions on Smart Grid*, 2019. **10**(3): p. 2453–2462.

[109] Baronti, F., Zanaboni, W., Roncella, R., Saletti, R., and Spagnuo, G., Open-circuit voltage measurement of lithium–iron–phosphate batteries. *2015 IEEE International Instrumentation and Measurement Technology Conference (I2mtc)*, 2015: p. 1711–1716.

[110] Bellache, K., Camara, M.B., and Dakyo, B., Hybrid electric boat based on variable speed diesel generator and lithium-battery—using frequency approach for energy management. *2015 International Aegean Conference on Electrical Machines & Power Electronics (Acemp), 2015 International Conference on Optimization of Electrical & Electronic Equipment (Optim) & 2015 International Symposium on Advanced Electromechanical Motion Systems (Electromotion)*, 2015: p. 744–749.

[111] Ben Ali, J., Khelif, R., Saidi, L., Chebel-Morello, B., and Fnaiech, F., The use of nonlinear future reduction techniques as a trend parameter for state of health estimation of lithium-ion batteries. *2015 16th International Conference on Sciences and Techniques of Automatic Control and Computer Engineering (Sta)*, 2015: p. 245–250.

[112] Bergman, M., Bergfelt, A., Sun, B., Bowden, T., Brandell, D., and Johansson, P., Graft copolymer electrolytes for high temperature Li-battery applications, using poly(methyl methacrylate) grafted poly(ethylene glycol) methyl ether methacrylate and lithium bis(trifluoromethanesulfonimide). *Electrochimica Acta*, 2015. **175**: p. 96–103.

[113] Baronti, F., Saletti, R., and Zamboni, W., Open circuit voltage of lithium-ion batteries for energy storage in DC microgrids. *2015 IEEE First International Conference on DC Microgrids (ICDCM)*, 2015: p. 343–348.

[114] Richardson, R.R., Osborne, M.A., and Howey, D.A., Battery health prediction under generalized conditions using a Gaussian process transition model. *Journal of Energy Storage*, 2019. **23**: p. 320–328.

[115] Tang, X.P., Zou, C.F., Yao, K., Lu, J.Y., Xia, Y.X., and Gao, F.R., Aging trajectory prediction for lithium-ion batteries via model migration and Bayesian Monte Carlo method. *Applied Energy*, 2019. **254**(113591): p. 1–12.

[116] Barillas, J.K., Li, J.H., Gunther, C., and Danzer, M.A., A comparative study and validation of state estimation algorithms for Li-ion batteries in battery management systems. *Applied Energy*, 2015. **155**: p. 455–462.

[117] Tang, X.P., Wang, Y.J., Zou, C.F., Yao, K., Xia, Y.X., and Gao, F.R., A novel framework for lithium-ion battery modeling considering uncertainties of temperature and aging. *Energy Conversion and Management*, 2019. **180**(1): p. 162–170.

[118] Wang, F.K. and Mamo, T., A hybrid model based on support vector regression and differential evolution for remaining useful lifetime prediction of lithium-ion batteries. *Journal of Power Sources*, 2018. **401**: p. 49–54.

[119] Patil, M.A., Tagade, P., Hariharan, K.S., *et al.*, A novel multistage support vector machine based approach for Li ion battery remaining useful life estimation. *Applied Energy*, 2015. **159**: p. 285–297.

[120] Ben Ali, J., Azizi, C., Saidi, L., Bechhoefer, E., and Benbouzid, M., Reliable state of health condition monitoring of Li-ion batteries based on incremental support vector regression with parameters optimization. *Proceedings of the Institution of Mechanical Engineers Part I-Journal of Systems and Control Engineering*, 2020: p. 1–14.

[121] Du, J.C, Zhang, W.G., Zhang, C.P., and Zhou, X.Z., Battery remaining useful life prediction under coupling stress based on support vector regression. *Cleaner Energy for Cleaner Cities*, 2018. **152**: p. 538–543.

[122] Ma, G.J., Zhang, Y., Cheng, C., Zhou, B.T., Hu, P.C., and Yuan, Y., Remaining useful life prediction of lithium-ion batteries based on false nearest neighbors and a hybrid neural network. *Applied Energy*, 2019. **253** (113626): p. 1–13.

[123] Reddy, G.T., Priya, S.R.M. Parimala, M., *et al.*, A deep neural networks based model for uninterrupted marine environment monitoring. *Computer Communications*, 2020. **157**: p. 64–75.

[124] Joshi, U. and Kumar, R., A novel deep neural networks based path prediction. *Cluster Computing—The Journal of Networks Software Tools and Applications*, 2020: p. 1–14.

[125] Chen, Y.Z., Md, U.H., Deepjyoti, D., and Michael, C., Stochastic battery operations using deep neural networks. *2019 IEEE Power & Energy Society Innovative Smart Grid Technologies Conference (ISGT)*, 2019: p. 1–13.

[126] Li, X.Y., Zhang, L., Wang, Z.P., and Dong, P., Remaining useful life prediction for lithium-ion batteries based on a hybrid model combining the long short-term memory and Elman neural networks. *Journal of Energy Storage*, 2019. **21**: p. 510–518.

[127] Zhang, Y.Z., Xiong, R., He, H.W., and Pecht, M.G., Long short-term memory recurrent neural network for remaining useful life prediction of lithium-ion batteries. *IEEE Transactions on Vehicular Technology*, 2018. **67**(7): p. 5695–5705.

[128] Li, P.H., Zhang, Z.J., Xiong, Q.Y., *et al.*, State of health estimation and remaining useful life prediction for the lithium-ion battery based on a

variant long short term memory neural network. *Journal of Power Sources*, 2020. **459**(228069): p. 1–12.

[129] Bhayo, B.A., Al-Kayiem, H.H., Gilani, S.I.U., and Ismail, F.B., Power management optimization of hybrid solar photovoltaic-battery integrated with pumped-hydro-storage system for standalone electricity generation. *Energy Conversion and Management*, 2020. **215**: p. 1–14.

[130] Bhowmik, P., Chandak, S., and Rout, P.K., State of charge and state of power management among the energy storage systems by the fuzzy tuned dynamic exponent and the dynamic PI controller. *Journal of Energy Storage*, 2018. **19**: p. 348–363.

[131] Nejad, S. and Gladwin, D.T., Online battery state of power prediction using PRBS and extended Kalman filter. *IEEE Transactions on Industrial Electronics*, 2020. **67**(5): p. 3747–3755.

[132] Guo, N.Y., Shen, J.W., Xiao, R.X., Yan, W.S., and Chen, Z., Energy management for plug-in hybrid electric vehicles considering optimal engine on/off control and fast state of charge trajectory planning. *Energy*, 2018. **163**: p. 457–474.

[133] Hahn, S.L., Storch, M., Swaminathan, R., Obry, B., Bandlow, J., and Birke, K.P., Quantitative validation of calendar aging models for lithium-ion batteries. *Journal of Power Sources*, 2018. **400**: p. 402–414.

[134] Hao, M.L., Li, J., Park, S., Moura, S., and Dames, C., Efficient thermal management of Li-ion batteries with a passive interfacial thermal regulator based on a shape memory alloy. *Nature Energy*, 2018. **3**(10): p. 899–906.

[135] Hong, D.H., Chen, L., Kong, Q.G., and Cao, H., First principles probing of photo-generated intermolecular charge transfer state in conjugated oligomers. *Chinese Journal of Chemical Physics*, 2018. **31**(2): p. 171–176.

[136] Baek, K.W., Hong, E.S., and Cha, S.W., Capacity fade modeling of a lithium-ion battery for electric vehicles. *International Journal of Automotive Technology*, 2015. **16**(2): p. 309–315.

[137] Baba, A. and Adachi, S., Simultaneous state of charge and parameter estimation of lithium-ion battery using log-normalized unscented Kalman filter. *2015 American Control Conference (ACC)*, 2015: p. 311–316.

[138] Wang, S.L., Yu, C.M., Fernandez, C., Chen, M.J., Li, G.L., and Liu, X.H., Adaptive state of charge estimation method for an aeronautical lithium-ion battery pack based on a reduced particle-unscented Kalman filter. *Journal of Power Electronics*, 2018. **18**(4): p. 1127–1139.

[139] Aung, H. and Low, K.S., Temperature dependent state of charge estimation of lithium ion battery using dual spherical unscented Kalman filter. *IET Power Electronics*, 2015. **8**(10): p. 2026–2033.

[140] Amiribavandpour, P., Shen, W.X., Mu, D.B., and Kapoor, A., An improved theoretical electrochemical–thermal modelling of lithium-ion battery packs in electric vehicles. *Journal of Power Sources*, 2015. **284**: p. 328–338.

[141] Yang, F.F., Zhang, S.H., Li, W.H., and Miao, Q., State of charge estimation of lithium-ion batteries using LSTM and UKF. *Energy*, 2020. **201**: p. 1–14.

[142] Xiong, R., Li, L.L., Yu, Q.Q., Jin, Q., and Yang, R.X., A set membership theory based parameter and state of charge co-estimation method for all-climate batteries. *Journal of Cleaner Production*, 2020. **249**: p. 1–14.

[143] Ren, H.B., Zhao, Y.Z., Chen, S.Z., and Yang, L., A comparative study of lumped equivalent circuit models of a lithium battery for state of charge prediction. *International Journal of Energy Research*, 2019: p. 1–12.

[144] Li, Y., Vilathgamuwa, M., Farrell, T., Choi, S.S., Tran, N.T., and Teague, J., A physics-based distributed-parameter equivalent circuit model for lithium-ion batteries. *Electrochimica Acta*, 2019. **299**: p. 451–469.

[145] Tran, N.T., Farrell, T., Vilathgamuwa, M., Choi, S.S., and Li, Y., A computationally efficient coupled electrochemical–thermal model for large format cylindrical lithium ion batteries. *Journal of the Electrochemical Society*, 2019. **166**(13): p. A3059–A3071.

[146] Chen, L., Xu, R.Y., Rao, W.N., *et al.*, Electrochemical model parameter identification of lithium-ion battery with temperature and current dependence. *International Journal of Electrochemical Science*, 2019. **14**(5): p. 4124–4143.

[147] Liu, X.Y., Li, W.L., and Zhou, A.G., PNGV equivalent circuit model and SOC estimation algorithm for lithium battery pack adopted in AGV vehicle. *IEEE Access*, 2018. **6**: p. 23639–23647.

[148] Zheng, Y.J., Gao, W.K., Ouyang, M.G., Lu, L.G., Zhou, L., and Han, X.B., State of charge inconsistency estimation of lithium-ion battery pack using mean-difference model and extended Kalman filter. *Journal of Power Sources*, 2018. **383**: p. 50–58.

[149] Yao, S.G., Liao, P., Xiao, M., Cheng, J., and Xu, L., Study on Thevenin equivalent circuit modeling of zinc–nickel single-flow battery. *International Journal of Electrochemical Science*, 2018. **13**(5): p. 4455–4465.

[150] Zhao, Y., Stein, P., Bai, Y., Al-Siraj, M., Yang, Y.Y.W., and Xu, B.X., A review on modeling of electro-chemo-mechanics in lithium-ion batteries. *Journal of Power Sources*, 2019. **413**(1): p. 259–283.

[151] Wang, S.L., Xu, Z.Y., Wu, X.L., *et al.*, Analyses and optimization of electrolyte concentration on the electrochemical performance of iron–chromium flow battery. *Applied Energy*, 2020. **271**: p. 1–14.

[152] Tian, Y., Lai, R.C., Li, X.Y., Xiang, L.J., and Tian, J.D., A combined method for state of charge estimation for lithium-ion batteries using a long short-term memory network and an adaptive cubature Kalman filter. *Applied Energy*, 2020. **265**: p. 1–14.

[153] Tremblay, O. and Dessaint, L.A., Experimental validation of a battery dynamic model for EV applications. *World Electric Vehicle Journal*, 2009. **3**(2): p. 289–298.

[154] Pals, C.R. and John, N., Thermal modeling of the lithium/polymer battery: II. Temperature profiles in a cell stack. *Journal of the Electrochemical Society*, 1995. **142**(10): p. 3282–3288.

[155] Xu, Y.D., Hu, M.H., Fu, C.Y., Cao, K.B., Su, Z., and Yang, Z., State of charge estimation for lithium-ion batteries based on temperature-dependent second-order RC model. *Electronics*, 2019. **8**(9): p. 1–14.

[156] Zhao, R.X., Kollmeyer, P.J., Lorenz, R.D., and Jahns, T.M., A compact methodology via a recurrent neural network for accurate equivalent circuit type modeling of lithium-ion batteries. *IEEE Transactions on Industry Applications*, 2019. **55**(2): p. 1922–1931.

[157] Yang, Q.X., Xu, J., Li, X.Q., Xu, D., and Cao, B.G., State of health estimation of lithium-ion battery based on fractional impedance model and interval capacity. *International Journal of Electrical Power & Energy Systems*, 2020. **119**: p. 1–14.

[158] Yang, L., Cai, Y.S., Yang, Y.X., and Deng, Z.W., Supervisory long-term prediction of state of available power for lithium-ion batteries in electric vehicles. *Applied Energy*, 2020. **257**: p. 1–14.

[159] Tian, Y., Huang, Z.J., Long, T., Tian, J.D., and Li, X.Y., Performance analysis and modeling of three energy storage devices for electric vehicle applications over a wide temperature range. *Electrochimica Acta*, 2019: p. 135317.

[160] Saldana, G., Martin, J.I.S., Zamora, I., Asensio, F.J., and Onederra, O., Analysis of the current electric battery models for electric vehicle simulation. *Energies*, 2019. **12**(14).

[161] Liu, X.Y., Li, W.L., and Zhou, A.G., PNGV equivalent circuit model and SOC estimation algorithm for lithium battery pack adopted in AGV vehicle. *IEEE Access*, 2018. **6**: p. 23639–23647.

[162] Zheng, L.F., Zhu, J.G., Lu, D.D.C., Wang, G.X., and He, T.T., Incremental capacity analysis and differential voltage analysis based state of charge and capacity estimation for lithium-ion batteries. *Energy*, 2018. **150**: p. 759–769.

[163] Lyu, Z.Q. and Gao, R.J., Li-ion battery state of health estimation through Gaussian process regression with Thevenin model. *International Journal of Energy Research*, 2020: p. 1–14.

[164] Hageman, S.C., Simple pspice models let you simulate common battery types. *Electron Design News*, 1993. **38**, 117–129.

[165] Chen, M. and Rincon-Mora, G. A., Accurate electrical battery model capable of predicting runtime and I–V performance. *IEEE Transactions on Energy Conversion*. **21**(2), 504–511.

[166] Wang, H., Wang, S.L., Zhang, E.Y., and Lu, L.X., An energy balanced and lifetime extended routing protocol for underwater sensor networks. *Sensors*, 2018. **18**(5): p. 1–13.

[167] Huo, Y.T., Hu, W., Li, Z., and Rao, Z.H., Research on parameter identification and state of charge estimation of improved equivalent circuit model of Li-ion battery based on temperature effects for battery thermal management. *International Journal of Energy Research*, 2020: p. 1–14.

[168] Xiong, R., Zhang, Y.Z., Wang, J., He, H.W., Peng, S.M., and Pecht, M., Lithium-ion battery health prognosis based on a real battery management

system used in electric vehicles. *IEEE Transactions on Vehicular Technology*, 2019. **68**(5): p. 4110–4121.

[169] Yang, Y., Yuan, W., Zhang, X.Q., *et al.*, A review on structuralized current collectors for high-performance lithium-ion battery anodes. *Applied Energy*, 2020. 276.

[170] Campestrini, C., Kosch, S., and Jossen, A., Influence of change in open circuit voltage on the state of charge estimation with an extended Kalman filter. *Journal of Energy Storage*, 2017. **12**: p. 149–156.

[171] Wang, M.S., Mu, Y.F., Li, F.X., *et al.*, State space model of aggregated electric vehicles for frequency regulation. *IEEE Transactions on Smart Grid*, 2020. **11**(2): p. 981–994.

[172] Shrivastava, P., Soon, T.K., Bin Idris, M.Y.I., and Mekhilef, S., Overview of model-based online state of charge estimation using Kalman filter family for lithium-ion batteries. *Renewable & Sustainable Energy Reviews*, 2019. **113**: p. 1–12.

[173] Guha, A. and Patra, A., Online estimation of the electrochemical impedance spectrum and remaining useful life of lithium-ion batteries. *IEEE Transactions on Instrumentation and Measurement*, 2018. **67**(8): p. 1836–1849.

[174] Guha, A. and Patra, A., State of health estimation of lithium-ion batteries using capacity fade and internal resistance growth models. *IEEE Transactions on Transportation Electrification*, 2018. **4**(1): p. 135–146.

[175] Ran, X.K., Chen, H.L., Liu, Z.M., and Chen, J.S., Delivering deep learning to mobile devices via offloading. *Vr/ArNetwork'17: Proceedings of the 2017 Workshop on Virtual Reality and Augmented Reality Network*, 2017: p. 42–47.

[176] Wang, L.M., Lu, D., Liu, Q., Liu, L., and Zhao, X.L., State of charge estimation for LiFePO4 battery via dual extended kalman filter and charging voltage curve. *Electrochimica Acta*, 2019. **296**: p. 1009–1017.

[177] Qian, L.J., Xin, F.L., Bai, X.X., and Wereley, N.M., State observation-based control algorithm for dynamic vibration absorbing systems featuring magnetorheological elastomers: Principle and analysis. *Journal of Intelligent Material Systems and Structures*, 2017. **28**(18): p. 2539–2556.

[178] Wang, S.L., Fernandez, C., Liu, X.H., Su, J., and Xie, Y.X., The parameter identification method study of the splice equivalent circuit model for the aerial lithium-ion battery pack. *Measurement & Control*, 2018. **51**(5–6): p. 125–137.

[179] Wang, S.L., Fernandez, C., Xie, Z.W., Li, X.X., Lou, C.Y., and Li, Q., A novel weight coefficient calculation method for the real-time state monitoring of the lithium-ion battery packs under the complex current variation working conditions. *Energy Science & Engineering*, 2019. **7**(6): p. 3038–3057.

[180] Propp, K., Auger, D.J., Fotouhi, A., Longo, S., and Knap, V., Kalman-variant estimators for state of charge in lithium–sulfur batteries. *Journal of Power Sources*, 2017. **343**: p. 254–267.

[181] Linghu, J.Q., Kang, L.Y., Liu, M., Luo, X., Feng, Y.B., and Lu, C.S., Estimation for state of charge of lithium-ion battery based on an adaptive high-degree cubature Kalman filter. *Energy*, 2019. **189**: p. 1–13.

[182] Tang, X.P., Wang, Y.J., Yao, K., He, Z.W., and Gao, F.R., Model migration based battery power capability evaluation considering uncertainties of temperature and aging. *Journal of Power Sources*, 2019. **440**: p. 1–14.

[183] Singh, K.V., Bansal, H.O., and Singh, D., Hardware-in-the-loop implementation of ANFIS based adaptive SoC estimation of lithium-ion battery for hybrid vehicle applications. *Journal of Energy Storage*, 2020. **27**: p. 1–14.

[184] Perea, A., Dontigny, M., and Zaghib, K., Safety of solid-state Li metal battery: Solid polymer versus liquid electrolyte. *Journal of Power Sources*, 2017. **359**: p. 182–185.

[185] Xu, X.M., Wu, D., Yang, L., Zhang, H., and Liu, G.J., State estimation of lithium batteries for energy storage based on dual extended Kalman filter. *Mathematical Problems in Engineering*, 2020. **2020**: p. 1–14.

[186] Pastor-Fernandez, C., Uddin, K., Chouchelamane, G.H., Widanage, W.D., and Marco, J., A comparison between electrochemical impedance spectroscopy and incremental capacity-differential voltage as Li-ion diagnostic techniques to identify and quantify the effects of degradation modes within battery management systems. *Journal of Power Sources*, 2017. **360**: p. 301–318.

[187] Parra, D., Swierczynski, M., Stroe, D.I., *et al.*, An interdisciplinary review of energy storage for communities: Challenges and perspectives. *Renewable & Sustainable Energy Reviews*, 2017. **79**: p. 730–749.

[188] Parker, J.F., Chervin, C.N., Pala, I.R., *et al.*, Rechargeable nickel-3D zinc batteries: An energy-dense, safer alternative to lithium-ion. *Science*, 2017. **356**(6336): p. 414–417.

[189] Shu, X., Li, G., Shen, J.W., Yan, W.S., Chen, Z., and Liu, Y.G., An adaptive fusion estimation algorithm for state of charge of lithium-ion batteries considering wide operating temperature and degradation. *Journal of Power Sources*, 2020. **462**: p. 1–14.

[190] Wang, S.L., Fernandez, C., Yu, C.M., Fan, Y.C., Cao, W., and Stroe, D.I., A novel charged state prediction method of the lithium ion battery packs based on the composite equivalent modeling and improved splice Kalman filtering algorithm. *Journal of Power Sources*, 2020. **471**(228450): p. 1–13.

[191] Yang, F.F., Song, X.B., Dong, G.Z., and Tsui, K.L., A coulombic efficiency-based model for prognostics and health estimation of lithium-ion batteries. *Energy*, 2019. **171**: p. 1173–1182.

[192] Yang, N.X., Fu, Y.H., Yue, H.Y., *et al.*, An improved semi-empirical model for thermal analysis of lithium-ion batteries. *Electrochimica Acta*, 2019. **311**: p. 8–20.

[193] Yin, Y.L. and Choe, S.Y., Actively temperature controlled health-aware fast charging method for lithium-ion battery using nonlinear model predictive control. *Applied Energy*, 2020. **271**: p. 1–14.

[194]   Pan, H.H., Lu, Z.Q., Lin, W.L., Li, J.Z., and Chen, L., State of charge estimation of lithium-ion batteries using a grey extended Kalman filter and a novel open-circuit voltage model. *Energy*, 2017. **138**: p. 764–775.

[195]   Oh, K.Y. and Epureanu, B.I., A phenomenological force model of Li-ion battery packs for enhanced performance and health management. *Journal of Power Sources*, 2017. **365**: p. 220–229.

[196]   Thanagasundram, S.A., Arunachala, R., Makinejad, K., Teutsch, T., and Jossen, A., A cell level model for battery simulation. *EEVC European Electric Vehicle Congress*, 2012: p. 2–13.

[197]   Liu, D.T., Li, L., Song, Y.C., Wu, L.F., and Peng, Y., Hybrid state of charge estimation for lithium-ion battery under dynamic operating conditions. *International Journal of Electrical Power & Energy Systems*, 2019. **110**: p. 48–61.

[198]   Lu, J.H., Chen, Z.Y., Yang, Y., and Lv, M., Online estimation of state of power for lithium-ion batteries in electric vehicles using genetic algorithm. *IEEE Access*, 2018. **6**: p. 20868–20880.

[199]   Xu, W.H., Wang, S.L., Fernandez, C., Yu, C.M., Fan, Y.C., and Cao, W., Novel reduced-order modeling method combined with three-particle nonlinear transform unscented Kalman filtering for the battery state of charge estimation. *Journal of Power Electronics*, 2020. **20**(6): p. 1541–1549.

[200]   Chen, Z.H., Sun, H., Dong, G.Z., Wei, J.W., and Wu, J., Particle filter-based state of charge estimation and remaining-dischargeable-time prediction method for lithium-ion batteries. *Journal of Power Sources*, 2019. **414**: p. 158–166.

[201]   Li, J., Landers, R.G., and Park, J., A comprehensive single-particle-degradation model for battery state of health prediction. *Journal of Power Sources*, 2020. **456**: p. 1–14.

[202]   Naseri, F., Farjah, E., Ghanbari, T., Kazemi, Z., Schaltz, E., and Schanen, J.L., Online parameter estimation for supercapacitor state of energy and state of health determination in vehicular applications. *IEEE Transactions on Industrial Electronics*, 2020. **67**(9): p. 7963–7972.

[203]   Parvini, Y., Vahidi, A., and Fayazi, S.A., Heuristic versus optimal charging of supercapacitors, lithium-ion, and lead-acid batteries: An efficiency point of view. *IEEE Transactions on Control Systems Technology*, 2018. **26**(1): p. 167–180.

[204]   Nayak, P.K., Grinblat, J., Levi, E., Levi, M., Markovsky, B., and Aurbach, D., Understanding the influence of Mg doping for the stabilization of capacity and higher discharge voltage of Li- and Mn-rich cathodes for Li-ion batteries. *Physical Chemistry Chemical Physics*, 2017. **19**(8): p. 6142–6152.

[205]   Mussa, A.S., Klett, M., Behm, M., Lindbergh, G., and Lindstrom, R.W., Fast-charging to a partial state of charge in lithium-ion batteries: A comparative ageing study. *Journal of Energy Storage*, 2017. **13**: p. 325–333.

[206]   Verbrugge, M. and Tate, E., Adaptive state of charge algorithm for nickel metal hydride batteries including hysteresis phenomena. *Journal of Power Sources*, 2004. **126**(1): p. 236–249.

[207] Huria, T., Ludovici, G., and Lutzemberger, G., State of charge estimation of high power lithium iron phosphate cells. *Journal of Power Sources*, 2014. **249**: p. 92–102.

[208] Xing, Y.J., He, W., Pecht, M., and Tsui, K.L., State of charge estimation of lithium-ion batteries using the open-circuit voltage at various ambient temperatures. *Applied Energy*, 2014. **113**: p. 106–115.

[209] Jiménez-Bermejo, D., Fraile-Ar, J., Castao-Solis, S., Merino, J., and Lvaro-Hermana, R., Using dynamic neural networks for battery state of charge estimation in electric vehicles. *Procedia Computer Science*, 2018. **130**: p. 533–540.

[210] El Mejdoubi, A., Oukaour, A., Chaoui, H., Gualous, H., Sabor, J., and Slamani, Y., State of charge and state of health lithium-ion batteries' diagnosis according to surface temperature variation. *IEEE Transactions on Industrial Electronics*, 2016. **63**(4): p. 2391–2402.

[211] Zhang, C., Li, K., Deng, J., and Song, S.J., Improved realtime state of charge estimation of LiFePO4 battery based on a novel thermoelectric model. *IEEE Transactions on Industrial Electronics*, 2017. **64**(1): p. 654–663.

[212] Bartlett, A., Marcicki, J., Onori, S., Rizzoni, G., Yang, X.G., and Miller, T., Electrochemical model-based state of charge and capacity estimation for a composite electrode lithium-ion battery. *IEEE Transactions on Control Systems Technology*, 2016. **24**(2): p. 384–399.

[213] Mukai, K., Inoue, T., Kato, Y., and Shirai, S., Superior low-temperature power and cycle performances of Na-ion battery over Li-ion battery. *ACS Omega*, 2017. **2**(3): p. 864–872.

[214] Mu, H., Xiong, R., Zheng, H.F., Chang, Y.H., and Chen, Z.Y., A novel fractional order model based state of charge estimation method for lithium-ion battery. *Applied Energy*, 2017. **207**: p. 384–393.

[215] Motapon, S.N., Lupien-Bedard, A., Dessaint, L.A., Fortin-Blanchette, H., and Al-Haddad, K., A generic electrothermal Li-ion battery model for rapid evaluation of cell temperature temporal evolution. *IEEE Transactions on Industrial Electronics*, 2017. **64**(2): p. 998–1008.

[216] Yang, P., Sant, G., and Neithalath, N., A refined, self-consistent Poisson–Nernst–Planck (PNP) model for electrically induced transport of multiple ionic species through concrete. *Cement & Concrete Composites*, 2017. **82**: p. 80–94.

[217] Yang, J.F., Xia, B., Shang, Y.L., Huang, W.X., and Mi, C.C., Adaptive state of charge estimation based on a split battery model for electric vehicle applications. *IEEE Transactions on Vehicular Technology*, 2017. **66**(12): p. 10889–10898.

[218] Yu, Q.Q., Xiong, R., Lin, C., Shen, W.X., and Deng, J.J., Lithium-ion battery parameters and state of charge joint estimation based on H-infinity and unscented kalman filters. *IEEE Transactions on Vehicular Technology*, 2017. **66**(10): p. 8693–8701.

[219] Zhao, W., Li, P., Han, K., *et al.*, Low-temperature and high-voltage Zn-based liquid metal batteries based on multiple redox mechanism. *Journal of Power Sources*, 2020. **463**: p. 1–14.

[220] Zhao, J.B., Zhang, X., Zhao, Q., Wang, L.J., and Wang, Y.Z., Enhanced cyclability and dynamic properties of P2-type $Na_{0.59}Co_{0.20}Mn_{0.80}O_2$ cathode by B-doping for sodium storage. *Chemical Physics*, 2020. **529**: p. 1–14.

[221] Zhao, C., Yin, H., and Ma, C.B., Equivalent series resistance-based real-time control of battery-ultracapacitor hybrid energy storage systems. *IEEE Transactions on Industrial Electronics*, 2020. **67**(3): p. 1999–2008.

[222] Zhang, X., Wang, Y.J., Liu, C., and Chen, Z.H., A novel approach of remaining discharge energy prediction for large format lithium-ion battery pack. *Journal of Power Sources*, 2017. **343**: p. 216–225.

[223] Xu, C.S., Zhang, F.S., Feng, X.N., *et al.*, Experimental study on thermal runaway propagation of lithium-ion battery modules with different parallel-series hybrid connections. *Journal of Cleaner Production*, 2021. **284** (124749): p. 1–13.

[224] Dong, Z., Dey, S., Perez, H., and Moura, S.J., Remaining useful life estimation of lithium-ion batteries based on thermal dynamics. *2017 American Control Conference (ACC)*, 2017: p. 4042–4047.

[225] Zhang, D., Cadet, C., Yousfi-Steiner, N., Druart, F., and Bérenguer, C., PHM-oriented degradation indicators for batteries and fuel cells. *Fuel Cells*, 2017. **17**(2): p. 268–276.

[226] Fan, X.L., Chen, L., Ji, X., *et al.*, Highly fluorinated interphases enable high-voltage Li-metal batteries. *Chem*, 2018. **4**(1): p. 174–185.

[227] Zhang, Y.L., Du, X.Y., and Salman, M., Battery state estimation with a self-evolving electrochemical ageing model. *International Journal of Electrical Power & Energy Systems*, 2017. **85**: p. 178–189.

[228] Geronimo, I., Payne, C.M., and Sandgren, M., Hydrolysis and transglycosylation transition states of glycoside hydrolase family 3 beta-glucosidases differ in charge and puckering conformation. *Journal of Physical Chemistry B*, 2018. **122**(41): p. 9452–9459.

[229] Zhang, Y.Z., Xiong, R., He, H.W., and Liu, Z., A LSTM-RNN method for the lithuim-ion battery remaining useful life prediction. *2017 Prognostics and System Health Management Conference (Phm-Harbin)*, 2017: p. 1059–1062.

[230] Douaa, F., Julia, M., Henrik, E., Kristina, E., Daniel, L., and Fouad, G., Towards high-voltage Li-ion batteries: Reversible cycling of graphite anodes and Li-ion batteries in adiponitrile-based electrolytes. *Electrochimica Acta*, 2018. **281**: p. 299–311.

[231] Zhang, Z.X., Hu, C.H., Si, X.S., Zhang, J.X., and Quan, S., A prognostic approach for systems subject to Wiener degradation process with cumulative-type random shocks. *2017 6th Data Driven Control and Learning Systems (Ddcls)*, 2017: p. 694–698.

[232] Shen, D.X., Wu, L.F., Kang, G.Q., Guan, Y., and Peng, Z., A novel online method for predicting the remaining useful life of lithium-ion batteries considering random variable discharge current. *Energy*, 2021. **218** (119490): p. 1–17.

[233] Sadabadi, K.K., Jin, X., and Rizzoni, G., Prediction of remaining useful life for a composite electrode lithium ion battery cell using an electrochemical

model to estimate the state of health. *Journal of Power Sources*, 2021. **481** (228861): p. 1–10.

[234] Zhang, Z. X., Si, X.S., Hu, C.H., and Pecht, M.G., A prognostic model for stochastic degrading systems with state recovery: Application to Li-ion batteries. *IEEE Transactions on Reliability*, 2017. **66**(4): p. 1293–1308.

[235] Ferraresi, G., Kazzi, M. E., Czornomaz, L., Tsai, C. L., Uhlenbruck, S., and Villevieille, C., Electrochemical performance of all-solid-state Li-ion batteries based on garnet electrolyte using silicon as a model electrode. *ACS Energy Letters*, 2018. **3**(4): p. 1006–1012.

[236] Zhao, S., Duncan, S.R., and Howey, D.A., Observability analysis and state estimation of lithium-ion batteries in the presence of sensor biases. *IEEE Transactions on Control Systems Technology*, 2017. **25**(1): p. 326–333.

[237] Lipu, M., Hannan, M.A., Karim, T.F., Hussain, A., and Mahlia, T., Intelligent algorithms and control strategies for battery management system in electric vehicles: Progress, challenges and future outlook. *Journal of Cleaner Production*, 2021. **292**(126044): p. 1–27.

[238] Zhong, Y.R., Yang, K.R., Liu, W., He, P., Batista, V., and Wang, H.L., Mechanistic insights into surface chemical interactions between lithium polysulfides and transition metal oxides. *Journal of Physical Chemistry C*, 2017. **121**(26): p. 14222–14227.

[239] Fujita, T., Alam, M.K., and Hoshi, T., Thousand-atom ab initio calculations of excited states at organic/organic interfaces: Toward first-principles investigations of charge photogeneration. *Physical Chemistry Chemical Physics*, 2018. **20**(41): p. 26443–26452.

[240] Zhou, X., Stein, J.L., and Ersal, T., Battery state of health monitoring by estimation of the number of cyclable Li-ions. *Control Engineering Practice*, 2017. **66**: p. 51–63.

[241] Zhou, X., Hsieh, S.J., Peng, B., and Hsieh, D., Cycle life estimation of lithium-ion polymer batteries using artificial neural network and support vector machine with time-resolved thermography. *Microelectronics Reliability*, 2017. **79**: p. 48–58.

[242] Arora, S., Kapoor, A., and Shen, W.X., A novel thermal management system for improving discharge/charge performance of Li-ion battery packs under abuse. *Journal of Power Sources*, 2018. **378**: p. 759–775.

[243] Sun, F.C., Xiong, R., and He, H.W., A systematic state of charge estimation framework for multi-cell battery pack in electric vehicles using bias correction technique. *Applied Energy*, 2016. **162**: p. 1399–1409.

[244] Gao, W.F., Song, J.L., Cao, H.B., *et al.*, Selective recovery of valuable metals from spent lithium-ion batteries—Process development and kinetics evaluation. *Journal of Cleaner Production*, 2018. **178**: p. 833–845.

[245] Yang, D., Zhang, X., Pan, R., Wang, Y.J., and Chen, Z.H., A novel Gaussian process regression model for state of health estimation of lithium-ion battery using charging curve. *Journal of Power Sources*, 2018. **384**: p. 387–395.

[246]   Assat, G. and Tarascon, J.M., Fundamental understanding and practical challenges of anionic redox activity in Li-ion batteries. *Nature Energy*, 2018. **3**(5): p. 373–386.

[247]   Yang, J., Peng, Z., Wang, H.M., Yuan, H.M., and Wu, L.F., The remaining useful life estimation of lithium-ion battery based on improved extreme learning machine algorithm. *International Journal of Electrochemical Science*, 2018. **13**: p. 4991–5004.

[248]   Diao, W.P., Xue, N., Bhattacharjee, V., Jiang, J.C., Karabasoglu, O., and Pecht, M., Active battery cell equalization based on residual available energy maximization. *Applied Energy*, 2018. **210**: p. 690–698.

[249]   Williard, N., He, W., Hendricks, C., and Pecht, M., Lessons learned from the 787 dreamliner issue on lithium-ion battery reliability. *Energies*, 2018. **6**(9): p. 4682–4695.

[250]   Assefi, M., Hooshmand, A., Hosseini, H., and Sharma, R., Battery degradation temporal modeling using LSTM networks. *2018 17th IEEE International Conference on Machine Learning and Applications (ICMLA)*, 2018: p. 853–858.

[251]   Astaneh, M., Dufo-Lopez, R., Roshandel, R., Golzar, F., and Bernal-Agustin, J.L., A computationally efficient Li-ion electrochemical battery model for long-term analysis of stand-alone renewable energy systems. *Journal of Energy Storage*, 2018. **17**: p. 93–101.

[252]   Azzollini, I.A., Di Felice, V., Fraboni, F., *et al.*, Lead-acid battery modeling over full state of charge and discharge range. *IEEE Transactions on Power Systems*, 2018. **33**(6): p. 6422–6429.

[253]   Bahaloo-Horeh, N., Mousavi, S.M., and Baniasadi, M., Use of adapted metal tolerant *Aspergillus niger* to enhance bioleaching efficiency of valuable metals from spent lithium-ion mobile phone batteries. *Journal of Cleaner Production*, 2018. **197**: p. 1546–1557.

[254]   Bartlett, T. and Cookson, J., 21st International conference on solid state ionics highlights of the latest developments in solid-state batteries for energy storage. *Johnson Matthey Technology Review*, 2018. **62**(2): p. 204–207.

[255]   Barz, T., Seliger, D., Marx, K., *et al.*, State and state of charge estimation for a latent heat storage. *Control Engineering Practice*, 2018. **72**: p. 151–166.

[256]   Bayir, R. and Soylu, E., Real time determination of rechargeable batteries' type and the state of charge via cascade correlation neural network. *Elektronika Ir Elektrotechnika*, 2018. **24**(1): p. 25–30.

[257]   Beckner, W., Mao, C.M., and Pfaendtner, J., Statistical models are able to predict ionic liquid viscosity across a wide range of chemical functionalities and experimental conditions. *Molecular Systems Design & Engineering*, 2018. **3**(1): p. 253–263.

[258]   Ben Ali, J. and Saidi, L., A new suitable feature selection and regression procedure for lithium-ion battery prognostics. *International Journal of Computer Applications in Technology*, 2018. **58**(2): p. 102–115.

[259] Smith, K., Saxon, A., Keyser, M., Lundstrom, B., and Roc, A., Life prediction model for grid-connected Li-ion battery energy storage system. *2017 American Control Conference (ACC)*, 2017.

[260] Alberto, B., Michael, H., Martin, J., Alfredo, U., and Pablo, S., Combined dynamic programming and region-elimination technique algorithm for optimal sizing and management of lithium-ion batteries for photovoltaic plants. *Applied Energy*, 2018. **228**: p. 1–11.

[261] Sarasketa-Zabala, E., Gandiaga, I., Rodriguez-Martinez, L.M., and Villarreal, I., Calendar ageing analysis of a LiFePO$_4$/graphite cell with dynamic model validations: Towards realistic lifetime predictions. *Journal of Power Sources*, 2014. **272**: p. 45–57.

[262] Beuse, M., Battery economics death by a thousand charges. *Nature Energy*, 2018. **3**(5): p. 363–364.

[263] Beuse, M., Schmidt, T.S., and Wood V., A "technology-smart" battery policy strategy for Europe. *Science*, 2018. **361**(6407): p. 1075–1077.

[264] Miao, Q., Xie, L., Cui, H.J., Liang, W., and Pecht, M., Remaining useful life prediction of lithium-ion battery with unscented particle filter technique. *Microelectronics Reliability*, 2019. **53**(6): p. 805–810.

[265] Bezha, M. and Nagaoka, N., Predicting voltage characteristic of charging model for Li-Ion battery with ANN for real time diagnosis. *2018 International Power Electronics Conference (IPEC-Niigata 2018-ECCE Asia)*, 2018: p. 3170–3175.

[266] Biswas, A., Gu, R., Kollmeyer, P., Ahmed, R., and Emadi, A., Simultaneous state and parameter estimation of Li-ion battery with one state hysteresis model using augmented unscented Kalman filter. *2018 IEEE Transportation and Electrification Conference and Expo (ITEC)*, 2018: p. 1065–1070.

[267] Bobrikov, I.A., Samoylova, N.Y., Ivanshina, O.Y., *et al.*, Abnormal phase-separated state of Li$_x$Ni$_{0.8}$Co$_{0.15}$Al$_{0.05}$O$_2$ in the first charge: Effect of electrode compaction. *Electrochimica Acta*, 2018. **265**: p. 726–735.

[268] Cai, W.J., Qi, K., Chen, Z.Y., Guo, X.P., and Qiu, Y.B., Effect of graphene oxide with different oxygenated groups on the high-rate partial-state of charge performance of lead-acid batteries. *Journal of Energy Storage*, 2018. **18**: p. 414–420.

[269] Cano, Z.P., Banham, D., Ye, S.Y., *et al.*, Batteries and fuel cells for emerging electric vehicle markets. *Nature Energy*, 2018. **3**(4): p. 279–289.

[270] Abe, K., Yoshitake, H., Kitakura, T., Hattori, T., Wang, H., and Yoshio, M., Additives-containing functional electrolytes for suppressing electrolyte decomposition in lithium-ion batteries. *Electrochimica Acta*, 2018. **49**(26): p. 4613–4622.

[271] Lee, Y.K., Park, J., vLu, W., A comprehensive experimental and modeling study on dissolution in Li-ion batteries. *Journal of the Electrochemical Society*, 2019. **166**(8): p. A1340–A1354.

[272] Rashid, M. and Gupta, A., Mathematical model for combined effect of SEI formation and gas evolution in Li-ion batteries. *ECS Electrochemistry Letters*, 2018. **3**(10): p. A95–A98.

[273] Carkhuff, B.G., Demirev, P.A., and Srinivasan, R., Impedance-based battery management system for safety monitoring of lithium-ion batteries. *IEEE Transactions on Industrial Electronics*, 2018. **65**(8): p. 6497–6504.

[274] Chakraborty, I., Nandanoori, S.P., and Kundu, S., Virtual battery parameter identification using transfer learning based stacked autoencoder. *2018 17th IEEE International Conference on Machine Learning and Applications (ICMLA)*, 2018: p. 1269–1274.

[275] Chemali, E., Kollmeyer, P.J., Preindl, M., Ahmed, R., and Emadi, A., Long short-term memory networks for accurate state of charge estimation of Li-ion batteries. *IEEE Transactions on Industrial Electronics*, 2018. **65**(8): p. 6730–6739.

[276] Cai, L., Meng, J.H., Stroe, D.I., Peng, J.C., Luo, G.Z., and Teodorescu, R., Multi-objective optimization of data-driven model for lithium-ion battery SOH estimation with short-term feature. *IEEE Transactions on Power Electronics*, 2020. 1.

[277] Chen, Y.T., Biookaghazadeh, S., and Zhao, M., Exploring the capabilities of mobile devices supporting deep learning. *HPDC'18: Proceedings of the 27th International Symposium on High-Performance Parallel and Distributed Computing: Posters/Doctoral Consortium*, 2018: p. 17–18.

[278] Chen, W. Li, G.D., Pei, A., *et al.*, A manganese–hydrogen battery with potential for grid-scale energy storage. *Nature Energy*, 2018. **3**(5): p. 428–435.

[279] Chingin, K. and Barylyuk, K., Charge-state-dependent variation of signal intensity ratio between unbound protein and protein–ligand complex in electrospray ionization mass spectrometry: The role of solvent-accessible surface area. *Analytical Chemistry*, 2018. **90**(9): p. 5521–5528.

[280] Liaw, B.Y., Roth, E.P., Jungst, R. G., Nagasubramanian, G., Case, H.L., and Doughty, D.H., Correlation of Arrhenius behaviors in power and capacity fades with cell impedance and heat generation in cylindrical lithium-ion cells. *Journal of Power Sources*, 2018. **119–121**: p. 874–886.

[281] Ecker, M., Gerschler, J.B., Vogel, J., *et al.*, Development of a lifetime prediction model for lithium-ion batteries based on extended accelerated aging test data. *Journal of Power Sources*, 2017. **215**: p. 248–257.

[282] Belt, J., Utgikar, V., and Bloom, I., Calendar and PHEV cycle life aging of high-energy, lithium-ion cells containing blended spinel and layered-oxide cathodes. *Journal of Power Sources*, 2018. **196**(23): p. 10213–10221.

[283] Zhou, C.K., Qian, K.J., Allan, M., and Zhou, W.J., Modeling of the cost of EV battery wear due to V2G application in power systems. *IEEE Transactions on Energy Conversion*, 2017. **26**(4): p. 1041–1050.

[284] Li, N., Gao, F., Hao, T.Q., Ma, Z., and Zhang, C.H., SOH balancing control method for the mmc battery energy storage system. *IEEE Transactions on Industrial Electronics*, 2018. **65**(8): p. 6581–6591.

[285] Neubauer, J. and Pesaran, A., The ability of battery second use strategies to impact plug-in electric vehicle prices and serve utility energy storage applications. *Journal of Power Sources*, 2011. **196**(23): p. 10351–10358.

[286] Yang, F.F., Wang, D., Xing, Y.J., and Tsui, K.L., Prognostics of Li (NiMnCo)O$_2$-based lithium-ion batteries using a novel battery degradation model. *Microelectronics Reliability*, 2017. **70**, 70–78.

[287] Ecker, M., Nieto, N., Kabitz, S., *et al.*, Calendar and cycle life study of Li (NiMnCo)O$_2$-based 18650 lithium-ion batteries. *Journal of Power Sources*, 2018. **248**: p. 839–851.

[288] Schuster, S.F., Bach, T., Fleder, E., *et al.*, Nonlinear aging characteristics of lithium-ion cells under different operational conditions. *Journal of Energy Storage*, 2019. **1**: p. 44–53.

[289] Ng, K.S., Moo, C.S., Chen, Y.P., and Hsieh, Y.C., Enhanced coulomb counting method for estimating state of charge and state of health of lithium-ion batteries. *Applied Energy*, 2018. **86**(9): p. 1506–1511.

[290] Eom, S.W., Kim, M.K., Kim, I.J., Moon, S.I., Sun, Y.K., and Kim, H.S., Life prediction and reliability assessment of lithium secondary batteries. *Journal of Power Sources*, 2007. **174**(2): p. 954–958.

[291] Andre, D., Appel, C., Soczka-Guth, T., and Sauer, D.U., Advanced mathematical methods of SOC and SOH estimation for lithium-ion batteries. *Journal of Power Sources*, 2013. **224**: p. 20–27.

[292] Eddahech, A., Briat, O., Woirgard, E., and Vinassa, J.M., Remaining useful life prediction of lithium batteries in calendar ageing for automotive applications. *Microelectronics Reliability*, 2019. **52**(9–10): p. 2438–2442.

[293] Eddahech, A., Briat, O., Henry, H., Deletage, J.Y., Woirgard, E., and Vinassa, J.M., Ageing monitoring of lithium-ion cell during power cycling tests. *Microelectronics Reliability*, 2018. **51**(9–11): p. 1968–1971.

[294] Cho, H.M., Choi, W.S., Go, J.Y., Bae, S.E., and Shin, H.C., A study on time-dependent low temperature power performance of a lithium-ion battery. *Journal of Power Sources*, 2017. **198**: p. 273–280.

[295] Koltypin, M., Aurbach, D., Nazar, L., and Ellis, B., More on the performance of LiFePO$_4$ electrodes—The effect of synthesis route, solution composition, aging, and temperature. *Journal of Power Sources*, 2017. **174** (2): p. 1241–1250.

[296] Xu, J., Wang, X., Yuan, N.Y., Hu, B.Q., Ding, J.N., and Ge, S.H., Graphite-based lithium ion battery with ultrafast charging and discharging and excellent low temperature performance. *Journal of Power Sources*, 2019. **430**, 74–79.

[297] Madani, S., Schaltz, E., and Knudsen Kær, S., An electrical equivalent circuit model of a lithium titanate oxide battery. *Batteries*, 2019. **5**(1): p. 1–14.

[298] Zhu, R., Duan, B., Zhang, J.M., Zhang, Q., and Zhang, C.H., Co-estimation of model parameters and state of charge for lithium-ion batteries with recursive restricted total least squares and unscented Kalman filter. *Applied Energy*, 2020. **277**: p. 1–14.

[299] Drozhzhin, O.A., Shevchenko, V.A., Zakharkin, M.V., *et al.*, Improving salt-to-solvent ratio to enable high-voltage electrolyte stability for advanced Li-ion batteries. *Electrochimica Acta*, 2018. **263**: p. 127–133.

[300]   Zhang, K., Zhao, P., Sun, C.F., Wang, Y.R., and Chen, Z.W., Remaining useful life prediction of aircraft lithium-ion batteries based on F-distribution particle filter and kernel smoothing algorithm. *Chinese Journal of Aeronautics*, 2020. **33**(5): p. 1517–1531.

[301]   Du, R.L., Liu, J.Q., Zhou, D., and Meng, G., Adaptive Kalman filter enhanced with spectrum analysis to estimate guidance law parameters with unknown prior statistics. Proceedings of the Institution of Mechanical Engineers Part G—*Journal of Aerospace Engineering*, 2018. **232**(16): p. 3078–3099.

[302]   Du, J.Y., Zhang, X.B., Wang, T.Z., *et al.*, Battery degradation minimization oriented energy management strategy for plug-in hybrid electric bus with multi-energy storage system. *Energy*, 2018. **165**: p. 153–163.

[303]   Cui, Y.Z., Zuo, P.J., Du, C.Y., *et al.*, State of health diagnosis model for lithium ion batteries based on real-time impedance and open circuit voltage parameters identification method. *Energy*, 2018. **144**: p. 647–656.

[304]   Du, X.P., Qian, Z., Chen, Z.L., and Rao, Z.H., Experimental investigation on mini-channel cooling-based thermal management for Li-ion battery module under different cooling schemes. *International Journal of Energy Research*, 2018. **42**(8): p. 2781–2788.

[305]   Deng, J., Bae, C., Marcicki, J., Masias, A., and Miller, T., Safety modelling and testing of lithium-ion batteries in electrified vehicles. *Nature Energy*, 2018. **3**(4): p. 261–266.

[306]   Erisen, N., Emerce, N.B., Erensoy, S.C., and Eroglu, D., Modeling the effect of key cathode design parameters on the electrochemical performance of a lithium-sulfur battery. *International Journal of Energy Research*, 2018. **42**(8): p. 2631–2642.

[307]   Xu, W., Xu, J.L., Liu, B.L., Liu, J.J., and Yan, X.F., A multi-timescale adaptive dual particle filter for state of charge estimation of lithium-ion batteries considering temperature effect. *Energy Science & Engineering*, 2020: p. 1–14.

[308]   Wang, Y.J. and Chen, Z.H., A framework for state of charge and remaining discharge time prediction using unscented particle filter. *Applied Energy*, 2020. **260**: p. 1–14.

[309]   Dewangga, B.R., Herdjunanto, S., and Cahyadi, A., Unknown input observer for battery open circuit voltage estimation: An LMI approach. *Proceedings of 2018 the 10th International Conference on Information Technology and Electrical Engineering (ICITEE)*, 2018: p. 471–475.

[310]   Bi, Y., Yin, Y., and Choe, S.Y., Online state of health and aging parameter estimation using a physics-based life model with a particle filter. *Journal of Power Sources*, 2020. **476**: p. 1–14.

[311]   Ding, W., Qiu, D.F., Bolgar, M.S., and Miller, S.A., Improving mass spectral quality of monoclonal antibody middle-up LC-MS analysis by shifting the protein charge state distribution. *Analytical Chemistry*, 2018. **90**(3): p. 1560–1565.

[312]   Ahwiadi, M. and Wang, W., An adaptive particle filter technique for system state estimation and prognosis. *IEEE Transactions on Instrumentation and Measurement*, 2020. **69**(9): p. 6756–6765.

[313] Tang, X., Liu, K., Wang, X., Liu, B., Gao, F., and Widanage, W. D., Real-time aging trajectory prediction using a base model-oriented gradient-correction particle filter for lithium-ion batteries. *Journal of Power Sources*, 2019. **440**(227118): p. 1–12.

[314] Shi, E.W., Xia, F., Peng, D.G., Li, L., Wang, X.K., and Yu, B.L., State of health estimation for lithium battery in electric vehicles based on improved unscented particle filter. *Journal of Renewable and Sustainable Energy*, 2019. **11**(2): p. 1–12.

[315] Sangwan, V., Kumar, R., and Rathore, A.K., State of charge estimation of Li-ion battery at different temperatures using particle filter. *Journal of Engineering*, 2019(18): p. 5320–5324.

[316] Dong, L., Liang, F.X., Wang, D., Zhu, C.Z., and Liu, J.H., Safe ionic liquid-sulfolane/LiDFOB electrolytes for high voltage Li-1.15(Ni0.36Mn0.64) (0.85)O-2 lithium ion battery at elevated temperatures. *Electrochimica Acta*, 2018. **270**: p. 426–433.

[317] Feng, F., Teng, S.L., Liu, K.L., *et al.*, Co-estimation of lithium-ion battery state of charge and state of temperature based on a hybrid electrochemical–thermal–neural-network model. *Journal of Power Sources*, 2020. **455**: p. 1–14.

[318] Li, Y., Liu, K.L., Foley, A.M., *et al.*, Data-driven health estimation and lifetime prediction of lithium-ion batteries: A review. *Renewable and Sustainable Energy Reviews*, 2019. **113**(109254): p. 1–13.

[319] Liu, K.L., Li, K., Peng, Q., Guo, Y.J., and Zhang, L., Data-driven hybrid internal temperature estimation approach for battery thermal management. *Complexity*, 2018. **2018**(9642892): p. 1–13.

[320] Guo, J., Li, Z.J., and Pecht, M., A Bayesian approach for Li-ion battery capacity fade modeling and cycles to failure prognostics. *Journal of Power Sources*, 2017. **281**: p. 173–184.

[321] Diao, W.P., Saxena, S., Han, B., and Pecht, M., Algorithm to determine the knee point on capacity fade curves of lithium-ion cells. *Energies*, 2019. **2019**;12:2910.

[322] Bloom, I., Cole, B.W., Sohn, J.J., *et al.*, An accelerated calendar and cycle life study of Li-ion cells. *Journal of Power Sources*, 2001. **165**(2), 566–572.

[323] Ramadass, P., Haran, B., White, R., and Popov, B. N., Mathematical modeling of the capacity fade of Li-ion cells. *Journal of Power Sources*, 2003. **123**(2): p. 230–240.

[324] Kumar, A., Das, S., and Mallipeddi, R., A reference vector-based simpli-fied covariance matrix adaptation evolution strategy for constrained global optimization. *IEEE Transactions on Cybernetics*, 2020. 1–14.

[325] Liu, F., Ma, J., and Su, W.X., Unscented particle filter for SOC estimation algorithm based on a dynamic parameter identification. *Mathematical Problems in Engineering*, 2019. **2019**: p. 1–13.

[326] Cadini, F., Sbarufatti, C., Cancelliere, F., and Giglio, M., State of life prognosis and diagnosis of lithium-ion batteries by data-driven particle filters. *Applied Energy*, 2019. **235**: p. 661–672.

[327] Tian, J.P., Xiong, R., Shen, W.X., Wang, J., and Yang, R.X., Online simultaneous identification of parameters and order of a fractional order battery model. *Journal of Cleaner Production*, 2020. **247**: p. 1–14.

[328] Wang, R.R. and Feng, H.L., Lithium-ion batteries remaining useful life prediction using Wiener process and unscented particle filter. *Journal of Power Electronics*, 2020. **20**(1): p. 270–278.

[329] Weber, N., Landgraf, S., Mushtaq, K., *et al.*, Modeling discontinuous potential distributions using the finite volume method, and application to liquid metal batteries. *Electrochimica Acta*, 2019. **318**: p. 857–864.

[330] Wang, Z., Ouyang, D.X., Chen, M.Y., Wang, X.H., Zhang, Z., and Wang, J., Fire behavior of lithium-ion battery with different states of charge induced by high incident heat fluxes. *Journal of Thermal Analysis and Calorimetry*, 2019. **136**(6): p. 2239–2247.

[331] Tang, X.P., Gao, F.R., Zou, C.F., Yao, K., Hu, W.G., and Wik, T., Load-responsive model switching estimation for state of charge of lithium-ion batteries. *Applied Energy*, 2019. **238**: p. 423–434.

[332] Zheng, Y.J., Gao, W.K., Han, X.B., Ouyang, M.G., Lu, L.G., and Guo, D.X., An accurate parameters extraction method for a novel on-board battery model considering electrochemical properties. *Journal of Energy Storage*, 2019. **24**: p. 1–14.

[333] Vijayaraghavan, V., Shui, L., Garg, A., Peng, X.B., and Singh, V.P., Crash analysis of lithium-ion batteries using finite element based neural search analytical models. *Engineering with Computers*, 2019. **35**(1): p. 115–125.

Index

www.ingramcontent.com/pod-product-compliance
Lightning Source LLC
Chambersburg PA
CBHW050512190326
41458CB00005B/1507